SPRINGER
LABMANUAL

G. M. Rothe

Electrophoresis of Enzymes

Laboratory Methods

With 59 Figures

Springer Verlag

Berlin Heidelberg New York
London Paris Tokyo
Hong Kong Barcelona Budapest

Professor Dr. GUNTER M. ROTHE

Johannes Gutenberg-Universität
Institut für Allgemeine Botanik
Fachbereich Biologie
D-55099 Mainz

ISBN-13:978-3-642-79071-3   e-ISBN-13:978-3-642-79069-0
DOI: 10.1007/978-3-642-79069-0

Library of Congress Cataloging-in-Publication Data

Rothe, G. (Gunter)
Electrophoresis of enzymes : laboratory methods / Gunter M. Rothe.
    (Springer laboratory)
Includes bibliographical references.
ISBN-13:978-3-642-79071-3
1. Enzymes – Separation – Laboratory manuals. 2. Electrophoresis – Laboratory manuals.
I. Title. II. Series.
QP601.R786 1994
574.19'25 – dc20

Production: PRODUserv Springer Produktions-Gesellschaft, Berlin
Typesetting: Dataconversion by Fotosatz-Service Köhler OHG, Würzburg
Cover-layout: Struve & Partner, Heidelberg

SPIN 10123478   52/3020-543210   Printed on acid-free paper

To my parents

Preface

This book is addressed to both new and experienced workers interested in the
properties of enzymes or using enzymes as gene markers in areas such as biochemistry, clinical pathology, zoology, botany, systematics, ecology, population genetics or
natural resources management.

The first chapter introduces into the classification, structure and size of enzymes
and isozymes. The second chapter provides protocols to extract (iso)enzymes from
microbial, plant, animal and human tissue and cells-respectively. Methods to extract
enzymes from cell-organelles of plant and animal tissues are also presented. The third
chapter is dedicated to reliable electrophoretic procedures being in use to separate
native enzymes. Some quantitative electrophoretic methods are also given to determine the size and net charge of (iso)enzymes. Chapter four summarizes methods to
renature enzymes after sodium dodecylsulphate electrophoresis. Chapter five gives
an overview on the various "histochemical" reactions being in use to visualize
enzymes following electrophoresis. Specific staining protocols together with suitable
electrophoretic systems for more than 100 different enzymes are listed in chapter six.
Finally, methods of data acquisition from enzyme patterns produced by electrophoresis are presented and various examples given to calculate populational genetic
statistics.

Acknowledgement

I am very grateful to Prof. Dr. Bertold J. Radola who encouraged me to write this book. I am also much indebted to all my students but especially to Renate Bohrmann, Ute Does, Dr. Werner D. Maurer, Brigitte Niethard, Monika Noll, Dr. Irene Novotny, Ina Pappe, Dr. Huschang Pukhanbaba, Anke Richert and Beate Schellenberg. I thank PD Dr. Klaus Honomichl and Hans Weidmann for their help in establishing the computer program MOL-MASS and I wish to thank Peter Enders and the staff of Springer-Verlag for their assistance and patience.

Mainz, March 1994   Gunter M. Rothe

Abbreviations

ADP:	Adenosine-5′-diphosphate
AOL:	Agar overlay
ATP:	Adenosine-5′-triphosphate
BIS:	N,N′-methylenebisacrylamide
CG:	Cellogel
DNA:	Deoxyribonucleic acid
EDTA:	Ethylenediaminetetraacetic acid
FAD:	Flavin adenine dinucleotide
FM:	Flow method
IEF:	Isoelectric focussing
MOL:	Membrane overlay
M_r:	molecular mass [g/mol]
MTT:	3-(4,5-Dimethylthiazol-2-yl)-2,5-diphenyltetrazolium bromide
NAD:	β-Nicotinamide adenine dinucleotide
NADH:	β-Nicotinamide adenine dinucleotide, reduced form
NADP:	β-Nicotinamide adenine dinucleotide phosphate
NADPH:	β-Nicotinamide adenine dinucleotide phosphate, reduced form
PAA:	Polyacrylamide
PAGE:	Polyacrylamide gel electrophoresis
Pi:	Orthophosphate
PMS:	Phenazine methosulphate
POL:	Paper overlay
SDS:	Sodium dodecylsulphate
SGE:	starch gel electrophoresis
% T:	(g Acrylamide + g BIS)/100 ml
Temed:	N,N,N′,N′-Tetramethylethylenediamine
Tris:	Tris (hydroxymethyl) aminomethane
U:	Units
UTL:	Ultrathin layer

Contents

1 Introduction

Enzymes are protein molecules with catalytic abilities. To date about 1500–2000 enzymes have been described and catalogued [1].

Bacteria such as *Escherichia coli* are estimated to possess about 3000 different proteins per cell while higher eukaryote cells may synthesize about 50 000 proteins of which the majority are enzymes.

1.1 Enzyme Classification

1.1.1 The EC Nomenclature

The presently accepted nomenclature of enzymes is that recommended by the Enzyme Commission (set up in 1955 by the International Union of Biochemistry in consultation with the International Union of Pure and Applied Chemistry) [1]. This system is based on the specific chemical reaction catalyzed by an enzyme. All enzymes known so far can be grouped into six main groups (with several sub-groups specifying the reaction more precisely).

The system for classification of enzymes also serves as a basis for assigning code numbers to them. The code numbers are prefixed by EC (Enzyme Commission) and contain four numbers separated by points, with the following meaning [1]:

(a) the first number shows to which of the six main classes an enzyme belongs to;
(b) the second figure indicates the sub-class;
(c) the third figure gives the sub-sub-class; and
(d) the fourth figure is the serial number of the enzyme in its sub-sub-class:

1.1.1.1 Oxidoreductases

To this group belong all enzymes catalyzing oxidation-reduction reactions. The substrate (AH_2) that is oxidized is regarded as a hydrogen donor ($AH_2 + B = A + BH_2$). The recommended name is dehydrogenase but, as an alternative, reductase is used. The name oxidase is restricted to enzymes which exclusively use O_2 as the hydrogen acceptor. The second figure in the code number of oxidoreductases indicates the group in the hydrogen donor which undergoes oxidation (e. g. CH–OH, CHO, CH–CH,

CHNH$_2$, NAD(P)H). The third figure indicates the type of acceptor involved: 1 denotes (NAD(P)), 2 a cytochrome, 3 O$_2$, 4 a disulphide, 5 a quinone etc [1].

1.1.1.2 Transferases

These are enzymes which catalyze the transfer of a group, e.g. a methyl or glycosyl group, from one compound to another. In many cases the donor is a cofactor (coenzyme) carrying the group to be transferred. The second figure in the code number of transferases indicates the group transferred: a one carbon group (2.1), a carbonyl group (aldehyde or ketone) (2.2), a glycosyl group in 2.3 and so on. The third figure informs on the group transferred: e.g. subclass 2.1 is subdivided into methyltransferases (2.1.1), hydroxymethyl and formyltransferases (2.1.2) and so on.

1.1.1.3 Hydrolases

These enzymes catalyze the hydrolytic cleavage of C–O, C–N, C–C and some other bonds, including phosphoric anhydride bonds. Their trivial names are formed by adding the suffix … ase to the substrate which they hydrolyze. A number of hydrolases acting on ester, glycosyl, peptide, amide, or other bonds are known to catalyze not only the hydrolytic removal of a particular group from their substrates, but also the transfer of this group to a suitable acceptor molecule. Yet they are not grouped as transferases because the transfer of a specific group to water as the acceptor molecule is considered to be their main physiological function. The second figure in the code number of hydrolases indicates the nature of the bond hydrolyzed: e.g. esterases (3.1), glycosidases (3.2) and so on. The third figure generally specifies the nature of the substrate: e.g. carboxylic esters (3.1.1), thiol esters (3.1.2), phosphoric monoesters (3.1.3), O-glycosides (3.2.1), N-glycosides (3.2.2) and so on.

1.1.1.4 Lyases

These enzymes cleave C–C, C–O, C–N, and other bonds by elimination, forming double bonds, or conversely adding groups to double bonds. Recommended names are decarboxylase, aldolase, "dehydratase" (if water is eliminated) or "hydro-lyase" (if the reverse reaction is more important or the only one which can be demonstrated). "Synthase" but not synthetase may be used as in tryptophan synthase or cystathionine β-synthase. The second figure in the EC number indicates the bond being cleaved: e.g. C–C-lyases (4.1), C–O-lyases (4.2) and so on. The third figure informs on the group that is eliminated: e.g. CO$_2$ (4.1.1) or H$_2$O (4.2.1).

1.1.1.5 Isomerases

These enzymes catalyze geometric or structural changes within a molecule. According to the type of isomerism, they may be called racemases, epimerases, cis-transisomerases, isomerases, tautomerases, mutases, or cyclo-isomerases.

1.1.1.6 Ligases (Synthetases)

These enzymes catalyze the linkage of two molecules coupled with the hydrolytic breakdown of a pyrophosphate bond in ATP or an analogous compound. The bonds formed are often high energy bonds. The second figure in the code number indicates the bond formed: e.g. C–O (6.1), C–S (6.2) etc. Sub-sub-classes are only in use in the group of C–N-ligases.

The system of the nomenclature and the classification of enzymes is based exclusively on the reaction that is catalyzed and does not consider their origin or multiplicity. Enzymes catalyzing the same reaction but isolated from different species will have varying amino acid sequences so that they may be distinguished by electrophoretic methods. They may have different sizes and net negative charges and they may even differ in their catalytic behaviour.

1.1.2 Multienzyme Systems

Where more than a single catalytic activity is performed by a protein molecule the EC's recommendation is that it should be referred to as a system. The enzyme shikimate dehydrogenase of higher plants may be taken as a simple example since, by chromatographic methods, it cannot be separated from the preceding enzyme of the prechorismate pathway which is 3-dehydroquinate dehydratase [2]. Although both enzymes form a system they appear under distinct EC numbers (EC 1.1.1.25 and EC 4.2.1.10). In bacteria all enzymes of the pre-chorismate pathway occur separately [3] while they aggregate to a multienzyme complex in *Neurospora crassa* [4]. Similarly, homoserine dehydrogenase and aspartate kinase activities are associated with a single polypeptide chain but are numbered EC 1.1.1.3 and EC 2.7.2.4. The debranching enzyme gives another example. It acts both as amylo-1,6-glucosidase and as 4-α-D-glucanotransferase and thus appears as EC 3.2.1.33 and EC 2.4.1.25 in the classification scheme [5]. When multifunctional enzymes are localized in electrophoretic support media it is trivial that they will appear as one band. However, if two different enzyme activities appear at the same location after electrophoresis this does not prove the existence of an enzyme system. In most instances the two activities may be separated when altering the electrophoretic conditions such as the buffer, the pH-value and the support medium.

Shikimate dehydrogenase, dehydroquinate dehydratase

Homoserine dehydrogenase, aspartate kinase

Debranching enzyme

1.2 Enzyme Subunit Composition

Few enzymes consist of a single peptide chain only (Table 1.1). Most are polymeric which means that several polypeptides, named subunits, are forming an active enzyme unit. If an "oligomeric" enzyme is separated into subunits, the catalytic function is generally lost. It is, however, possible to separate subunits of polymeric enzymes and recombine them into an active enzyme as has been shown for mammalian lactate dehydrogenase. If the two homopolymeric lactate dehydrogenase forms LDH_1 (AAAA) and LDH_5 (BBBB) from beef heart are frozen in molar sodium chloride

Monomeric enzymes

Table 1.1. Monomeric enzymes (data taken from [7] and [11])

Enzyme (EC number)	Source	Locus (man)	Mol mass [g/mol]
Acid phosphatase (3.1.3.2)	human	ACP_1	15 000
Adenosine deaminase (3.5.4.4)	human	ADA	34 000
Adenylate kinase (2.7.4.3)	human human cow	AK_1 AK_2 	22 000 40 000 31 000
α-Amylase (3.2.1.1)	human	AMY_1 AMY_2	55 000 55 000
Carbonate dehydratase (4.2.1.1)	rat human	Kidney Erythrocyte RBC–B CA_1, CA_2	26 000 24 000 29 000
Cathepsin B (3.4.22.1)	rat	liver	22 500
Esterase (3.1.1.1)	human	ESA_1 ESB_4 ESD	55 000 22 000 28 000
α-Frucosidase (3.2.1.51)	human	α–FUC	50 000
β-Galactosidase (3.2.1.23)	human	βGal_A	72 000
Guanylate kinase (2.7.4.8)	human	GUK_1	22 000
Mannose phosphate isomerase (5.3.1.8)	human	MPI	43 000
NADH diaphorase (1.6.2.2)	human	DIA	30 000
Nucleosidetriphosphate adenylate kinase (2.7.4.10)	human	AK_3	24 000
Pepsin (ogen) (3.4.23.*)	human	Pg	40 000
Phosphoglucomutase (2.7.5.1)	human	PGM_1 PGM_2 PGM_3	51 000 61 000 53 000
Phophoglycerate kinase (2.7.2.3)	human	PGK	50 000
Phosphoglyceromutase (2.7.5.3)	human	$PGAM_M$ $PGAM_B$	30 000 30 000
Ribonuclease (3.1.26.1)	human	Urine	21 500
Uridine phosphorylase (2.4.2.3)	human	UMPK	28 000

* not further specified

and thawed after several hours the four subunits of each homomer recombine to the homopolymeric (LDH$_1$, LDH$_5$) forms, but also the heteromeric forms LDH$_2$ (AAAB), LDH$_3$ (AABB) and LDH$_4$ (ABBB). All five isoenzymes will appear in the calculated proportions of 1:4:6:4:1 which would be expected upon random reassociation of the subunits. Electrophoresis after similar but separate treatment of LDH$_1$ or LDH$_5$ showed the presence of the single isoenzyme only [6].

    Most enzymes investigated are dimeric in structure [7, 8] (Table 1.2). Only a few enzymes are trimers (Table 1.3), several are tetramers (Table 1.4) and very few are composed of more than four subunits [7, 8] (Table 1.5). Most enzymes are composed of identical subunits (Table 1.2 – 1.5) but in some enzymes the subunits are of different size (Table 1.6) and function [7, 8].

(margin note: LDH)

(margin note: Oligomeric enzymes)

Table 1.2. Dimeric enzymes composed of equally sized subunits (data taken from a compilation of [7] and [8])

Enzyme (EC number)	Source (location)	Mol mass [g/mol]	Isozymes; pI
Acetylcholinesterase (3.1.1.7)	cobra *(Naja naja atra)*	144000	> 10; 6.25 – 6.4
	Bungarus fasciatus (venom)	126000	10; 4.3 – 5.3
Acid phosphatase (3.1.3.2)	human (prostate gland)	104000	> 8; 4.47, 5.62, 6.02, 6.78, 7.12, 7.83
Adenosine deaminase (3.5.4.4)	rabbit (kidney)	215000	4; 4.15, 4.50, 5.05, 5.65
Adenosylhomocysteinase (3.3.1.1)	*Lupinus lutens* (seeds)	110000	1; 4.9
Adenylate kinase (2.7.4.3)	rat (liver)	46000	2; 7.5, 8.0
Alcohol dehydrogenase (1.1.1.1)	*Rhodopseudomonas acidophila*	120000	1; 9.3
	horse (liver)	80000	12; 8.08, 8.28, 8.51, 8.7, 9.29
	rat (liver)	68000	1; 9.7
Aldolase (4.1.2.13)	yeast	80000	3; 5.1, 5.2, 5.3
Alkaline phospatase (3.1.3.1)	human	116000	2; 4.6
	calf	140000	1; 4.4
	(placenta)	80000	5; 5.06, 5.17, 5.20, 5.26, 5.38
	E. coli		
2-Amino-adipate aminotransferase (2.6.1.39)	rat (kidney)	85000	1; 6.56
L-Aminoacid oxidase (1.4.3.2)	*Crotalus adamanteus* (venom)	135000	18; 5.71, 5.87

Table 1.2 (continued)

Enzyme (EC number)	Source (location)	Mol mass [g/mol]	Isozymes; pI
Aminobutyrate transaminase (I, II) (2.6.1.19)	pig (liver)	110 000	4; 6.10, 6.30 (I), 5.90, 6.34 (II)
δ-Amino-laevulinate synthase (2.3.1.37)	rat (liver, mitochondria)	120 000	1; 4.5
Aspartate aminotransferase (2.6.1.1)	pig (heart)	82 000	1; 5.68
Carbonate dehydratase III (4.2.1.1)	rabbit (skeletal muscle)	58 000	1; 8.41 (monomer) 9.34 (dimer)
Carboxypeptidase G$_1$ (3.4.22.12)	*Pseudomonas stutzeri*	92 000	4; 7.1
Cellulase (GB-2 component) (3.2.1.4)	*Pyricularia oryzae*	240 000	1; 4.05
Creatine kinase (2.7.3.2)	rabbit	81 000	3; 6.6, 6.7, 6.9
Cytochrome c oxidase (1.9.3.1)	*Pseudomonas sp.*	120 000	1; 6.9
Dihydropteridine reductase (1.6.99.7)	rat (liver) sheep (liver)	51 000 52 000	1; 6.35 1; 5.4
Enolase (4.2.1.11)	yeast;	88 000	7; 5.4, 5.6, 6.0, 6.6, 6.7, 6.8, 7.1
	rabbit (muscle)	85 000	3; 7.7, 8.4, 8.8
Formaldehyde dehydrogenase (1.2.1.1)	human (liver)	81 000	1; 6.35
Fructokinase (2.7.1.4)	cow (liver)	56 000	1; 5.7
Glucose-6-phosphate dehydrogenase (1.1.1.49)	*Candida utilis*	110 000	3; 5.5, 5.87, 6.54
Glucose-6-phosphate isomerase (5.3.1.9)	human, wild type, Singh variant (erythrocytes)	131 000	1; 9.25 3; 9.25, 9.40, 9.57
β-D-Glucosidase (3.2.1.21)	almond	135 000	1; 7.3
Glutamate decarboxylase (4.1.1.15)	human (brain)	140 000	4; 5.0, 5.1, 5.2, 5.4
Glutathione transferase (2.5.1.18)	human (erythrocytes)	47 500	1; 4.5

Table 1.2 (continued)

Enzyme (EC number)	Source (location)	Mol mass [g/mol]	Isozymes; pI
Glycerol-3-phosphate dehydrogenase (1.1.1.8)	rabbit (skeletal muscle)	78 000	1; 6.45
sn-Glycerol-3-phosphate dehdrogenase (1.1.1.94)	E. coli	51 000	1; 6.0
Guanine deaminase (3.5.4.3)	rabbit (liver)	112 000	1; 4.78
Hexokinase (PI, PII) (2.7.1.1)	yeast	104 000	2; 5.0 (P II), 5.3 (P I)
Histidyl-t-RNA synthetase (6.1.1.21)	rabbit (reticulocytes)	122 000	1; 5.0
Homoserine dehydrogenase (1.1.1.3)	Rhodospirillum rubrum	110 000	4; 5.0, 5.3, 5.7, 6.1
Hydrogenase (1.18.3.1)	Chromatium E. coli (membrane bound)	100 000 113 000	2; 4.2, 4.4 2; 4.2
3-Hydroxyacyl-CoA dehydrogenase (1.1.1.35)	pig (heart)	65 000	1; 8.95
Leucyl-t-RNA synthethase (6.1.1.4)	yeast	120 000	1; 4.9
Lipoxydase (lipoxygenase) (1.13.11.12)	soybean	180 000	1; 5.65 3; 5.68, 6.15, 6.25
Luciferase	firefly	100 000	2; 5.7, 6.4
Malate dehydrogenase (1.1.1.37)	Saccharomyces cerevisiae (mitochondria, cytoplasma) bovine (mitochondria)	68 000 75 000 70 000	1; 6.8 2; 6.75, 7.1 2; 8.0 – 8.5
α-D-Mannosidase I, II (3.2.1.24)	Phaseolus vulgaris	220 000	2; 5.1 (I), 6.1 (II)
Myrosinase C (thioglucosidase) (3.2.3.1)	rapeseed; white mustard	135 000 151 000	3; 4.96, 4.99, 5.06 1; 5.08
Nitrogenase (1.18.2.1)	Klebsiella	67 000	3; 4.0, 5.9
Pantothenase (3.5.1.22)	Pseudomonas	100 000	1; 4.7
Phenol sulphotransferase I (acryl sulphotransferase) (2.8.2.1)	rat (liver)		1; 8.5

Table 1.2 (continued)

Enzyme (EC number)	Source (location)	Mol mass [g/mol]	Isozymes; pI
Phenylalanine (histidine) aminotransferase (2.6.1.58)	mouse (liver, mitochondria)	80 000	5; 5.6, 6.0, 6.2, 6.5, 6.7
6-Phosphogluconate dehydrogenase (1.1.1.43)	*Neurospora crassa*	115 000	2; 4.93, 5.50
Phosphorylase (muscle phosphorylase a) (2.4.1.1)	honey bee (venom)	40 000	1; 10.5
Phosphorylase (2.4.1.1)	*Klebsiella pneumoniae*	180 000	1; 5.3
Postproline dipeptidyl aminopeptidase (dipeptidyl peptidase IV) (3.4.14.2)	lamb (kidney)	230 000	1; 4.9
Protease (3.4.24.4)	*Bacteroides amylophilus*	60 000	2; 4.25
Protein kinase, cGMP-dependent (2.7.1.37)	cow (lung) dog (heart)	150 000 69 000	1; 5.4 1; 4.0
Protocollagen hydroxylase (1.14.11.2)	chicken (embryo)	113 000	1; 4.4
Serinepyruvate aminotransferase (2.6.1.51)	mouse (liver) dog (liver) cat (liver)	80 000 80 000 80 000	4; 6.1, 6.3, 6.6, 6.9 2; 6.6, 6.9 2; 6.6, 6,9
Seryl *t*-RNA synthetase (6.1.1.11)	*E. coli* (strain K12)	100 000	1; 3.9
Sulphatase A (3.1.6.1)	ox (liver)	107 000	1; 3.6
Superoxide dismutase (1.15.1.1)	*Porphyridium cruentum*	40 000	1; 4.2
Thioredoxin reductase (1.6.4.5)	rat (Novikoff tumor)	116 000	1; 5.1
Triosephosphate isomerase (5.3.1.1)	rabbit (muscle)	53 000	5; 5.8, 6.2, 7.0, 7.7, 8.0
Trytophanyl *t*-RNA synthetase (6.1.1.2)	*E. coli* B human (placenta)	74 000 120 000	3; 6.2 1; 5.8
Xanthin dehydrogenase (1.2.1.37)	*Streptomyces cyanogenus*	125 000	1; 4.4

Table 1.3. Trimeric enzymes composed of equally sized subunits (data taken from a compilation of [7] and [8])

Enzyme (EC number)	Source (location)	Mol mass [g/mol]	Isozymes
β-N-Acetyl-D-hexosaminidase (3.2.1.52)	*Trigonella foenum graecum* (seeds)	84000 72000	1; 6.78 (I) 1; 6.30 (II) 4.90 dimer of II 4.65 dimer of I
Adenine phosphoribosyl transferase (2.4.2.7)	human (fibroblast)	34000	1; 4.48
Carboxylesterase E₁ (3.1.1.1)	rat (liver microsomes)	177000	1; 5.65
Hypoxanthine phosphoribo-syltransferase (2.4.2.8)	Chinese hamster (liver, V79 tissue culture cells)	78000	6; 6.2, 6.3, 6.6
	human (erythrocytes)	81000	3; 5.6, 5.7, 5.9
Nucleoside phosphoacyl-hydrolase (3.6.1.24)	human (placenta)	93000	3; 5.64, 5.74, 5.86
Ornithine transcarbamylase (2.1.3.3)	human (liver) ox (liver)	110000 108000	1; 7.5 7; 6.19, 6.36, 6.44, 6.49, 6.59, 6.77, 6.95
Purine nucleoside phosphorylase (2.4.2.1)	human (erythrocytes)	90000	6; 5.85, 5.92, 6.02, 6.08, 6.14, 6.25
	cow (spleen)	90000	1; 5.4

Table 1.4. Tetrameric enzymes composed of equally sized monomers (data taken from a compilation of [7] and [8])

Enzyme (EC number)	Source (location)	Mol mass [g/mol]	Isozymes; pI
Acetylcholinesterase (3.1.1.7)	Electrophorus electricus (electric eel tissue)	280000	5 major, 3 minor; 5.5 – 6.0
Acetyl-CoA acetyltransferase (I, A, B) (2.3.1.9)	cow (liver, mitochondria)	152000	3; 6.9, 7.5, 8.8
α-N-Acetyl-D-galactosamidase (3.2.1.49)	limpet (*Patella vulgata*)	200000	1; 5.5
Adenosylhomocysteinase (3.3.1.1)	calf (liver)	237500	2; 5.8, 6.0

Table 1.4 (continued)

Enzyme (EC number)	Source (location)	Mol mass [g/mol]	Isozymes; pI
Aldehyde dehydrogenase (1.2.1.3)	cow (liver) horse (liver)	220 000 245 000	1: 5.4 2; 5.05, 4.80
Aldolase A C (4.1.2.13)	human (erythrocytes) rat (brain)	158 000 148 000	1; 8.9 1; 4.28
Alkaline phosphatase I variant (3.1.3.1)	human (liver) human (placenta)	136 000 120 000	1; 4.5 6; 3.4, 4.3, 4.6, 5.4
5'-AMP aminohydrolase (3.5.4.6)	human (erythrocytes)	285 000	1; 5.5
Aspartate-semialdehyde dehydrogenase (1.2.1.11)	yeast	156 000	1; 6.17
Catalase (1.11.1.6)	*Neurospora crassa* human	320 000 263 000	1; 5.0 1; 6.7
DNase V (3.1.21.1)	calf (thymus)	53 000	1; 10.3
DDT dehydrochlorinase (4.5.1.1)	house fly	120 000	3; 6.3, 6.9, 7.4
Diacetyl reductase (1.1.1.5)	*Aerobacter aerogens*	100 000	4; 5.75, 5.9, 6.55, 6.8
Fructose-bis-phosphatase (3.1.3.11)	mouse (liver)	143 000	1; 6.1
L-Fucose dehydrogenase (1.1.1.122)	sheep (liver)	123 000	1; 5.8
α-L-Fucosidase (3.2.1.51)	human (liver)	200 000	6; 5.2, 5.4, 5.6, 5.9, 6.2, 6.4
Fumarase (4.2.1.2)	pig (heart)	194 000	11; 5.9 – 8.2
Galactonate dehydratase (4.2.1.6)	*Pseudomonas*	240 000	1; 4.5
α-D-Galactosidase (3.2.1.22)	*E. coli* (strain K 12)	329 000	1; 5.1
β-D-Glucoronidase (3.2.1.31)	mouse (urine)	280 000	4; 5.58, 5.78, 5.95, 6.02
Glutaminase (I, II, III) (3.5.1.2)	Pseudomonas	146 000	3; 7.8 (III), 8.05 (II), 8.35 (I)

Table 1.4 (continued)

Enzyme (EC number)	Source (location)	Mol mass [g/mol]	Isozymes; pI
Glutathione peroxidase (1.11.1.9)	human (placenta)		1; 4.8
Glyceraldehyde 3-phosphate dehydrogenase (1.2.1.9)	fish (muscle)	160 000	3; 7.9, 8.25, 8.42
Hydroxymethylglutaryl-CoA reductase (1.1.1.88)	rat (liver, microsomes)	200 000	1; 6.2
4-Hydroxyphenylpyruvate dioxygenase (1.13.11.27)	*Pseudomonas;* sp. P. J. 874	150 000	1; 4.8
3 (or 17) β-Hydroxysteroid dehydrogenase (1.1.1.51)	*Pseudomonas testosteroni*	98 500	6; unknown
Lactate dehydrogenase (1.1.1.27)	*Ambystoma mexicanum*	140 000	9; 5.24 (LDH1), 5.58 (LDH2), 5.62, 5.74 (LDH3), 5.80, 6.07 (LDH4), 6.14, 6.52 (LDH5), 6.60
Neuraminidase (3.2.1.18)	influenza virus; A_2/1957	200 000	6; 5.2, 5.35, 5.5, 5.8, 6.2, 6.5
Nucleosidediphosphatase (3.6.1.6)	rat (liver, cytosol)	120 000	2; 4.7, 5.0
Phosphoenolpyruvate carboxylase (4.1.1.31)	*E. coli*	402 000	1; 4.92
6-Phosphofructokinase (2.7.1.11)	*Lactobacillus acidophilus*	154 000	4.9 – 5.1
Phosphoglycerate dehydrogenase (1.1.1.95)	chicken (liver)	165 000	1; 8.95
Phosphoglycolate phosphatase (3.1.3.18)	tobacco (leaves)	86 300	1; 3.85
Pyrophosphatase (inorganic) (3.6.1.1)	*Thiobacillus*	88 000	1; 5.05
Pyruvate kinase (2.7.1.40)	yeast	220 000	1; 6.6
	chicken (skeletal muscle)	212 000	2; 8.45, 8.77
type A	pig (kidney)	249 000	1; 5.6
type L	human (liver)	240 000	2; 5.85, 6.28

Table 1.4 (continued)

Enzyme (EC number)	Source (location)	Mol mass [g/mol]	Isozymes; pI
Pyruvate oxidase (1.2.3.3)	*E. coli*	240 000	1; 5.6
Tyrosinase (1.10.3.1)	*Porcellia laevis* (cuticle)	122 000	2; 6.1, 7.1
Urease (3.5.1.5)	Jack bean	485 000	1; 4.88

Table 1.5. Enzymes composed of more than four subunits (data taken from a compilation of [7] and [8])

Enzyme (EC number)	Source (location)	Mol mass [g/mol]	Number of subunits (M_r)	Isozymes; pI
Aminolevulinate dehydratase (4.2.1.24)	human (erythrocytes)	252 000	8 (31 000)	1; 4.9
Bilirubinglucuronoside glucuronosyl transferase (2.4.1.95)	rat (liver)	160 000	6 (28 000)	1; 7.9
Creatinine amidohydrolase (creatininase) (3.5.2.10)	*Pseudomonas putida* (strain C-83)	175 000	8 (22 000)	1; 4.7
Glutamine synthetase (6.3.1.12)	*Azotobacter vinelandii*	640 000	12 (53 000)	1; 4.6
Isocitrate dehydrogenase (1.1.1.41)	baker's yeast	375 000	10 (40 000)	1; 5.5
L-Lactate dehydrogenase (1.1.1.27)	*E. coli* (membranes)	480 000	12 (43 000)	1; 8.3

Table 1.6. Enzymes with differently sized subunits (data taken from a compilation of [7] and [8])

Enzyme (EC number)	Source (location)	Mol mass [g/mol]	Subunit No; M_r	Isozymes; pI
Acetyl-CoA choline O-acetyltransferase (2.3.1.6)	squid (head) ganglia	93 000	1; 37 000 1; 56 000	6; 5.0 – 6.2
Alkaline phosphatase (3.1.3.1)	human (Nasopharyngeal tumor, KB cells)	136 000	1; 64 000 1; 72 000	1; 4.3
α-Amylases (1A, 1B, 2A, 2B) (3.2.1.1)	human (subman-dibular salvia)	220 000	2; 57 000 (1A, 1B) 2; 54 000 (2A, 2B)	4; 5.9 (1A, 2A) 6.4 (1B, 2B)
	human (salvia)	125 000	1; 61 000 (A) 1; 64 000 (B)	2; 5.9, 6.4 (A) 2; 5.9 6.4 (B)
Anthranilate synthase (4.1.3.27)	*Serratia*	150 000	2; 21 000 2; 60 000	1; 4.6
γ-Butyrobetaine 2-oxoglutarate dioxygenase (1.14.11.1)	*Pseudomonas* Sp. AK 1	90 000	1; 39 000 1; 37 000	1; 5.1
Cathepsin B forms I, II, III (3.4.22.1)	pig (liver)	29 000 (I, II) 29 000 (III)	1; 25 000 (I) 1; 4 000 5.8 (III)	3; 5.2 5.4 (II)
Cystathionine β-synthase (4.2.1.22)	human (skin fibroblasts)	123 000	1; 53 000 1; 70 000	1; 5.7
DNA polymerase (2.7.7.7)	calf (thymus, coto)	160 000	1; 90 000 1; 60 000	3; 5.3, 5,8, 6.3
DNA polymerase-a	human (KB cells)	140 000	1; 76 000 1; 66 000	1; 5.1
β-D-Glucosidase (3.2.1.21)	*Cicer arietinum*	110 000	1; 63 000 1; 43 000	3; 9.0, 9.3 10.0
γ-Glutamyltransferase (2.3.2.2)	beef (colostrum)	80 000	1; 25 000 1; 55 000	1; 3.85
	rat (kidney)	68 000	1; 46 000 1; 22 000	12; 6.12 – 9.20
Hydrogenase (1.18.3.1)	*Desulfovibrio vulgaris*	89 000	1; 59 000 1; 28 000	2; 5.8, 6.2
Nitrogenase (1.18.2.1)	*Klebsiella pneumonia*	218 000	2; 50 000 2; 60 000	1; un-known

Table 1.6 (continued)

Enzyme (EC number)	Source (location)	Mol mass [g/mol]	Subunit No; M_r	Isozymes; pI
L-Prolylpeptide hydrolase (3.4.11.5)	chicken (embryo)	248 000	2; 64 000 (α) 2; 60 000 (β)	2; un- known
UMP: pyrophosphate phosphoribosyltransferase (2.4.2.9)	yeast	80 000	1; 58 000 1; 25 000	2; 5.27, 5.35
Urokinase (3.4.21.31)	human (urine)	47 000	1; 33 000 1; 18 000	2; 8.6, 8.9

1.3 Isozyme Classification

About 50 % of all enzymes investigated so far exist in multiple molecular forms, isozymes. These usually differ in electrophoretic mobility. Besides, they may have slightly different catalytic abilities. Differences in electrophoretic mobility may result from charge and (or) size variabilities. Enzyme multiplicity can depend on genetic factors (a) directly or (b) indirectly. The first group of isozymes may be further subdivided into two classes [9 – 12]: (1) isoenzymes and (2) allozymes. Enzymes of class (1) are also known as primary isozymes, those of class (2) as alloenzymes while those which evolve by post-translational modifications (3) are named secondary isozymes. The classification given is of practical use, and it is recommended to use the term isozyme in a more general sense since very few multiple molecular forms can definitely be separated into one or the other type of class [9 – 11].

(margin note: Isoenzymes, allozymes, isozymes)

1.3.1 Multiple Loci Determining Isoenzymes

Isoenzymes may be distributed between different cell compartments, e.g. the cytoplasm, mitochondria, chloroplasts or some other cellular component (Table 1.7). Such enzymes are encoded in at least two different genes (Table 1.7). If they are multimeric their subunits will not combine to hybrid forms [11]. The gene loci of these enzymes are often located on different chromosomes, as is the case with the human enzymes malate dehydrogenase (where the soluble form is coded by a gene on chromosome 2, whereas the mitochondrial form is encoded in chromosome 7), isocitrate dehydrogenase (soluble form: chromosome 2; mitochondrial form: chromosome 15) [13, 14], and superoxide dismutase (soluble form: chromosome 2; mitochondrial form: chromosome 6) [15, 16].

(margin note: Enzyme loci, enzyme locations)

Multiple loci coding for enzymes of identical substrate specificity are usually attributed to the occurrence of gene duplications which occurred in the course of evolution. As a result of point mutation the duplicated genes subsequently diverged in amino acid composition leading to different enzyme forms separable by electrophoresis.

Table 1.7. Number of isozyme loci, subunit number and subcellular localization of enzymes commonly assayed in animals and plants

Enzyme	Loci	Subunits	Localization[b]	Origin[c]	
AAT (GOT)[a]	2 (4)	2	c, mt	a	
Aspartate aminotransferase	2 (4)	2[b]	c, p, mb, mt[c]		p
ACO	2	1	c, mt	a	
Aconitase	1 – 3	1	c, mt		p
ACP	many	1 (2)		a	
Acid phosphatase	2 – 4	1 (2)	varies		p
ADA	1	1		a	
Adenosine deaminase					p
ADH		2	c	a	
Alcohol dehydrogenase	1 – 3	2	c		p
AK	3	1	c, mt	a	
Adenylate kinase	1 – 2	1	c, p		p
AKP	many	1		a	
Alkaline phosphatase					p
ALD		4		a	
Aldolase	2	4	c, p		p
CAR	2	1		a	
Carbonate dehydratase					p
CAT				a	
Catalase	1	4	mb		p
CK	2	1		a	
Creatine kinase					p
DIA		1		a	
Diaphorase	1 – 4	1, 2, 4	c, p, mt		p
EST	many	1 (2)		a	
Esterase	2 – 10	1, 2	c		p
FUM		2		a	
Fumerase	1	4	mt		p
PEP	many	1 (2)		a	
Peptidase					p
PGD		2	c	a	
6-Phosphogluconate dehydrogenase	2	2	c, p		p
PGM	3	1	c	a	
Phosphoglucomutase	2	1	c, p		p

Table 1.7 (continued)

Enzyme	Loci	Subunits	Localization[b]	Origin[c]	
PK	2	1		a	
Pyruvate kinase					p
SKD				a	
Shikimate dehydrogenase	2	1	c, p		p
SOD	2	2, 4	c, mt	a	
Superoxide dismutase	1 – 2	2, 4	c, p, mt		p
TPI				a	
Triosephosphate isomerase	2	2	c, p		p

[a] Letters decipt enzyme abbreviations.

[b] Localization: cytosol (c), plastid (p), mitochondria (mt) and microbody (mb).

[c] Animals (mostly mammalia) (a), plants (p).

[d] Cytosolic GAPD uses NAD, plastidic GAPD uses NADP, IDH is mostly cytosolic and NAD specific, MDH and ME are mostly NAD and NADP active in plants.

1.3.2 Multiple Allelism

Multiple loci are in general common to all members of a species while multiple alle-lism results in differences between individual members of a species with respect to the patterns of a certain isozyme system. At any given locus a number of different alleles may occur in a population of individuals. If each allele codes for a structurally distinct enzyme version these will differ from one individual to another. Such enzyme forms are called allozymes or alloenzymes. The number of electrophoretically de-tectable allozymes depends on (a) the number of differently charged polypeptides evolved and (b) the number of subunits which form a catalytically active enzyme unit. Individuals having homogeneous alleles with respect to a certain enzyme will pro-duce a single species of enzyme, but this enzyme will differ from one individual to another, according to the particular alleles they happen to carry at the locus in que-stion [11]. The inheritance of allozymes has been used to study the subunit structure of enzymes. If each of two homozygous individuals synthesize an enzyme form of

Number of isozymes resulting from subunit numbers different negative charge the number of isozymes in heterozygous individuals depend on the subunit composition of the enzyme. With monomeric enzymes two enzyme forms will occur in heterozygotes (A and B), with dimeric enzymes there are three forms (AA, AB, BB), with trimeric enzymes there are four forms (AAA, AAB, ABB, BBB) and with tetrameric enzymes there are five different forms (A_4, A_3B, A_2B_2, AB_3, B_4).

Every gene can mutate and so initiate the formation of a slightly modified protein. Therefore it can be assumed that every enzyme has sometime existed as an isozyme. The stable isozymes presently represent those forms that have proved profitable to the progress of evolution and are therefore established in the genes of a given species.

1.3.3 Secondary Isozymes

Secondary isozymes are generated by posttranslational modifications of a given protein structure [12]. Nine different mechanisms have been described leading to the formation of secondary isozymes: (1) aggregation and polymerization, (2) oxidation or reduction, (3) limited proteolysis, (4) differences in the carbohydrate contents, (5) deamidation, (6) aggregation of substrates or cosubstrates, (7) temperature effects, (8) pH effects, and (9) conformational isomerism [17]. An example to the formation of secondary isozymes is given by the enzyme catalase. In mouse kidney five major enzyme forms of catalase have been found which are either localized in the soluble fraction, red cells, or peroxisomes. They all are encoded in a single genetic locus and their multiplicity is caused by the progressive attachment of negatively charged sialic acid residues to each of its four subunits [18]. Form 1 is charge x, form 2 is x + (one negative charge (−1)), form 3 is x + (−2), form 4 is x + (−3), and form 5 is x + (−4) [19].

Catalase

1.3.4 Isozyme Numbering

It is recommended that iso(en)zymes are named according to the extent of migration in an electrophoretic support medium rather than on the basis of tissue distribution (e.g. brain type, muscle type, etc.) since this distribution can vary between different

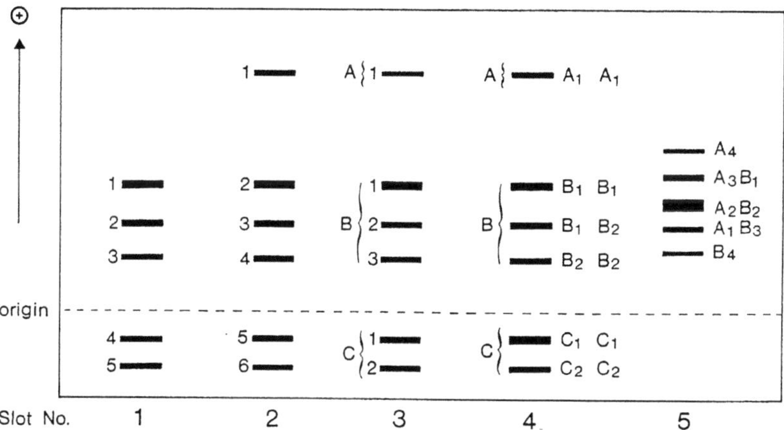

Fig. 1.1. Diagram illustrating the possible evolution of an isozyme nomenclatural system when using starch gel electrophoresis. *Slot 1* shows the numbers assigned to the bands observed in the first study. *Slot 2* shows a possible renumbering later when a faster band is observed. *Slot 3* shows a possible relabeling system based on a regional approach. *Slot 4* shows a regional labeling when knowing the genotype of visualized allozymes. *Different suffix* numbers indicate different alleles coding for different primary structures. In the example given the enzyme has a dimeric structure. In *slot 5* the labeling of the isoenzyme patterns of mammalian lactate dehydrogenase as occurring in human blood is shown. In this case the *suffix numbers* indicate the number of subunits with A_4 and B_4 being the homomeric forms of tetrameric LDH and A_3B_1, A_2B_2 and A_1B_3 being the heteromeric forms of LDH

species or in a single species or tissue with the stage of development. Iso(en)zymes are numbered starting with the species of highest mobility towards the anode or by grouping them into several classes with a number of sub-types as is illustrated in Fig. 1.1. Alternatively, a regional approach may be used, dividing the patterns into regions, with letters designating the regions, and numbers designating the bands within a region. Both systems are flexible enough to adapt labelling to future requirements [20]. Once the genetics of an isozyme or allozyme system have been established the phenotypic designations should be permanent so that a more explanative labeling system can be used.

1.4 References

1. Karlson P, Bielka H, Horecker BL, Jacoby WB, Keil B, Liébecq C, Lindberg B, Webb EC (1979) Enzyme Nomenclature, Recommendations of the Nomenclature Committee of the International Union of Biochemistry, Academic Press, New York, San Francisco, London
2. Boudet AM, Lécussan R (1974) Planta 119: 71–79
3. Berlyn MB, Giles NH (1969) J Bacteriol 99: 222–230
4. Lumsden J, Coggins JR (1977) Biochem J 161: 599–607
5. Price C, Stevens L (1982) Fundamentals of Enzymology. Oxford University Press, Oxford New York
6. Markert CL (1963) Science 140: 1329–1330
7. Righetti PG, Tudor G (1981) J Chromatogr 220: 115–194
8. Righetti PG, Caravaggio T (1976) J Chromatogr 127: 1–28
9. Markert CL (1975) Biology of isozymes. In: Markert CL (ed) Isozymes, vol. I (Molecular Structure). Academic Press, New York San Francisco London, pp 1–9
10. Markert CL (1977) Isozymes: the development of a concept and its application. In: Rattazzi MC, Scandalios JG, Whitt GS (eds) Isozymes: Current tropics in biological and medical research, vol. I. Alan R Liss, New York, pp 1–16
11. Harris H, Hopkinson DA (1976) Handbook of enzyme electrophoresis in human genetics. North-Holland Publ Comp, Amsterdam Oxford; Amer. Elsévier Pub Comp Inc, New York
12. Shaw CR (1969) Int Rev Cytol 25: 297–332
13. Shows TB (1972) Biochem Genet 7: 193–204
14. Turner BM, Fisher RA, Harris H (1974) Ann Hum Genet 37: 455–467
15. Van Someren H, Van Henegouwen HB, Los-Würzer-Figurelli E, Doppert B, Veruloet M, Meera Khan P (1974) Humangenetik 25: 189–201
16. Creagan R, Tischfield J, Ricciuti F, Ruddle FH (1973) Humangenetik 20: 203–209
17. Rothe GM (1980) Hum Genet 56: 129–155
18. Jones GL, Masters CL (1972) FEBS Lett. 21: 207–210
19. Holmes RS, Duley JA (1975) Biochemical and genetic studies of peroxisomal multiple enzyme systems: α-hydroxyacid oxidase and catalase. In: Markert CL (ed) Isozymes vol. I (Molecular Structure). Academic Press, New York San Francisco London, pp 191–121
20. Brewer GJ, Sing CF (1970) An introduction to isozyme techniques. Academic Press, New York San Francisco London
21. Ferguson A (1980) Biochemical systematics and evolution. Blackie & Son Ltd, Glasgow
22. Pasteur N, Pasteur G, Bonhomme F, Catalan J, Britton-Davidian J (1988) Practical isozyme genetics. Ellis Horwood Ltd Publishers, Chichester
23. Kephart SR (1990) Amer J Bot 77: 693–712

2 Extraction of Enzymes from Tissues, Cells and Cell-Organelles

2.1 Methods to Extract Enzymes from Microorganisms

Microorganisms, such as bacteria, algae, moulds and others, are ruptured by sonication, by passage through a French press [1, 2] (Fig. 2.1) or a Manton-Gaulin homogenizer [3], by blending with glass beads [4], or by digesting the cell walls enzymically [5]. Extract preparation is preferably performed in the cold ($+4\,°C$).

In any case, 100 – 500 mg of microorganism-rich material may be suspended in a small volume (1 – 2 ml) of buffer and then ruptured. As extraction buffer 100 mmol l^{-1} phosphate, pH 7.0, containing a reducing agent (0.1 – 1 mmol l^{-1} 2-mercaptoethanol, or 0.05 – 0.1 mmol l^{-1} dithiothreitol (or dithioerythritol) or ascorbic acid) and in some cases one or several proteinase inhibitors (see Table 2.1) may be used. But the separation buffer system used in electrophoresis may also be taken to homogenize the cells [10]. Cell rupture is easily monitored by microscopic examination. If the ruptured cell suspension contains a gelatinous aggregate of nucleic acids, it is advisable to sonicate it before centrifugation. By definition, the crude extract is the clear though mostly opalescent liquid resulting after centrifugation of ruptured cells. In case a fatty overlayer is formed on the fluid obtained after centrifugation, a syringe with a long needle may be used to withdraw carefully the crude extract enclosed between the sediment and the overlayer. After preparing the crude extract, its contamination by nucleic acids can be estimated by measuring its absorbances at 260 and 280 nm [11]. Ratios of these absorbances close to one indicate a significant contamination with nucleic acids. These may be removed by precipitation with protamine sulphate (0.2 – 0.4 wt/vol%, final concentration), streptomycin (1 – 2 wt/vol%, final concentration), $MnCl_2$ (50 mmol l^{-1}), lysozyme (12 mg/ml) [12] or 6,9-diamino-2-ethoxyacridine (Ethodin or Rivanol) [13]. An increased ionic strength (0.2 M $(NH_4)_2SO_4$) decreases the strength of the nucleic acid-protein interaction and may improve the efficiency of the separation [6]. If this step is necessary the remaining ammonium sulphate must be removed before the extract can be used for electrophoresis. Since none of the precipitating agents is universally applicable it may be necessary to try several approaches before a largely nucleic acid free enzyme extract is obtainable.

Yeast cells are somewhat difficult to rupture. Sonication and other less vigorous methods cannot be applied. There are three different methods available to set enzymes free from yeast cells: (a) pressure homogenization, (b) autolysis with toluene

Extraction
medium for
microorganisms

Removal
of nucleic acids
from crude
extracts

Rupture
of yeast cells

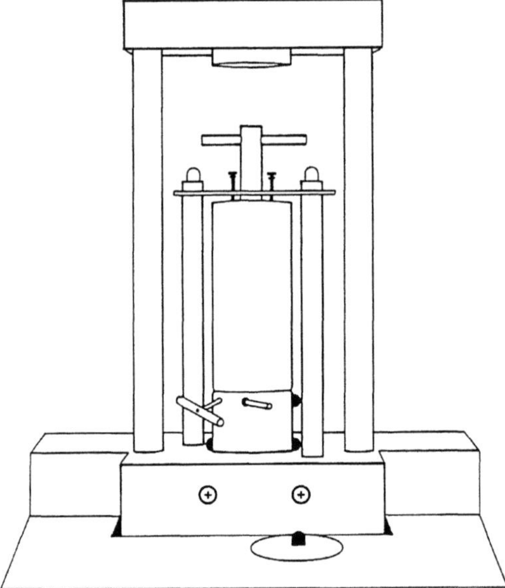

Fig. 2.1. French pressure cell for the disintegration of blood cells, unicellular organisms, animal tissue and other biological material. Pressure cells may be used to disrupt the cellular walls of plants while leaving the cell nucleus undisturbed. The high pressure capacity reaches a maximum of 40 000 psi disintegrating even the most durable cell material, such as spores, cellulose, etc. Pressure cells are available for 40 ml (20 000 psi), 35 ml (40 000 psi) and 1.4 – 3.7 ml 20 000 psi)

Table 2.1. Methods to rupture yeast cells [6]

Pressure homogenization

Yeast cells can be homogenized effectively by pressure (French press, Fig. 2.2; Manton-Gaulin homogenizer). Releasing protein from yeast cells is more rapid at 30 °C than at 5 °C and at higher pressures [6, 7]. One part of yeast cake may be suspended in two parts (w/v) of buffer and passed through a homogenizer [8]. As extraction buffer a 5 mmol l^{-1} Tris–HCL buffer, containing 10 mmol l^{-1} MgCl$_2$, 1 mmol l^{-1} dithiothreitol, pH 8.1 and a proteinase inhibitor (Table 2.1) may be used [8]. At 30 °C and 550 kg/cm^2, 62 % of the protein is released in one pass, 75 % in two and 95 % in four passes [3]. About 100 mg protein is obtained from 1 g of completely ruptured yeast cells

Blending frozen cells

One part of crumbled yeast cells are frozen in one and a half parts (w/v) of liquid nitrogen. Then liquid N$_2$ and frozen yeast are poured into a stainless steel Waring blender and homogenized for 4 min at 1-min intervals. After each minute, the frozen yeast powder is scraped off the inner surface of the container. The fine frozen powder is then suspended in 20 mmol l^{-1} sodium phosphate, pH 7.5, containing 2-mercaptoethanol and a proteinase inhibitor (Table 2.1), allowed to thaw, and stirred for 1 h [9]. Afterwards it is centrifuged to remove cell wall debris.

Table 2.2. Selected proteinase inhibitors (1)

Inhibitor	Toxicity	Properties
Diazoacetyl-norleucine methyl ester plus Cu^{2+}		Inactivates acid (carboxyl) proteinases and other proteins, particularly sulfhydryl enzymes [15]. [a] Efc: 1 mM reagent and 1 mM Cu^{2+}
Diisopropyl-fluoro-phosphate (DFP)	Very toxic[b]	Inhibits serine proteinases. Store 100 mmol l^{-1} stock solution in dry isopropanol at -20 °C in small aliquots. Dilute ten-fold before use. Efc: 1 mmol l^{-1}
Dimethyl-dichlorovinyl phosphate (DDVP) (syn. Dichlorvos, Vapona)	Non toxic [16] but may be carcinogenic [18]	Inhibits serine proteinases. Relatively stable, [a] HL at pH 7.6, 37 °C: 32 h [8]. Efc: 1 mmol l^{-1}
Ethylenediaminetetra-acetic acid (EDTA)	Non toxic	Inhibits metalloproteinases and other metalloenzymes. Efc: 1–10 mmol l^{-1}
p-Hydroxy-mercuri-benzoate (PMB)		Inactivates thiol proteinases with pH optimum at 4–7 and other sulfhydryl enzymes. Efc: 1 mmol l^{-1}
Pepstatin		Inactivates acid (carboxyl) proteinases with pH optimum 2–5. Reversible weak inhibitor above pH 6.0 [19]. Efc: 0.1–1 μmol l^{-1} (10 μg/ml)
o-Phenanthroline		Inactivates metalloproteinases and other metalloenzymes. Efc: 1 mmol l^{-1}
Phenylmethylsulphonyl fluoride (PMSF)	Relatively non toxic	Inhibits serine proteinases. Stable for months in isopropanol. HL at pH 7, 25 °C: 110 min [20]. Reacts with other proteins [21]. Use with caution. Efc: 1 mmol l^{-1}
Trasylol	Non toxic	Stable in neutral to acid media (mol mass 11 600), identical to the pancreatic trypsin inhibitor [22]. Use 24–50 Kallikrein units/ml [6].

[a] Many other synthetic [23, 24], microbial [25] or plant [26] proteinase inhibitors have been described; HL: half live; Efc: effective concentration.

[b] All operations with pure DFP and solutions exceeding 1 mmol l^{-1} concentrations are to be done in a hood with good air flow. It is recommended to wear polyvinyl gloves and not to contaminate clothing. Immediate access to atropine is strongly recommended as a precaution against accidental exposure to DFP. Aqueous DFP solutions < 1 mmol l^{-1} may be used outside of hood, but contact with skin has to be prevented. For complete hydrolysis of DFP, contaminated glassware is placed in 500 mmol l^{-1} NaOH for at least 24 h [6].

and 2-mercaptoethanol and (c) blending deep-frozen cells. Each of these methods has its advantages and disadvantages. Method (a) is rapid and effective but requires a special equipment (French press). Autolysis is not recommended as an extraction procedure. Workinq with liquid nitrogen needs several facilities which are not available in all laboratories, but it is rapid and effective [9] (see Table 2.1).

Proteinase inhibitors

The main problem when using yeast cells results from their high proteinase activities [14]. To stop these activities one or several proteinase inhibitors (see Table 2.2) are added to the yeast suspension before pressure homogenization. As an alternative the proteinase inhibitors are mixed with the buffer which is added after freezing the cells in liquid nitrogen. In any case the time in which the crude extract is in contact with the proteinases should be as short as possible. Upon electrophoresis, proteinases and other enzymes are expected to be separated from each other. But comigration is also possible. Several methods can be used to test whether an enzyme

Modification of enzymes by proteinases

was modified by proteinases prior to electrophoresis: (a) an antibody prepared against the pure enzyme exhibits a single fused precipitin line in the Ouchterlony double diffusion test against both the homogeneous enzyme and the enzyme in the cell free extract; (b) immunoelectrophoresis of the cell free extract produces one precipitin arc with the same mobility as the pure enzyme; (c) immunoprecipitates of both the pure enzyme and the enzyme in the crude extract yield protein bands with the same electrophoretic mobility in sodium dodecyl sulphate electrophoresis; and (d) the pure enzyme and the enzyme in the cell free extract have the same isoelectric point [6].

2.2 Methods to Extract Enzymes from Animal Soft Tissue

Tissue samples should be obtained as fresh as possible.

If crude extracts cannot be prepared at once the tissues must be stored in ice if the delay is short, but they should be frozen, preferably at liquid nitrogen temperatures (–196 °C) if the period between obtaining the sample and carrying out the assay is long (> 6 h). However, it will be necessary to investigate the effect of freezing on enzyme activity. The tissue samples should be washed free of blood to avoid contamination by blood enzymes. Blotting of tissues on filter paper helps to remove adherent blood. The vascular system of organs should be freed from blood with 1.8 % saline before the organ is disrupted. It is better to store the tissue as a block rather than after homogenization.

At present the most widely used instrument for the homogenization of animal soft tissue is the ground-glass homogenizer fitted with a tight-fitting ground-glass (or Teflon) pestle (Potter-Eveljhem homogenizer) (Fig. 2.2). The tight-fitting pestle (free space approximately 0.2 mm) is rapidly rotated (around 1000 revolutions per min) by a stirrer motor. About 1 – 2 g of fresh tissue is cut into small pieces with scissors and suspended in 8 – 20 ml of medium (water) to give a one in ten dilution. Homogenization is achieved by pushing the ground glass tube up and down the rotating pestle for a few minutes. The ground-glass tube is cooled to prevent overheating of the sample. The extent of the homogenization depends on how tightly the

Preparation of tissues

pestle fits into the glass tube.

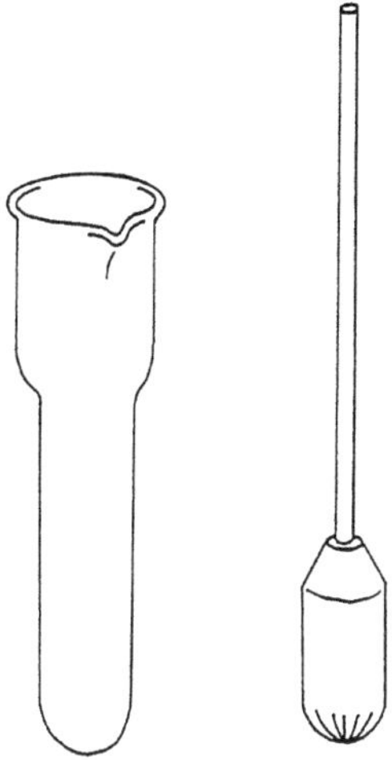

Fig. 2.2. Ground-glass tube and pestle of a Potter-Eveljhem homogenizer

Even tough tissues (e.g. mammary gland and some skeletal muscles) can be completely homogenized when using small amounts of tissue (100 mg). The glass homogenizer can be replaced by the Polytron homogenizer (Northern Medie Supply Ltd, Crosslands Lane, Newport Proad, North Cave, Brough HU15 2PG, England). The tissue is cut into small pieces and is then homogenized using the Polytron set at position 3/4 for 10 s at 0 °C. The Polytron is particularly useful for the homogenization of muscle from vertebrates and non-insect invertebrates. Alternatively, the muscle can be partially homogenized using a Silverson homogenizer (Silverson Machines Ltd, London), an Omni-Mixer (Ivan Sorvall, Newton, Conn. 06470, USA) or an Ultra-Turrax homogenizer (Janke & Kunkel, IKA-Werk, Staufen, Germany) (Fig. 2.3) followed by complete homogenization in a ground-glass homogenizer. Other methods of tissue disintegration are listed in Table 2.3.

Muscle homogenization

Homogenizers

Even after thorough homogenization in a ground-glass homogenizer, some enzymes may not be released from tissues or cells. Mitochondria, for example, may not have been completely disrupted, or vesicles may have been formed from intracellular membranes with enzymes trapped within them. This problem can be overcome in several ways. The homogenate may be subjected to ultrasonic vibrations (sonications) usually for periods of up to 1 min. Sonication is a rapid technique and, provided that the homogenate is cooled during the process (by keeping it on ice), an enzyme may not be adversely affected. On the other hand, sonication may not be completely effective for some tissues such as mammary gland and brain tissue. Non-ionic detergents can be used to solve this problem: deoxycholate or Triton X-100 is mixed with the homogenate; e.g. 10 μl of a 10 wt/vol.% solution of deoxycholate is added to each ml of homogenate [28]. A concentration of 0.1 wt/vol.% Triton X-100 increases the activity of 5-nucleotidase and adenosine kinase from many tissues. However, detergents can inactivate enzymes and therefore their effect on the activity and electrophoretic mobility of enzymes must be carefully checked.

Formation and solubilization of vesicles

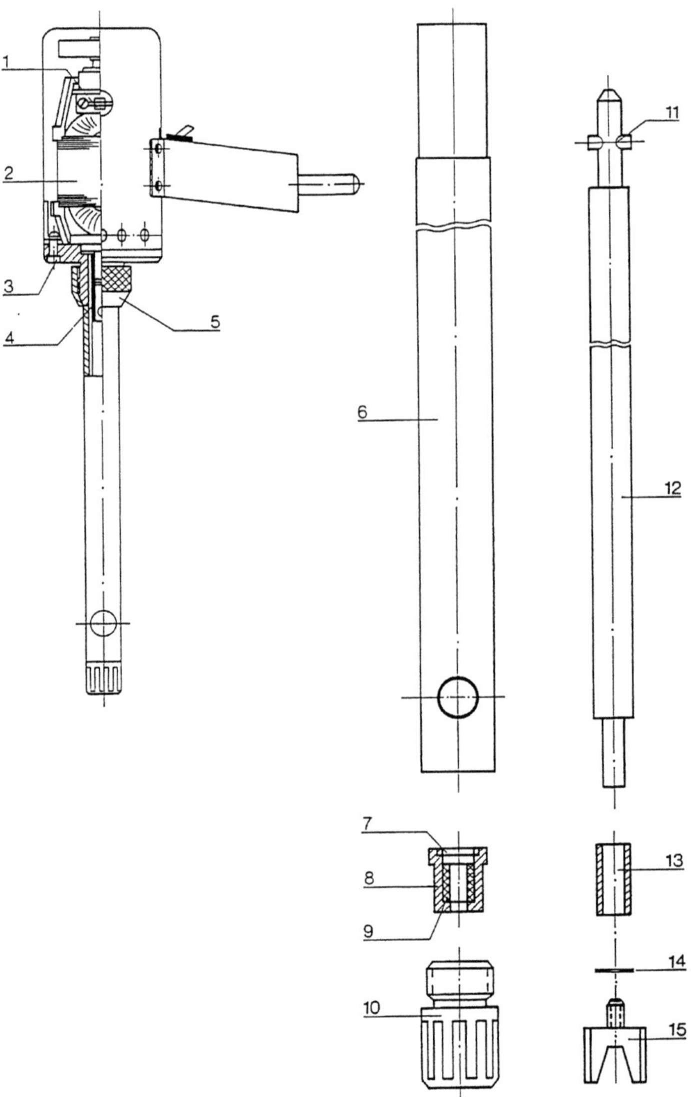

Fig. 2.3. Detailed view of an Ultraturrax homogenizer; *1:* carbon brush, *2:* motor, *3:* screw, *4:* clutch, *5:* nut, *6:* shank tube, *7:* run disc, *8:* bearing bush, *9:* slotted ring bearing, *10:* stator, *11:* clutch pin, *12:* axle, *13:* running tube, *14:* disc, *15:* rotor (by curtesy of Janke & Kunkel, Staufen, Germany)

Table 2.3. Methods to homogenate or disintegrate cells of animal tissues

Method	Tissue
Mechanical homogenisation Pestle homogenizer	Parenchymatous tissue, brain, heart
Grinding with sand or glass beads Glass bed homogenizer	Muscle, frozen tissue Ascites tumor cells, muscle
Blade homogenizer (blender)	Muscle
Sonic homogenization	Universal application
Thermal disintegration freezing and thawing [a]	Universal application
Chemical disintegration digitonin, Triton X-100	Mitochondria, muscle
Biological-enzymatic disintegration autolysis, maceration; bacterial proteases	Muscle, heart

[a] Freezing mixtures: acetone-CO_2; alcohol-CO_2; liquid air; liquid nitrogen. Data taken from [27].

Table 2.4. Buffer salts used in protein extraction

Buffer	pK-values
Sodium acetate	4.75
Sodium bicarbonate	6.50; 10.25
Sodium citrate	3.09; 4.75; 5.4 [a]
Tris-chloride [b]	8.21
Sodium phosphate	1.5; 7.5; 12.0
Tris phosphate [b]	7.5; 8.21

The buffer concentration is usually $20-100$ mmol l^{-1}.
[a] Sodium citrate binds Ca^{2+} [29].
[b] Tris: Tris(hydroxymethyl)aminomethane.

When a tissue is disrupted, acids may be set free. Consequently, it is necessary to buffer the homogenizing medium to achieve a pH at which the respective enzyme is most stable. This pH-value can be found by trial and error. The buffering capacity of a buffer is best at the pK-value of the buffer salt. Therefore, the buffer pH should be as close as possible to the pKa-value, and not more than one pH unit from the pKa-value (Table 2.4). Many enzymes are most stable in the region of neutral pH, but some retain activity at a pH different from neutrality (e.g. phosphofructokinase is best extracted at pH 8.2) [28].

pH of extraction buffer

Tissue homogenization causes a dilution of enzymes which can inactivate them. This problem can be reduced by the addition of bovine serum albumin ($0.1-1$ wt/vol.%) or up to 5 mmol l^{-1} glycerol to the homogenization medium [28]. Addition of substrate, cofactors or allosteric activators can stabilize some enzymes during homogenization (e.g., dehydrogenases) [28].

Dilution of enzymes

Heavy metal ions (Fe, Zn, Cu, Pb, Hg) inhibit many enzymes. They can be introduced into the homogenate with the tissue, glass-ware, distilled water or as contami-

nants with reagents. Usually, 1 – 2 mmol l⁻¹ of the chelating agent ethylenediaminote-
traacetic acid (EDTA) are used to complex ions to hinder their reaction with proteins.
Sometimes it is necessary to lower the concentration of Ca^{2+} to concentrations below
10^{-7} mol l⁻¹ since higher concentrations of Ca^{2+} (10^{-7} – 10^{-5} mol l⁻¹) can modulate the
activity of some enzymes (e.g. mitochondrial glycerol phosphate dehydrogenase)
[28]. The calcium concentration can be reduced by addition of a chelating agent which

Effect of Ca^{2+}-ions on enzymes has a more specific affinity towards Ca^{2+} such as ethanedioxybis (ethylamine) tetra-
acetic acid (EGTA). The use of EGTA is recommended if other metal ions (e.g. Mg^{2+})
must not be complexed. The concentration of EGTA in the homogenization medium
is usually about 1 mmol.

Protection of thiol groups Oxidation of thiol groups of enzymes can cause inactivation; –SH groups are oxidized to –S–S–groups. Oxidation of –SH groups can be prevented by inclusion of a
thiol reagent into the homogenizing medium. These reagents operate according to the
following reaction:

$$\text{Enzyme}\begin{matrix}S\\|\\S\end{matrix} + 2R-SH = \text{Enzyme}\begin{matrix}SH\\\\SH\end{matrix} + R-S-S-R.$$

Mercaptoethanol, Dithiothreitol Mercaptoethanol and Cleland's reagent (dithiothreitol) are mostly used as thiol
reagents. 2-Mercaptoethanol is cheap but has an unpleasant odour; it is slowly oxidiz-
ed in solution and must be used in relatively high concentrations (10 – 30 mmol l⁻¹)
which may interfere with subsequent assays. Dithiothreitol is almost odourless, and
can be used at low concentrations (0.1 – 2 mmol l⁻¹).

Composition of an extraction medium for animal tissues The following medium has been suggested for the extraction of enzymes from
animal tissues such as muscle, brain, liver and mammary gland [28]: 50 mmol l⁻¹
triethanolamine-HCl, 2 mmol l⁻¹ $MgCl_2$, 1 mmol l⁻¹ EDTA, 2 mmol l⁻¹ dithiothreitol,
adjusted with KOH to pH 7.5.

A number of enzymes exist in enzymatically interconvertible forms, one of which
is much less active than the other. In such cases a way must be found to obtain the
enzymatically active form upon extraction. Glycogen phosphorylase, for example,
can be obtained in the active "a" form if extracted (at pH 6.2) in the presence of ATP-

Interconvertible enzyme forms Mg^{2+}, Ca^{2+} and F⁻ [28]. Under these conditions the phosphatase is inhibited which
would otherwise convert phosphorylase a into the inactive phosphorylase b. Besides,
the kinase is stimulated which converts inactive phosphorylase b into active phosphorylase a. Phosphorylase b may be extracted in the presence of the allosteric acti-
vator AMP. Another example is pyruvate dehydrogenase which is activated by a
phosphatase and inhibited by a kinase. Suitable extraction conditions favour activation of the phosphatase and inhibit the kinase [30].

2.3 Differential Extraction of Cytosolic and Mitochondrial Enzymes from Animal Soft Tissues

By successive extraction with buffered media of increasing ionic strength, cytosolic
and mitochondrial enzymes can easily be liberated from uniform animal tissues like
liver or muscle. The following procedure has been suggested [31].

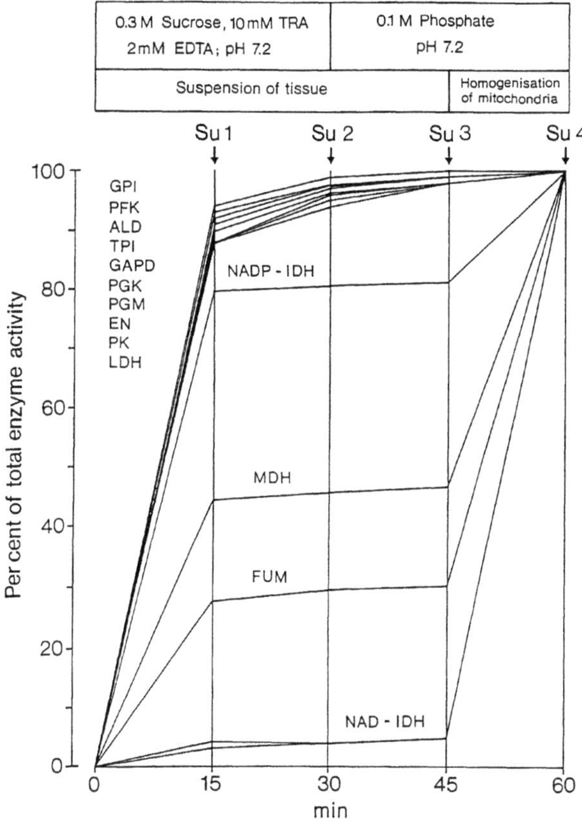

Fig. 2.4. Stepwise extraction of extra- and intramitochondrial enzymes from rat liver. The activities of the glycolytic enzymes and those of the citrate cycle are expressed as percent of the total activities and plotted against the extraction time [32]. *Su 1* to *Su 4*: supernatants of extraction steps 1 to 4 as described in the text (TRA: triethamolamine, EDTA: ethylendiaminetetraacetic acid-Na$_2$. Of the total activities of the glycolytic enzymes, 90% are in Su 1. Enzymes located in the extramitochondrial space are extracted in *Su 1* to *Su 3* using media of increasing ionic strength. Mitochondria are not broken before the last extraction step (*Su 4*). Abbreviations: *ALD*, aldolase (4.1.2.13); *EN*, enolase (4.2.1.11); *FUM*, fumarate hydratase (4.2.1.2); *GAPD*, glycerinaldehyde phosphate dehydrogenase (1.2.1.12); *GPI*, glucose phosphate isomerase (5.3.1.9); *LDH*, lactate dehydrogenase (1.1.1.27); *MDH*, malate dehydrogenase (1.1.1.37); *NAD-IDH*, NAD-isocitrate dehydrogenase (1.1.1.41); *NADP-IDH*, NADP-isocitrate dehydrogenase (1.1.1.42); *PFK*, phosphofructokinase (2.7.1.11); *PGK*, phosphoglycerate kinase (2.7.2.3); *PGM*, phosphoglyceromutase (2.7.5.3); *PK*, pyruvate kinase (2.7.1.40); *TPI*, triosephosphate isomerase (5.1.1.1)

A procedure to extract cytosolic and mitochondrial enzymes

Immediately after removal of a muscle from an animal the tissue is carefully cut into small pieces in the cold. Small muscles like those of insects may be disrupted with a needle. Care must be taken not to squeeze the tissue while mincing it. Approximately 450 mg of minced tissue are put in a centrifuge tube and suspended with 8.55 ml of extraction medium I which has the following composition: 300 mmol l^{-1} sucrose, 10 mmol l^{-1} triethanolamine-HCl, 2 mmol l^{-1} ethylenediaminetetraacetate (EDTA), adjusted to pH 7.2 with 2 N NaOH.

Homogenization of muscle tissue

During extraction the pH-value of the medium is kept constant using a dilute solution of NaOH. Extraction is performed with a small rectangular glass or plastic spatula or by using a motor driven plunger at a frequency of around 80 lifts per minute. The plunger comprises a disc in its center rectangularly fitted to a rod. The diameter of the disc is 4 mm smaller than the inner diameter of the centrifuge tube in which the minced tissue is kept. During the extraction procedure the centrifuge tube is cooled to +4 °C and care must be taken not to grind or to crush the tissue pieces. After precisely 15 min the extraction is stopped and the suspension is centrifuged for 20 min at 100 000 g. The clear supernatant containing mostly cytosolic enzymes is separated and stored under the label Su 1 (Fig. 2.4 [31]).

The sediment is re-extracted with 8.55 ml of extraction medium I as described above and centrifuged for 20 min at 100 000g. The supernatant is decanted and stored under the label Su 2 (Fig. 2.4). The sediment is suspended in 9.0 ml of a 100 mmol l^{-1} KH$_2$PO$_4$/ Na$_2$HPO$_4$-buffer of pH 7.2 and continuously whirl mixed for 15 min. Afterwards the suspension is centrifuged and the supernatant (Su 3) is taken off (Fig. 2.4).

The sediment is completely transferred into a small glass vial and suspended in 4.5 ml of 100 mmol l^{-1} KH$_2$PO$_4$/NaHPO$_4$-buffer of pH 7.2. The suspension is intensively cooled with an ice-salt mixture and four-times homogenized for 30 s with one minute intervals in between. The cooled homogenate is re-extracted for 10 min by means of a magnetic stirrer and finally centrifuged. The resulting supernatant is labelled Su 4 (Fig. 2.4). The sediment is suspended in 6.75 ml of 50 mmol l^{-1} KH$_2$PO$_4$/Na$_2$HPO$_4$-buffer of pH 7.2 by using a fast rotating Teflon pestle. The resulting suspension can be used to estimate photometrically the activities of mitochondrial enzymes like succinate dehydrogenase. Non-ionic detergents like Triton X-100 may be used to solubilize such enzymes and to submit them to electrophoresis in the presence of Triton X-100 to preserve their catalytic activity [33].

2.4 Extraction of Insects

Extraction procedure

Insects may be immobilized with ether or carbon dioxide allowing the gas to dissipate before they are used. Individual insects such as *Drosophila* can be used for the determination of genetic variation between flies. The following extraction procedure has been described for this species [10].

One fly is put in a 5 ml plastic centrifuge tube, and a tiny drop of the buffer to be used, for example in the preparation of starch gels, is added from a Pasteur pipette. The fly is ground in the buffer with a glass stirring rod. Afterwards an additional 0.1 ml of buffer is added and then the homogenate is left to stand for a little while to sediment larger particles. The supernatant is used for electrophoresis. In a similar manner it is possible to study the individual larvae or pupae of *Drosophila*.

2.5 Extraction of Plant Tissues

Many plant tissues store considerable amounts of phenolics in their vacuols [34] while their contents in protein are quite low; e.g. *Picea abies* has 1 mg cm^{-3} [34, 35]. When these tissues are disrupted the phenols come into contact with the proteins and, unless protective agents are added, inactivate or denature them. This is also true for such compounds as terpenoids and resin acids [36, 37]. It cannot be expected that a certain extraction procedure will serve to extract all enzymes in an active state from a fully differentiated plant tissue. Different pH-levels may result in differential extraction of isozyme groups [37–39] and may affect lability of enzymes after extraction [40].

2.5.1 An Extraction Medium for Seeds and Herbs

In contrast to fully differentiated tissues of phenol-rich plants their seeds and vegetative buds are mostly free of large amounts of protein-interfering substances. The latter named tissues may be extracted with a simple acid (acetate-phosphate) or neutral (phosphate) buffer of a 100 mmol l^{-1} concentration [29] (Table 2.4). These media can also be used for herbs being substantially free of phenols like spinach [41] or pea leaves [42]. If the extracted enzymes are to be submitted to starchgel electrophoresis the gel buffer is commonly used as the extraction medium [10]. But phosphate buffers often retain the catalytic ability of enzymes better than buffers of Tris(hydroxymethyl)-aminomethane do. But phosphate buffers are to be avoided if metal ion-dependent enzymes are to be studied.

The action of phenols, terpenoids and resin acids

To extract seeds of *Camellia japonica* L., for example, embryos were removed and ground with a chilled mortar and pestle in 50 mmol l^{-1} Na-phosphate of pH 7.3, containing 5 wt/vol.% sucrose and 0.1% 2-mercaptoethanol [43].

Composition of extraction medium for herbs

2.5.2 Methods to Extract Woody Plants

A variety of phenol-binding components and non-ionic detergents have been used to extract active enzymes from woody plants, though with varying success depending on the tissue and species analyzed (Table 2.5).

The formation of complexes between proteins and phenolic compounds can be suppressed by a number of phenolic-adsorbents. These are, for example, polyvinylpyrrolidone [45], in a soluble from (PVP) or in an insoluble from (PVPP, Polyclar AT) [35, 37, 46, 47], casein [48], bovine serum albumin [49] or resins [34]. The phenol-scavenger PVP is frequently used but is not equally effective with all types of phenols [46, 47]. Phenol binding to exchange resins has been reported [34], but isozymes with

Phenolic adsorbents PVP, PVPP, casein, BSA, exchange resins

Table 2.5. Extraction procedures that were used to obtain enzymes in an active state from vegetative tissues of woody plants

Plant species: *Camellia japonica L.;*

Tissue: leaves;

Enzymes investigated (EC number in brackets):
aconitase (4.2.1.3), alcohol dehydrogenase (1.1.1.1), aldolase (4.1.2.13), esterase (3.1.1.6),
NAD(P)-glutamate dehydrogenase (1.4.1.3), glyceraldehyde-3-phosphate dehydrogenase (1.2.1.9),
isocitrate dehydrogenase (1.1.1.42), leucine aminopeptidase (3.4.11.1), NAD$^+$-malate dehydrogenase
(1.1.1.37), phosphoglucomutase (2.7.5.1), 6-phosphogluconate dehydrogenase (i.1.1.44), superoxide
dismutase (1.15.1.1), triosephosphate isomerase (5.3.1.1);

Extraction of enzymes:
leaves: leaves are diced and added to 5 ml of an extraction buffer containing 40 mmol l^{-1} Na-
phosphate, 200 mmol l^{-1} sucrose, 1 mmol l^{-1} EDTA, 3 mmol l^{-1} DTT, 5 mmol l^{-1} Na-ascorbate,
3 mmol l^{-1} Na-bisulfite, 6 mmol l^{-1} diethyldithiocarbamate, 5% PVP-40, and 0.1% 2-mercapto-
ethanol, final pH = 7.3. Homogenization is performed at 4 °C for 10 s. The resulting extract is
immediately poured into approximately 1.8 g of dry polyvinylpyrrolidone (insoluble grade),
yielding a paste that is electrophoresed without further processing.
Seeds: embryos are removed from their testae and ground with a chilled mortar and pestle in
50 mmol l^{-1} Na-phosphate, pH 7.3, containing 5 wt/vol.% sucrose and 0.1% 2-mercaptoethanol [43].

Plant species: *Picea abies (L.) Karst.;*

Tissue: needles;

Enzymes investigated (EC number in brackets):
alanine aminotransferase (2.6.1.2), aspartate aminotransferase (2.6.1.1), glucose 6-phosphate
dehydrogenase (1.1.1.49), glucose 6-phosphate isomerase (5.3.1.9), NAD-glutamate dehydrogenase
(1.4.1.2), NAD$^+$-malate dehydrogenase (1.1.1.37), 6-phosphogluconate dehydrogenase (1.1.1.44),
shikimate dehydrogenase (1.1.1.25);

Extraction of enzymes:
two g of needles are cut into pieces of about 1 mm and transferred into a 100 ml glass test tube
containing 20 ml of extraction medium. The mixture is ultrasonicated at 4 °C twice for 2 min each
and then flushed with N$_2$. Afterwards, the suspension is homogenized four times for 10 s with
pauses of 10 s, using an Ultraturrax blender at speed 7 – 8. Finally, the homogenate is filtered
through a nylon cloth, and the filtrate centrifuged at 38 000 g for 30 min at 4 – 10 °C. The clear fluid
between the sediment and the lipid overlayer is used as crude extract. The extraction medium
consists of a 100 mmol l^{-1} Na-phosphate buffer of pH 7.5, containing 5 wt/vol.% moistened insoluble polyvinylpyrrolidone and 0.5% Triton X-100. (Polyvinylpyrrolidone is washed intensively
with 100 mmol l^{-1} Na-phosphate, pH 7.5 and then added to the extraction medium.) Then 10 ml of
crude extract are loaded on a column of Fractogel TSK HW-40 (F) (Merck, Germany), previously
equilibrated with a 100 mmol l^{-1} Na-phosphate buffer of pH 7.5 having a hight of 8 cm and an i.d.
of 2.5 cm. High molecular weight compounds are eluted with equilibration buffer. In autumn
enzyme activities (nkat (g dry weight of current-year needles) $^{-1}$) were: alanine aminotransferase
(200 – 280), aspartate aminotransferase (500 – 900), glucose-6-phosphate dehydrogenase
(40 – 190), glucose-6-phosphate isomerase (300 – 600), glutamate dehydrogenase (7 – 50),
NAD$^+$-malate dehydrogenase (5200 – 11200), 6-phosphogluconate dehydrogenase (20 – 70) and
shikimate dehydrogenase (60 – 90) [34]. Chromatographically purified extracts were used
for gradient gel electrophoresis [44]

low isoelectric points may be selectively removed from extracts under conditions of low ionic strength [50]. Borate and germanate were also reported to complex phenolic compounds [51, 52] (Table 2.5). Thiol reagents such as 2-mercaptoethanol, cystein or dithiothreitol are added to extraction media to inhibit the enzymatic turnover of phenol oxidases which oxidize phenols to quinones in the presence of O_2. The quinones in turn tend to form covalent bonds with proteins thus irreversibly inactivating them [43, 53]. Another inhibitor of phenol oxidases is the compound diethyldithiocarbamate [54] but it is only effective at concentrations exceeding 10 mmol l^{-1}. Since the various phenol oxidases are active only in the presence of O_2 it is good practice to prepare crude extracts in the cold and under an atmosphere of N_2 (Table 2.5). Tissues of woody plants are frequently frozen in liquid nitrogen before rupture because this treatment makes them brittle and easy to grind. Maceration of tissues in liquid N_2 also provides a simple means of working at low temperature and in an inert atmosphere [55]. Most extraction media developed to extract enzymes from woody plants contain, besides PVP (or PVPP), a non-ionic detergent like polyethyleneglycol (PEG), Tween, Tergitol 15-S-9 or Triton X-100 (Table 2.5). These surfactants appear to act by overcoming phenol inhibiton of enzymes [50] and by liberating proteins from cell membranes and cell walls [55]. Triton X-100 is effective in the extraction medium at concentrations between 0.3 and 1.0 vol.% [37], Tween 80 at 1-2 vol.% [49, 55], polyethyleneglycol at 1 – 2 wt/vol.% and Tergitol 15-S-9 at 1 vol.% (Table 2.5). PVP is used at 5 – 8 wt/vol.% concentrations either in a soluble or insoluble form. The insoluble form sediments along with cell debris when crude extracts are centrifuged thus eliminating a considerable amount of soluble phenols. The polymer may be purified before use [37, 56]. In general a combination of polyvinylpyrrolidone and a non-ionic detergent results in a better yield of protein and enzyme activity than the exclusive use of one of these substances. In some cases it is necessary to purify the resulting crude enzyme extracts prior to use. When using reducing agents like dithiothreitol it should be kept in mind that these are potential inhibitors of enzymes with free SH-groups like glucose-6-phosphate dehydrogenase.

Phenol oxidases

Liquid N_2

Nonionic detergents PEG, Tween, Triton X-100, Tergitol

2.6 Concentration of Diluted Enzyme Extracts

There are a number of methods to concentrate protein solutions but only a few are simple and work with a large number of samples. The most popular ones are: (a) freezing tissues in liquid N_2, freeze-drying them and extracting them into a small volume of buffer, (b) the use of a stationary ultrafiltration system, (c) centrifuge filtration using microconcentrators, (d) affinity chromatography, and (e) electrophoretic concentration.

The advantages of freeze-drying are that enzyme activities can be stored for a long period of time and the ease with which proteins can be eluted from the dried material. The disadvantages are the high costs for a freeze-drier and the procurement of liquid N_2. Stationary disposal ultrafiltration devices are easy to use but expensive if a large number of samples are to be dealt with. Centrifuge filtration may also become quite expensive if each centrifuge unit is used only once, but it is easy to do and

Centrifuge filtration

can be performed in a relatively short time. Moreover, several systems can be used six to ten times if cleaned adequately (0.1 vol. % Triton X-100, stored in 0.1 vol. % glycerol). Two different systems are commercially available: (a) the system of Amicon and (b) the system of Sartorius (Fig. 2.5). Both work with low sample volumes (1 – 2.5 ml).

Concentration of protein samples is achieved in that, under the centrifugal force, part of the solute and any molecule which is smaller than the exclusion limit of the membrane which is 10 000, 30 000 or 50 000 (g/mol) is passed through it while the larger molecules are retained. The ultrafiltration membrane is sealed to the bottom of a flow so that the proteins concentrate in the centrifuge tube and not on the membrane (Fig. 2.5). The advantage of this system is that material which sediments upon centrifugation sediments on the bottom of the centrifuge tube and does not clog the ultrafiltration membrane. Protein denaturation is low in both devices and the time to concentrate 2.5 ml is 15 – 60 min depending on the concentration of the protein solution and the presence of other molecules which interfere with the filtration process.

Diatomaceous earth ("Celite") has been used to concentrate ovarian follicular cells from rabbits for two-dimensional gel electrophoresis [57].

Plasminogen activators which are proteolytic enzymes could be purified 700-fold by specific adsorption on a fibrin-Celite matrix. Washed Celite was suspended

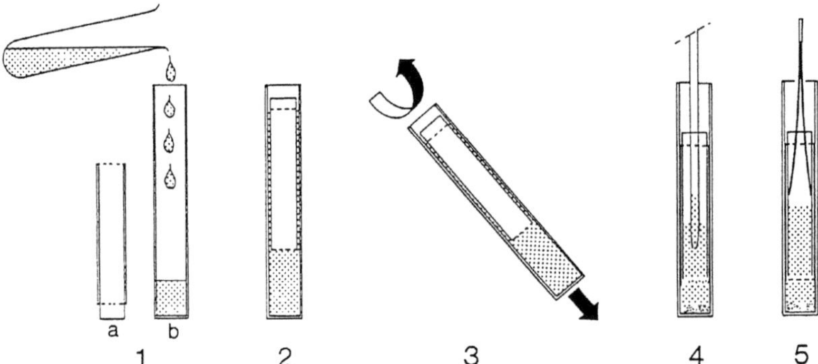

Fig. 2.5. Function of the micro ultrafiltration system Centrisart[R] of Sartorius GmbH, Göttingen, Germany.
The system consists of three parts: (1a) a float, (1b) a centrifuge tube, and a cap (not shown). The filtrating ultramembrane is sealed to the bottom of the float (1a). To operate the system, the cap is removed from the device and the float taken off. Care must be taken not to touch the membrane. After 0.1 – 2.5 ml sample are filled into the centrifuge tube (1b), the float is inserted, the system is capped and left untouched for 5 min to wet the membrane (2). Afterwards the system is put into a rotor made for a maximum of 17 × 100 mm centrifuge tubes (3). The maximum centrifugal force is 2500 g for swinging bucket rotors and 2000 g for fixed-angle rotors. When using concentrated protein solutions like serum, centrifugation is started at half-maximum speed. When lipid-rich solutions like milk are to be treated, only fixed-angle rotors should be used. At the end of centrifugation the ultrafiltrate is in the float and can be pipetted off (4) while the concentrated protein solution is in the centrifuge tube. Otherwise it may be removed after the float has been taken off with a spread pincette (5). The membrane contains about 2 mg glycerol which can be eluted from it by 5 min precentrifugation with distilled water, but then the membrane must not run dry!

in 50 mmol l⁻¹ sodium phosphate buffer of pH 7.4, containing 100 mmol l⁻¹ NaCl and 1 mmol l⁻¹ EDTA and mixed with a 2% fibrinogen solution (50 ml/20 g Celite). The protein was adsorbed by the diatomaceous earth and thus an "affinity column" was prepared for the chromatographic purification of plasminogen activators [58].

Concentration and purification of a wide variety of enzymes (and proteins) is possible if affinity chromatography with a group-specific ligand is used. Of the 2000 enzymes currently catalogued, 620 (31%) require a nucleotide coenzyme so that affinity chromatography using a nucleotide coenzyme-like ligand can be used to purify and to concentrate a large number of them. At least three ligands fulfil these requirements: (a) Blue Sepharose CL-6B, (b) 5'AMP-Sepharose 4B and (c) 2',5'ADP-Sepharose 4B. Blue Sepharose CL-6B contains the dye Cibacron Blue F3 G-A which is covalently attached to cross-linked agarose (by the triazine coupling method) [59]. Figure 2.6 shows the structure of the immobilized ligand. The blue dye Cibacron Blue F3G-A in Blue Sepharose CL-6B is capable of binding strongly to many proteins including kinases, dehydrogenases and most other enzymes requiring adenylyl-containing substances like NAD^+. It also binds several non-enzymic proteins such as albumin, lipoproteins, blood coagulating factors and interferon [60].

Affinity chromatography

Biospecifically adsorbed proteins can be eluted by low concentrations of the free cofactor whilst less specifically bound proteins require the use of much higher concentrations of cofactor or salt. The blue dye binds to the nucleotide-binding pocket of dehydrogenases. The use of Blue Sepharose CL-6B in place of immobilized nucleotides has the advantage that the ligand is stable so that the affinity gel may be repeatedly used. Table 2.6 shows some enzymes which were purified on agarose-bound Cibacron Blue F3G-A. When an enzyme binds specifically to Blue Sepharose CL-6B, the cofactor which has greatest affinity for the enzyme is most effective in elution [60]. A study of the desorption of human alcohol dehydrogenase from Blue Sepharose CL-6B by different concentrations of NADH indicated interlocus and interallelic differences in the affinities of the various isoenzymes [83].

Enzymes separated on Blue Sepharose

When using polyacrylamide gradient gels it is also possible to concentrate dilute enzyme extracts in situ. About 30 μl of sample can be applied to a sample slot of $5 \times 10 \times 1$ mm of a pore gradient gel used in a vertical position. After proteins have migrated into the gel for 10 min another aliquot of 30 μl of sample is applied to the gel. This procedure does not disturb the separation pattern but doubles the enzyme activity.

In situ concentration of enzyme solutions

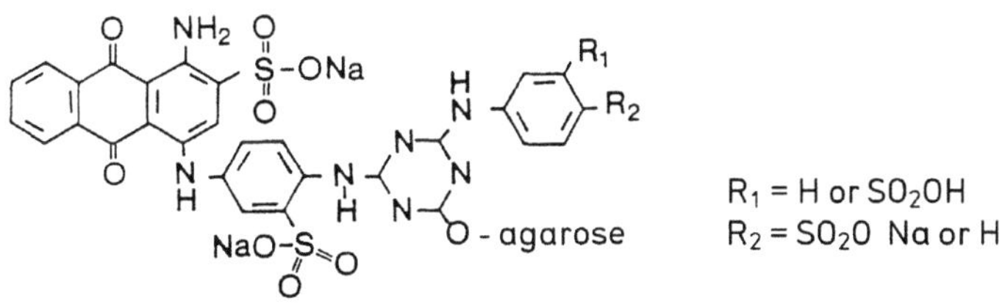

Fig. 2.6. Formula of the blue dye Cibacron Blue F 3G-A covalently attached to agarose (the product is commercially available under the trade name Blue-Sepharose from Pharmacia, Uppsala Sweden)

Table 2.6. Some enzymes with affinity for Cibacron Blue F3G-A

Enzyme	Ref.	Enzyme	Ref.
Adenylate cyclase	[61]	Hexokinase	[63]
Adenylate kinase	[62]	3-Hydroxy-3-methyl-glutaryl CoA reductase	[73]
Alcohol dehydrogenase	[63, 64]	Hydroxysteroid dehydrogenase	[74]
Aldolase	[63]	Isocitrate dehydrogenase	[75]
Amino acyl t-RNA synthetase	[65]	Malate dehydrogenase	[62]
AMP deaminase	[66]	Nitrate reductase	[76]
cyclic AMP-dependent protein kinase	[67]	Nucleoside kinase	[77]
Amylopectin-1,6-glucosidase	[68]	Phosphodiesterase	[78]
Aryl sulphatase	[69]	Phosphofructokinase	[62]
Creatine kinase	[63]	Phosphogluconate dehydrogenase	[62]
Deoxycytidine kinase	[70]	Phosphoglycerate kinase	[62]
DNA polymerase	[62]	Phosphoglyceromutase	[62]
Enolase	[63]	Polynucleotide kinase	[79]
Fructose diphosphatase	[62]	Protein kinase	[80]
Glucose-6-phosphate dehydrogenase	[63]	Pyruvate kinase	[63]
Glutathione reductase	[71]	Restriction endonucleases	[81]
Glyceraldehyde-3-phosphate dehydrogenase	[63]	Succinyl-CoA transferase	[82]
Glyoxalase	[72]		

2.7 Mammalian Blood

2.7.1 Erythrocytes

The following procedure may be used to study human erythrocytes as well as erythrocytes from a number of mammalian species such as sheep, goats, cattle, pigs, dogs, rats and mice [10].

2.7.1.1 Anticoagulants

Blood is usually anticoagulated for electrophoretic studies of erythrocyte enzymes, although this is not always necessary. Any one of several anticoagulants may be used **Heparin,** including heparin, ethylenediaminetraacetic acid (EDTA), and acid-citrate-dextrose **EDTA, ACD** (ACD). Anticoagulants which are metabolic inhibitors, such as fluoride and oxalate, affect various serum enzymes [27].

2.7.1.2 Blood Storage

Blood samples may be stored in a number of ways. Most enzymes can be preserved in an active state for at least one month if 0.15 ml of acid-citrate-dextrose (ACD) (0.8 g of citric acid, 2.2 g of trisodium citrate, and 2.45 g of dextrose per 100 ml of solution) is added to 1 ml of blood. This amount of ACD may be added to blood even though it has initially been drawn in heparin or EDTA. The blood is then stored at refrigerator temperatures (+2 to +6 °C). Many enzymes will be adequately preserved if hemolysates are stored frozen at − 20 to −70 °C. However, some will not.

After separation of serum or plasma from mammalia, cooling to + 4 °C is effective to stabilize enzymes. Freezing can lead to inactivation, for example, of lactate dehydrogenase. Serum creatine kinase can be stabilized by SH-reagents and acid phosphatase by acidification. Serum creatine kinase is sensitive to light. The addition of citrate binds Ca^{2+}-ions and prevents an activation of most coagulation proteases.

Composition of acid-citrate-dextrose (ACD)

Effect of freezing LDH, CK, ACP

2.7.1.3 Hemolysate Preparation

A simple hemolysate preparation is used by Grunbaum [84]. Red blood cells in whole blood are allowed to settle for several minutes. The supernatant plasma is transferred to a clean tube and may be used for serum protein analysis. For routine phentotyping, the cells need not to be washed. Whole blood cells (75 μl) are added to 200 μl of distilled water filled into tubes. The tubes are then capped and frozen (in liquid nitrogen if available). Afterwards they are thawed in warm water and again re-frozen, thawed and centrifuged. The resulting hemolysate may be stored at − 20 °C.

2.7.2 Serum

To obtain serum for electrophoretic studies, fresh whole blood is drawn without anticoagulant and placed in a glass tube. As soon as the clot forms the sample is centrifuged and the supernatant serum is used for electrophoresis. Several enzymes lose activity upon prolonged storage of serum (Table 2.7). Thus it is critically important that serum be very fresh, that heating during electrophoresis be prevented, and that the electrophoretic run be as short as possible. In this connection, cellulose acetate, acrylamide gel or agar gel electrophoreses are superior to starch gel electrophoresis [10]. If isoenzyme patterns of trace amounts of protein are to be investigated, it may be necessary to concentrate the serum proteins by one of the rapid methods described in Sec. 2.6.

2.7.3 Blood Cells

Blood from which blood cells are to be separated should always be used in the freshest possible condition. Leucocytes and platelets begin to clump and stick to surfaces after blood is drawn so that it is recommended to use plastic non-wetable equipment. Erythrocytes, leucocytes, platelets, and polymorphs may be prepared by rate zonal centrifugation while lymphocyte subfractionation, isolation of natural killer-cells or monocytes is best performed by isopycnic centrifugation.

Table 2.7. Stability of enzymes in human serum samples

Enzyme	Storage temperature[a]	
	2 – 8 °C	20 – 25 °C
Aldolase	loss of activity 5 days 8 %	loss of activity 5 days 15 %
Alkaline phosphatase	maximum 7 days	loss of activity 7 days 10 %
Acid phosphatase	no loss of activity during 7 days when stabilized by 5 mg NaHSO$_4$/ml	
α-Amylase	no loss of activity within 5 days	
Choline esterase	no loss of activity within 7 days	
Creatine kinase[b]	loss of activity 7 days 2 %	loss of activity 24 h 2 %
Creatine-MB isozyme[b]	24 h > 10 %	1 h < 10 %
Glutamate dehydrogenase	loss of activity 3 days 5 %	loss of activity 3 days 15 %
Aspartate aminotransferase	loss of activity 3 days 8 %	loss of activity 3 days 10 %
Alanine aminotransferase	loss of activity 3 days 10 %	loss of activity 3 days 17 %
γ-Glutamyltransferase	no loss of activity within 7 days	
Leucine aminopeptidase	no loss of activity within 7 days	
Lactate dehydrogenase	loss of activity 3 days 8 %	loss of activity 3 days 2 %
Lactate dehydrogenase-1-isozyme	loss of activity 7 days 5 %	loss of activity 7 days 5 %
Lipase	no loss of activity within 5 days	no loss of activity within 24 h

[a] Use well-stoppered containers .
[b] NAC activated.
Data taken from [27].

2.7.3.1 Preparation of Leucocytes and Platelets

Platelets may be obtained by centrifugation of 10-20 ml of anticoagulated whole blood at 500 g for 5 min. Heparin, ACD, or EDTA may be used as anticoagulants. The indicated conditions of centrifugation will sediment the erythrocytes and leucocytes but leave the platelets still in suspension. The suspension can be centrifuged at 2000 – 20 000 g for 10 – 20 min to precipitate the platelets. They may be disrupted by sonication, by repeated freezing and thawing, or by the addition of a detergent such as Triton X-100. If this detergent is used the pure solution is diluted tenfold and 1 volume of this solution is added to 0.25 to 1 volume of packed platelets. After disruption a 10 min centrifugation at 20 000 g clears the homogenate and reduces streaking of enzyme bands on the gel. The supernatant is used for electrophoresis [10].

Leucocyte preparation Leucocytes may be separated from 10 – 20 ml of whole blood by adding dextran of the average molecular weight of 235 000 so that a final concentration of 3 % is obtained. Dextran causes clumping of the erythrocytes which then settle more rapidly than the leucocytes do. Afterwards the mixture is allowed to stand undisturbed in the test tube for 30 min. The resulting supernatant contains the leucocytes. The leucocyte suspension is finally centrifuged, the supernatant discarded, and the sediment of leucocytes disrupted by sonication, repeated freezing and thawing, or the addition of Triton X-100 as described for the lysis of platelets [10]. Human peripheral blood mononuclear cells (PBMC) may also be separated on a discontinuous gradient of PVP-coated silica particles (Percoll[R]) [85]. PBMC are isolated from heparinized blood by density centrifugation on a Ficoll-Paque[R] cushion, from 1 ml blood about 1×10^6 – 2×10^6 PBMC are obtainable. This crude PBMC population consists of about 20 – 40 % monocytes, 60 – 80 % lymphocytes and 1 – 4 % polymorphonuclear cells. A discontinuous density gradient is prepared from 90 ml Percoll (d = 1.132 g/ml) and 8.965 ml Hanks'BSS ($10\times$) (Table 2.8), 1 ml Hepes (= N-2-hydroxyethylpiperazine-N'-2-ethanesulphonic acid) buffer (10 %) and 455 μl HCl (1 N); this stock solution is diluted with Hanks'BSS ($1\times$) supplemented with 1% Hepes buffer to obtain Percoll solutions of desities 1.062, 1.064, 1.066, 1.068 and 1.070 g/ml. Into a 50 ml conical disposable centrifuge tube are filled 3.75 ml of Hanks'BBS ($1\times$) supplemented with 1% Hepes buffer and underlayered successively with 7.5 ml portions of the Percoll solutions of densities 1.070 to 1.062. Finally the gradient is underlayered with about $150 – 200 \times 10^6$ PBMC suspended in 11.25 ml Percoll solution (d = 1.080 g/ml). Centrifugation is carried out at 390 g_{av} for 30 min at room temperature in a swing out rotor under careful acceleration and deceleration. After centrifugation the bottom of the tube may be pierced with a hollow needle and the gradient separated into fractions of 7.5 ml containing the cells banded at the interface between two different densities. The total recovery of cells is about 70 – 80 %. Fraction 2 contains a low number of erythrocytes. Most of the lymphocytes band in fractions 3, 4 and 5, in fraction 3 99% of the PBMC are lymphocytes. Monocytes are found in fraction 6. T and B lymphocytes are not separated [85]. Alternative methods to the one described here are also in use [89].

2.7.3.2 Isolation of Lymphocytes and Polymorphs

Lymphocytes from human blood may be isolated by centrifugation on the gradient medium Ficoll-Paque (Pharmacia, Uppsala, Sweden) [90]. Ficoll-Paque is a sterile, ready to use solution of 5.7 g Ficoll 400, 9 g sodium diatrizoate and Ca-Na$_2$-EDTA in 100 ml, the density of the solution is 1.077 ± 0.001 g/ml; sodium diatrizoate (Mr 635.92) is the sodium salt of 3,5-diacetamido-2,4,6-triiodobenzoic acid. It is included into the Ficoll solution to increase its density and osmolarity without increasing its viscosity so that lymphocytes can be effectively separated from other cells. The separation principle is based on the fact that red blood cells aggregate when they come into contact with Ficoll which increases their rate of sedimentation. They then, upon centrifugation, collect as a pellet at the bottom of the centrifuge tube where they are well separated from the lymphocytes [91 – 93]. Granulocytes also sediment through the Ficoll-Paque layer, because they increase in density when they come into contact with the slightly hypertonic Ficoll-Paque medium. Lymphocytes, monocytes and platelets do not enter the Ficoll-Paque solution. They collect as a band at the interface between the original blood sample and the separation medium.

Lymphocyte purification using Ficoll-Paque can be carried out according to the following procedure (Instruction manual, Pharmacia, Uppsala, Sweden). Fresh blood should be used and sample preparation should be performed at 18 – 20 °C (all glassware should be siliconized for 10 s with 1% dimethyldichlorosilane dissolved in an

Ficoll-Paque[R] medium

Collection of lymphocytes, monocytes and platelets

The Ficoll-Paque procedure to separate lymphocytes

organic solvent, and then thoroughly washed with distilled water; otherwise, tissue culture plasticware may be used).

To a 10 ml test-tube 2 ml of defibrinated or anticoagulated blood and 2 ml of balanced salt solution are added (final volume 4 ml of diluted sample). Both solutions are mixed by drawing the blood and buffer in and out of a Pasteur pipette.

At least 20 ml of balanced salt solution are needed for each sample. The balanced salt solution may be prepared from two stock solutions, A and B. Solution A: 1.0 g/l anhydrous D-glucose, 0.0074 g/l $CaCl_2 \cdot 2H_2O$, 0.1992 g/l $MgCl_2 \cdot 6H_2O$, 0.4026 g/l KCl and 17.565 g/l Tris; the substances are dissolved in 950 ml distilled water, concentrated hydrochloric acid is added until the pH is 7.6 and finally the volume is brought to 1 l. Solution B: 8.19 g/l NaCl. The balanced salt solution is prepared freshly each week by mixing one volumne of solution A with nine volumnes of solution B.

The 4 ml of diluted sample are carefully layered on to 3 ml Ficoll-Paque contained in a siliconized glass centrifuge tube (internal diameter 1.3 cm, volume 15 ml). When layering the sample care must be taken not to mix the diluted blood sample with the density medium. Then the centrifuge tube is spun at 400 g for 30 – 40 min at 18 – 20 °C. Finally the upper layer is removed by use of a siliconized Pasteur pipette, leaving the lymphocyte layer undisturbed at the interface. The upper layer which contains the plasma may be saved for later use. Finally, the lymphocytes are washed free from the platelets. For that purpose the lymphocyte layer is taken off with a Pasteur pipette and transferred to a clean centrifuge tube. (Removing excess Ficoll-Paque would cause contamination with granulocytes; removing excess supernatant would result in contamination with platelet.) Then 6 ml of balanced salt solution are added to the lymphocytes in the test-tube and the cells are suspended by gently drawing them in and out a Pasteur pipette. This procedure is followed by a 10 min centrifugation at 60 – 100 g at 18 – 20 °C. The supernatant is discarded and the washing procedure repeated once. After the supernatant has been discarded, the lymphocytes are suspended in an appropriate medium (e. g. medium 199; Table 2.9).

Ficoll-Paque, although with different densities, has also been used to isolate lymphocytes from mouse [94], dog [95], monkey [96], cow [97, 98], rabbit [99], horse [100], pig [100, 101] and fish [102].

Monocyte removal	Monocytes which remain in the lymphocyte fraction when the procedure given above is used may be removed by incubating the blood sample with iron (or iron carbonyl) particles before separation on Ficoll-Paque. The monocytes phagocytose the iron particles and become denser, with the result that they sediment through the Ficoll-Paque layer on centrifugation and are collected in the red blood cell pellet at the bottom of the centrifuge tube [93].

2.7.3.3 Lymphocyte Subfractionation

Separation of T and B lymphocytes on Percoll	T and B lymphocytes can be separated on a discontinuous silica sol (Percoll[R]) gradient [103, 104]. First, mononuclear cells are separated on Ficoll-Paque, e.g., as described above. The fraction containing lymphocytes and monocytes is then suspended in medium 199 (Table 2.9) containing 10 % heat inactivated fetal calf serum, not exceeding 80×10^6/ml. Aliquots of 2 ml are centrifuged in 12 ml plastic tubes and the supernatants discarded. Two ml volumes of each of Percoll[R] solutions "100", 70, 60, 50 and 40 % (4 ml of the 50 % concentration), are layered over the cell

Table 2.8. Composition of phosphate buffered saline (PBS) [86], Earle's [87] and Hanks'[88] medium

	Concentrations in mg/ml		
	PBS (Dulbecco)	Earle's salts	Hanks' BSS salts
NaCl	8000	6800	8000
KCl	200	400	400
Na_2HPO_4	1150	–	48
$NaH_2PO_4 \cdot H_2O$	–	140	–
KH_2PO_4	200	–	60
$MgCl \cdot 6H_2O$	100	–	– a)
$MgSO_4 \cdot 7H_2O$	–	200	200 a)
$CaCl_2$	100	200	140
glucose	–	1000	1000
phenol red	–	10	10
$NaHCO_3$	–	2200	350

Concentrations given in mg/ml.

a) Hanks'used 100 mg/l $MgCl_2 \cdot H_2O$ and 100 mg/l $MgSO_4 \cdot 7H_2O$. The concentrations indicated are those recommended by Biochrom KG, Berlin.

$CaCl_2 \cdot H_2O$ and phenol red are dissolved separately. To 1 l of solution are added 10 ml of phenol red 0.2%. The solution is autoclaved and to 10 ml amounts are added 0.25 ml of autoclaved 1.4% (isotonic) $NaHCO_3$. The solution is stored in the refrigerator for CO_2 equilibrium to pH 7.6 before final tightening of storage bottles

Table 2.9. Preparation of cell tissue culture medium 199 [105], slightly modified

The complete medium 199 is prepared from the stock solution A to Q [105]. But only L-amino acids instead of DL-amino acids are used and reducing their concentration to 50%. Besides, 1.0 mg l^{-1} ATP-Na_2 is taken instead of 10.0 mg l^{-1} ATP-Ba_2 [106].

Solution A: L-arginine · HCl (14 mg), L-histidine · HCl (4 mg), L-lysine · HCl (14 mg), L-tryptophane (4 mg), L-phenylalanine (10 mg), L-methionine (6 mg), L-serine (10 mg), L-threonine (12 mg), L-leucine (24 mg), L-isoleucine (8 mg), L-valine (10 mg), L-glutamic acid.H_2O (30 mg), L-aspartic acid (12 mg), L-α-alanine (10 mg), L-proline (8 mg), L-hydroxyproline (2 mg), glycine (10 mg), L-glutamine (20 mg), sodium acetate (16.3 mg). These ingredients are dissolved in 100 ml of double-strength Earle's solution.

Solution B: L-tryosine (200 mg), L-cystine (100 mg). These substances are dissloved with gentle heating in 100 ml of 0.075 N HCl.

Solution C: Nicotinic acid (25 mg), nicotin amide (25 mg), pyridoxine · HCl (25 mg), pyridoxal · HCl (25 mg), thiamin · HCl (10 mg), riboflavin (10 mg), Ca-pantothenate (10 mg), myoinositol (50 mg), p-aminobenzoic acid (50 mg), choline · HCl (500 mg), glass-distilled water to 200 ml.

Solution D: Cysteine · HCl (100 mg), glutathione (50 mg), ascorbic acid (50 mg), glass-distilled water to 100 ml. A fresh solution is prepared each month.

Solution E: d-biotin (10 mg), glass-distilled water to 100 ml.

Table 2.9 (continued)

Solution F:	Folic acid (10 mg), Earle's solution to 100 ml.
Solution G:	Vitamin A-acetate (10 mg), 1 ml Tween 80 (5% aqueous solution), 10 ml glass-distilled water to 100 ml. A fresh solution is prepared each month.
Solution H:	Calciferol (vit. D_2) (10 mg), Tween 80 (5% aqueous solution) (10 ml), glass-distilled water to 100 ml.
Solution I:	DL-α-tocopherol phosphate · Na_2 (10 mg), glass-distilled water to 100 ml.
Solution J:	Vitamin K3 (10 mg), glass-distilled water to 100 ml.
Solution K:	Cholesterol (10 mg), Tween 80 (5% aqueous solution) (10 ml), glass-distilled water to 100 ml.
Solution L:	Adenine sulphate (500 mg), glass distilled water to 100 ml.
Solution M:	Guanine · HCl (10 mg), xanthine (10 mg), hypoxanthine (10 mg), thymine (10 mg), uracil (10 mg), dissolved in 100 ml glass-distilled water with the addition of 0.35 ml concentrated NH_4OH.
Solution N:	d-ribose (100 mg), d-2-deoxyribose (100 mg), glass-distilled water to 100 ml. A 1.0 ml amount of this solution is incorporated in each liter of solution A.
Solution O:	Adenosine-5′-monophosphate (10 mg), glass-distilled water to 100 ml. A 4.0 ml portion of this solution is incorporated in each liter of solution A.
Solution P:	Adenosinetriphosphate · Na_2 (50 mg), glass-distilled water to 100 ml.
Solution Q:	D(+)glucose (5500 mg), glass-distilled water to 1000 ml.

Stock solutions A, B, K, L, M, P and Q are employed without further dilution. Solutions C, H and J are diluted 1:10 in glass-distilled water before use. Solutions D, E, F and I are diluted 1:100 in glass-distilled water before use. Afterwards the complete medium 199 is prepared from 25 ml solution A, 1 ml B, 0.1 ml C, 0.5 ml D, 0.5 ml E, 0.5 ml F, 0.5 ml G, 0.5 ml H, 0.5 ml I, 0.005 ml J, 0.10 ml K, 0.10 ml L, 0.15 ml M, 0.10 ml P, 10 ml Q and glass-distilled water to a total volume of 50 ml.
All solutions are sterilized by passage through UF fritted glass filters (corning). Stock solutions and the combined mixtures are stored in low-actinic glassware. The majority of these solutions are kept at 4 °C, but certain stock solutions, e.g., adenine, cholesterol, tyrosine and cystine, which tend to form precipitates in the refrigerator, are stored at room temperature [105]. Earle's solution [87] is composed as indicated in Table 2.7

pellets, starting with the "100%" and continuing with the decreasing concentrations. The "100%" Percoll^R concentration was made from 9 parts of Percoll^R (starting density 1.132 g/ml) and one part of 10x concentrated phosphate buffered saline (PBS 10x; Table 2.8). The lower concentrated Percoll^R solutions were made from the "100%" solution by dilution with PBS (1×). The gradient is spun at 450 g in a table-top centrifuge for 10 min at room temperature. The cells which will migrate to the interphase between 40% and 50% Percoll^R (d = 1.052/1.063 g/ml) are mostly B cells (plus macrophages) while the cells at the interphases between the steps 50% and 60% Percoll^R

(d = 1.063/1.075 g/ml) and 60 % and 70 % Percoll (d = 1.075/1.085 g/ml) will be mostly T cells [103].

They are carefully removed and washed twice in Hanks'BSS. Unseparated cells contained a mean of 10 % B lymphocytes and 75 % of T cells. After centrifugation 73 % of the cells at the interphase between 40 and 50 % Percoll[R] are B lymphocytes whereas at the interphases between 50/60 and 60/70 % of Percoll[R] 75 % of the T cells are found. Subfractionation of lymphocytes can also be performed on preformed self-generated gradients of Percoll[R] [104, 107].

2.7.3.4 Isolation of Natural Killer Cells

Isolation of natural killer lymphocytes using Percoll as density medium has been reported [108]. A lymphocyte preparation is incubated with cells of the K-562 cell line. This mixture is layered on top of a 17 % Percoll solution and centrifuged very gently (40 g, 7 min). The lymphocytes which form rosettes with the K-562 cells (together with excess K-562 cells) pass through the cushion leaving the other lymphocytes behind. The rosettes are then dissociated by agitation and layered on top of a 10 % solution of Percoll[R]. After a 7 min centrifugation at 40 g the K-562 cells will have passed through the cushion while the purified subfraction of the lymphocytes will remain above the Percoll[R] layer. These cells are natural killer (NK) effector cells [108].

2.7.3.5 Isolation of Monocytes

Density gradient centrifugation may also be used to separate monocytes from lymphocytes. Monocytes are somewhat less dense than lymphocytes, and although densities of both cell types overlap they can be separated with a high degree of purity. Cell separation is possible by use of preformed self-generated gradients or by use of discontinuous density gradients. Starting with a Percoll density of 1.07 g/ml, two distinct bands can be obtained after preforming a gradient between 1.08 and 1.04 g/ml [109]. The upper band will contain more than 95 % monocytes while the lower one will consist of more than 90 % lymphocytes.

Lymphocytes obtained by the standard Ficoll-Paque[R] procedure may be further subfractionated on a five-step gradient of Percoll of a density range of 1.062-1.070 g/ml [106]: 1.5 – 2.0 · 10⁷ peripheral blood mononuclear cells in 11.25 ml of Percoll in Hanks' BSS containing 1 % Hepes buffer of density 1.080 g/ml are underlayered below five solutions of Percoll of the following densities: 1.070, 1.068, 1.066, 1.064, and 1.062 g/ml. Of the lymphocytes, 90 % will band at the interphases of the densities 1.070/1.068, 1.068/1.066, and 1.006/1.064 g/ml, while monocytes will accumulate at a density of 1.064/1.062 g/ml [85]. One- and two-step procedures to isolate monocytes directly from whole blood in high yield have also been described [109].

It remains to mention that the entire spectrum of blood cells can be resolved on preformed gradients of Percoll (Table 2.10). Even erythrocytes have been subfractionated on step gradients [109] and on linear gradients [110] of Percoll.

Living cells can be separated from Percoll by washing with physiological saline (one volume of 1.8 % saline to one volume of cell suspension). The washing may be repeated two or three times and the cells collected between each washing step by centrifugation at 200 g for 2 – 3 min.

Table 2.10. Examples of mammalian cells separated in selfgenerating Percoll[R] gradients

Source	Density (g ml⁻¹)[a]	Centrifugation conditions
Rat liver cells		
Hepatocytes	1.07 - 1.10	30 000 g for 30 min
Kupffer cells	1.05 - 1.06	30 000 g for 30 min
Human blood cells		
Thrombocytes	1.04 - 1.06	[b]
Lymphocytes	1.06 - 1.08	
Granulocytes	1.08 - 1.09	
Erythrocytes	1.09 - 1.10	

[a] Density given as recorded density in a Percoll[R] gradient.

[b] Separation of blood cells is best carried out by pre-forming the gradient (starting density 1.090 g ml⁻¹) by centrifugation at 20 000 g for 20 min, then layering blood on top of the gradient. Blood cells are then separated by centrifugation at 1000 g for 5 min in a swinging-bucket rotor. The thrombocytes remain in the serum layer above the gradient; the serum layer can be removed with a pipette (rate-zonal separation). A further spin for 20 min at 1000 g separates the other cell types at their isopycnic densities [109].

2.8 Pancreatic Islets

Pancreatic islets may be isolated from the pancreas by the following method [111]. Adult mice are starved overnight and killed by decapitation. The pancreas is rapidly excised, transferred to ice-cold Hanks' medium (Table 2.8) and cut in small pieces with a pair of scissors. These and 1.5 ml Hanks' medium are added to a vial containing 2.5 mg collagenase. After a vigorous shaking for 18-20 min at 37 °C the digest is diluted with 10 ml ice-cold Hanks' medium and thoroughly mixed. It is then kept on ice and allowed to sediment for 2 min. The supernatant is discarded and 25 μl DNAase solution (10 $\mu g/\mu l$) are added.

A stock solution of Percoll[R] is prepared by mixing nine volumes of Percoll with one volumne of 10 × concentrated Hanks' medium (density = 1.045 g/ml). Then 3.5 volumes of this stock solution are mixed with 6.5 volumes of Hanks' medium and poured into a vessel 25 – 80 mm wide. The pancreas digest is layered on top of this medium by means of a pipette and the pancreatic islets are allowed to sediment. The pancreas digest is then removed by suction and the pancreatic islets collected from the bottom of the vessel. The density of (ob/ob mice) pancreas islets is between 1.065 –1.070 g/ml. The density of the corresponding exocrine tissue is between 1.015 – 1.045 g/ml.

2.9 Isolation of Subcellular Organelles

2.9.1 Isopycnic Centrifugation

Separation of subcellular particles by density gradient centrifugation became very popular after the development of an almost ideal separation medium. This medium (Percoll[R], Pharmacia, Uppsala, Sweden) consists of colloidal silica particles of 15 – 30 nm diameter with a non-dialyzable coat of polyvinylpyrrolidone (PVP). Percoll has the following specifications: density = 1.130 + 0.005 g ml⁻¹, conductivity = 1.0 [mS cm⁻¹], osmolality ≤ 25 mOs kg⁻¹ H_2O, viscosity = 10 ± 5 cP at 20 °C, pH = 9.0 ± 0.5 at 25 °C. With Percoll, gradients can be formed within the density range 1.0 – 1.3 g/ml and within the pH range 5.5 to 10.0 (Proton concentrations below pH 5.5 and divalent cations can cause gelling of the silica matrix).

Properties of Percoll

Density gradients suitable to separate subcellular organelles either increase steadily in density (linear or sigmoidal) or comprise at least two layers of different density (step gradients). "S"-shaped gradients with a fairly flat region occupying most of the centrifuge tube are useful to separate particles of similar densities. A self generating gradient procedure is used to produce such gradients: a solution of Percoll is placed in a centrifuge tube and centrifuged in a fixed angle rotor for 30 – 60 min at a centrifugational force of 50 000 – 100 000. Either the suspension containing the

Table 2.11. Preparation of subcellular particles from mammalian tissues by use of self generating Percoll[R] gradients

Tissue [Ref.]	Procedure	Subcellular particle		
		name	marker enzyme	density [g/ml]
Human liver (biopsy) [112]	Mix homogenate with Percoll in 0.25 mol l⁻¹ sucrose, generate an almost linear gradient from p = 1.035 to 1.15 [g/ml]. Generate gradient in situ by a 45 min centrifugation at 50 000 g	plasma membrane	alkaline phosphatase	1.04
		endoplasmic reticulum	neutral-α-glucosidase	1.055
		lysosomes	N-acetyl-β- glucosaminidase	1.04 – 1.07 1.08 – 1.11
		mitochondria	glutamate dehydrogenase	1.09 – 1.11
Rat liver [113]	Prepare 50 % Percoll in 0.25 mol l⁻¹ sucrose (starting density: 1.07 [g/ml]. Generate gradient in situ by a 30 min centrifugation at 63 000 g	peroxisomes	catalase	1.05 – 1.07

subcellular organelles is mixed with Percoll and the particles are separated upon gradient formation (Table 2.11) or the suspension of subcellular particles is layered on top of the gradient formed and the entire tube is inserted in a swinging-bucket rotor where it is centrifuged for a few minutes at a low centrifugational force (1000 to 5000 g).

Since Percoll is non-toxic to biological materials and does not adhere to membranes, it is usually unnecessary to remove it from purified preparations. If desired, subcellular particles can be separated from Percoll by centrifugation at 100 000 g for 2 h in a swing-out rotor. The biological material remains above the pellet formed [114, 115].

Many enzyme assays can be carried out in the presence of Percoll without interference. The enzymes 5′-nucleotidase (plasma membranes), glucose-6-phosphatase (microsomes), β-glucuronidase (lysosomes) and succinic dehydrogenase (mitochondria) from rat liver homogenates were analysed in the presence of Percoll and no activity losses observed [114]. Labile succinic dehydrogenase was stabilized by Percoll. Aryl sulphatase, alkaline phosphatase, acid phosphatase, β-galcosidase, N-acetyl-α-D-glucosaminidase and β-glucosaminidase have also been analyzed in the presence of Percoll without interference from the density medium [114]. Due to light scattering by Percoll, it is preferable to use enzyme assays based on fluorescence rather than absorbance [112 – 116].

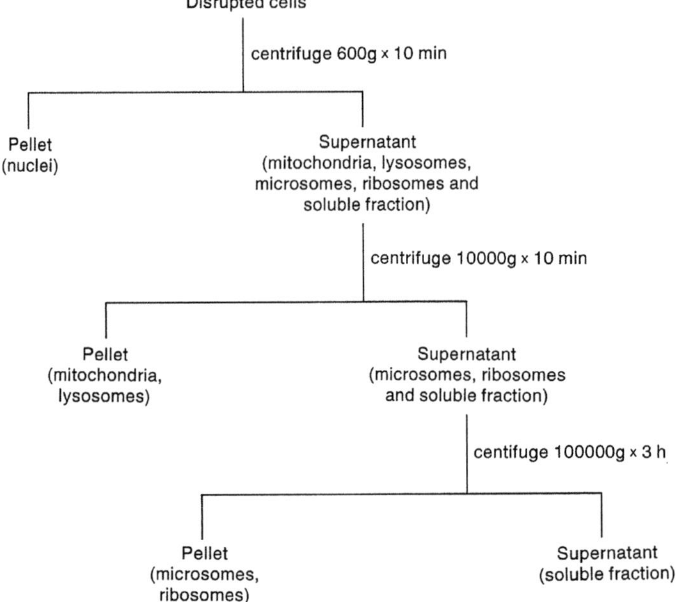

Fig. 2.7. Procedure for subcellular fractionation of mammalian cells by differential centrifugation [117]

2.9.2 Separation of Cell-Organelles from Mammalian Tissues

2.9.2.1 Separation of Nuclei, Mitochondria and Microsomes by Rate Zonal Centrifugation

Cell organelles from mammalian tissues may be separated at increasing centrifugational forces. If the procedures indicated in Fig. 2.7 are followed, the sedimented fractions are usually resuspended in buffer and then resedimented once or twice to obtain a fairly good separation of the main organelles nuclei, mitochondria and microsomes (disrupted endoplasmic reticulum). It is difficult to separate mitochondria from lysosomes by the procedure given in Fig. 2.7 because both have similar sizes and densities. Although various procedures have been described to separate mitochondria from lysosomes by rate zonal centrifugation it is recommended that the more elegant and simple methods of isopycnic centrifugation be used (see Sect. 2.9.2.3).

2.9.2.2 Separation of Golgi Apparatus from Rat Liver by Rate Zonal Centrifugation

Purification of the Golgi apparatus from rat liver was achieved by the following procedure [118].

    Male rats, 200 – 250 g, 50 days old (Holtzman strain), provided with standard diet and unlimited drinking water, are killed by decapitation and drained of blood. The livers are removed, weighed, and minced rapidly at room temperature. All other procedures are performed at 0 – 4 °C. The minced tissue, in lots of approximately 10 g, is mixed with 20 ml of cold homogenization medium. The homogenization medium is prepared by combining 50 ml of reagent a, 25 ml of reagent b, 10 ml each of reagents c and d and 5 ml of reagent e (reagent a: mix 50 ml of a 0.2 mol l⁻¹ Tris maleate solution (24.2 g Tris and 23.2 g maleic acid in 100 ml) and 37 ml of 0.2 N NaOH, dilute to a total of 200 ml, and adjust to pH 6.4 with NaOH, reagent b: 2 mol l⁻¹ sucrose, prepared in reagent a, reagent c: 10 % dextran (M_r = 225 000), reagent d: 10 mmol l⁻¹ $MgCl_2 \cdot 6H_2O$, reagent e: 0.1 mol l⁻¹ 2-mercaptoethanol). Homogenization is performed for 40 – 80 s at 5000 – 10 000 rpm using a Polytron 20 ST homogenizer. The homogenate is squeezed through a single layer of Miracloth to remove unbroken cells and connective tissue. The homogenate is centrifuged for 15 – 30 min at 2000 to 5000 g using a swing out rotor to concentrate the Golgi apparatus. The lipid on top of the tube and the supernatant fluid are removed by suction. The yellow-brown phase of the pellet (Golgi apparatus) which lies above the red to pink and dark brown layers containing whole cells, nuclei, and fragments of plasma membrane is resuspended in a portion of the lipid-free supernatant (final volume of about 6 ml for each liver homogenized), layered on 1.5 – 2 volumes of 1.25 mol l⁻¹ sucrose, and centrifuged for 30 min at 90 000 – 150 000 g using a rotor of the swingout type. The Golgi apparatus is concentrated at the 1.25 mol l⁻¹ sucrose-homogenate interface while mitochondria and endoplasmic reticulum enter the sucrose layer. The band containing the Golgi apparatus is removed from the top using a Pasteur pipette fitted with a rubber aspirator. This fraction is resuspended in the clear red supernatant fluid (for optimal preservation of morphology), in the homogenization medium or in distilled water (for highest fraction purity). The Golgi apparatus are collected by centrifugation at 2000 g for 30 min.

Table 2.12. Marker enzymes used in subcellular fractionation of animal cells [117]

Subcellular fraction	Enzyme
Nuclei	DNA nucleotidyltransferase
Nuclei	NMN adenylyltransferase
Mitochondria	Succinate dehydrogenase
Mitochondria	Cytochrome c oxidase
Endoplasmic reticulum	Glucose-6-phosphatase
Lysosomes	Acid phosphatase
Lysosomes	Ribonuclease
Peroxisomes	Catalase
Peroxisomes	Urate oxidase
Plasma membrane	5′-Nucleotidase
Cytosol	Glucose-6-phosphate dehydrogenase
Cytosol	Lactate dehydrogenase
Cytosol	6-Phosphofructokinase

The supernatant fluid is removed, and the surface of the pellet is carefully rinsed with distilled water. The pellets are finally resuspended in distilled water or an appropriate buffer solution at a concentration of 5 – 10 mg of Golgi apparatus protein per ml. The yield is 5 – 10 mg of Golgi apparatus protein per 10 g of liver with a fraction purity of at least 70 – 80 %.

Yield of Golgi apparatus

Further removal of contaminating cell components and a fraction purity of 90 % or more is achieved by repeated resuspension in distilled water or isolation medium and centrifugation at 2000 g for 30 min. Although dictyosomes remain intact, washing removes some of the peripheral tubules of the cisternae through vesiculation and reduces total yield of Golgi apparatus protein.

Levels of contaminating cell components are estimated from assays of marker enzymes (Table 2.12). Assays are carried out as soon as possible after preparation of the fractions. Enzyme assays are initiated by adding Golgi apparatus fractions diluted immediately before assay, and are performed at 37 °C with gentle shaking.

2.9.2.3 Separation of Mitochondria from Rat Brain by Isopycnic Centrifugation

A rapid isolation of metabolically active mitochondria from rat brain is possible by isopycnic centrifugation in Percoll gradients [119]. Two procedures were used for that purpose employing discontinuous Percoll gradients which yielded well-coupled mitochondria with high rates of respiratory activity and little residual contamination by synaptosomes or myelin. The procedures do not require an ultracentrifuge or swing-out rotor [114]:

Composition of extraction buffer

Rats (230 – 320 g) are fasted overnight prior to killing by decapitation. The brains are removed within 1 min into ice-cold "isolation buffer" which contains 0.32 mol l^{-1} sucrose, 1 mmol l^{-1} EDTA (K$^+$-salt), and 10 mmol l^{-1} Tris-HCl (pH 7.4). The cerebellum and underlying structures are removed and the remaining material is used as the

Fig. 2.8. Rapid isolation of metabolically active mitochondria from rat brain [117]

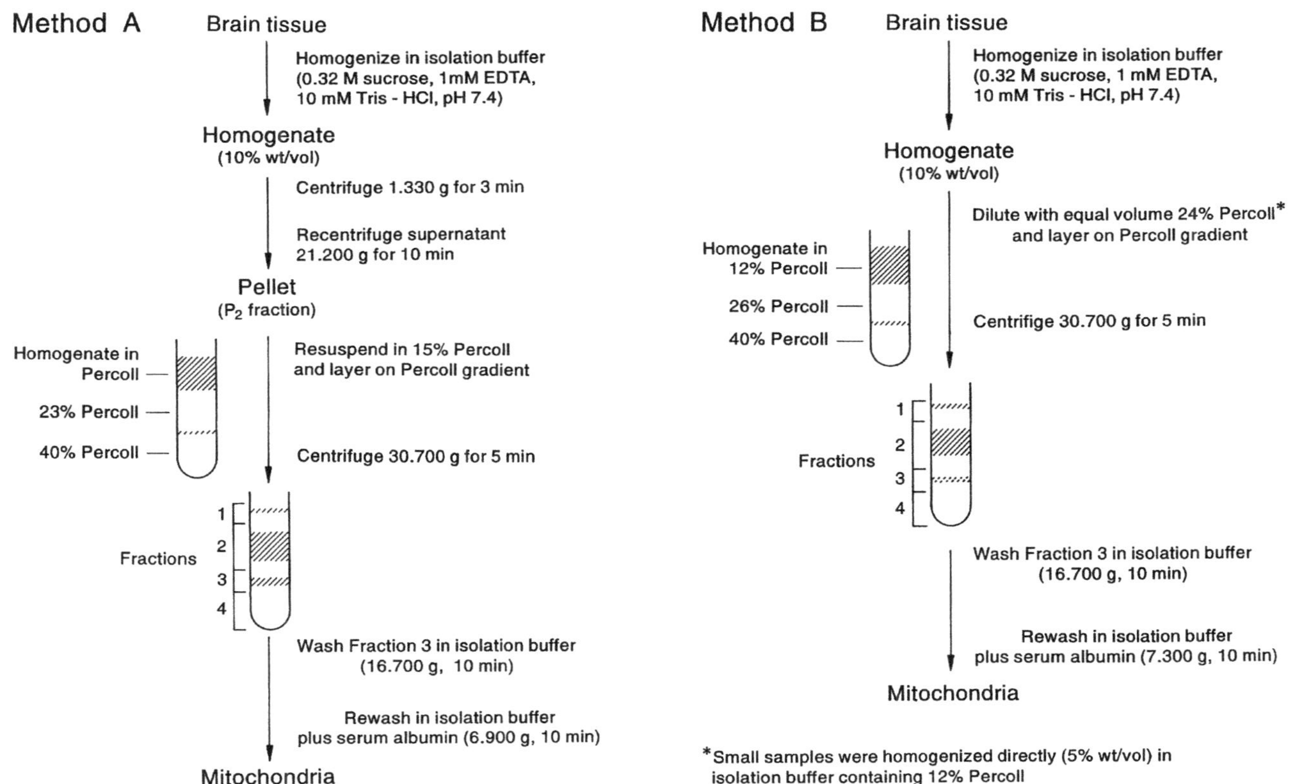

Method A
Brain tissue
Homogenize in isolation buffer (0.32 M sucrose, 1mM EDTA, 10 mM Tris - HCl, pH 7.4)
Homogenate (10% wt/vol)
Centrifuge 1.330 g for 3 min
Recentrifuge supernatant 21.200 g for 10 min
Pellet (P₂ fraction)
Homogenate in Percoll
23% Percoll
40% Percoll
Resuspend in 15% Percoll and layer on Percoll gradient
Centrifuge 30.700 g for 5 min
Fractions
1
2
3
4
Wash Fraction 3 in isolation buffer (16.700 g, 10 min)
Rewash in isolation buffer plus serum albumin (6.900 g, 10 min)
Mitochondria

Method B
Brain tissue
Homogenize in isolation buffer (0.32 M sucrose, 1 mM EDTA, 10 mM Tris - HCl, pH 7.4)
Homogenate (10% wt/vol)
Dilute with equal volume 24% Percoll* and layer on Percoll gradient
Homogenate in 12% Percoll
26% Percoll
40% Percoll
Centrifige 30.700 g for 5 min
Fractions
1
2
3
4
Wash Fraction 3 in isolation buffer (16.700 g, 10 min)
Rewash in isolation buffer plus serum albumin (7.300 g, 10 min)
Mitochondria

*Small samples were homogenized directly (5% wt/vol) in isolation buffer containing 12% Percoll

"forebrain". All solutions are ice-cold, and subsequent manipulations of the tissue are carried out on ice. The tissue is finely minced with scissors in a small amount of isolation buffer and washed three times with this buffer. The tissue in isolation buffer (10 wt/vol%) is homogenized by hand in an all-glass Dounce homogenizer (Kontes, Kineland, NJ, USA) using four up and down strokes with the A pestle (total clearance 0.12 mm) and eight strokes with the B pestle (total clearance 0.05 mm).

Centrifugation on a stepped gradient of Percoll

Method A: the homogenate is centrifuged at 1330 g for 3 min in a Beckman JA 20 rotor (Beckman centrifuge, model J2 – 21, Beckman, München, Germany) at 4 °C. The supernatant is carefully decanted and the pellet resuspended in half of the original volume using a Dounce homogenizer. This homogenate is recentrifuged as above, the supernatant retained, and the pellet discarded. The supernatants are pooled and centrifuged at 2000 g for 10 min in a JA 20 rotor. The decanted supernatant is discarded and the pellet resuspended in 15 % Percoll (10 ml/g of tissue originally homogenized). The discontinuous density gradient is prepared in 16 × 102 mm polyallomer centrifuge tubes by layering 3-ml fractions of the resuspended pellet on two preformed layers consisting of 3.5 ml of 23 % Percoll above 3.5 ml of 40 % Percoll. The lower layers are prepared manually immediately before to the start of the preparation by adding the 40 % Percoll below the 23 % Percoll. Tubes are centrifuged for 5 min at 30 700 g in a JA 21 rotor (acceleration time: 54 s, deceleration with brake: 100 s). Three major bands of material are obtained (Fig. 2.8), and the material banding near the interface of the lower two Percoll layers (fraction 3 in Fig. 2.8) are diluted 1:4 by gently mixing with isolation buffer. They are then centrifuged at 16 700 g for 10 min in a JA 21 rotor. The supernatant is removed with a Pasteur pipette to within a few millimeters of the base of the tube and the material is gently resuspended. Fatty acid-free bovine serum albumin (10 mg/ml) is added (0.5 ml per forebrain originally used) and the mixture is diluted with isolation buffer (final volume 3 ml per forebrain). After centrifugation at 6900 g for 10 min in a JA 21 rotor, the supernatant is rapidly decanted and the pellet gently resuspended in isolation buffer (0.3 ml per forebrain), using a fine Teflon stirring rod. This fraction is stored on ice for further investigations.

Fractionation of cerebral subregions

Method B: the initial homogenate is mixed 1:1 with a 24 % Percoll solution. For subregions (less than one cerebral hemisphere), a 5 wt/vol% homogenate is prepared directly in 12 % Percoll and no further dilution is necessary. Approximately 3.0-ml aliquots of homogenate in Percoll are layered over two preformed layers (prepared as in method A) consisting of 3.5 ml of 26 % Percoll on top of 3.5 ml of 40 % Percoll in 16 × 102 mm polyallomer centrifuge tubes. Centrifugation of the gradient is performed at 30 700 g for 5 min. The fraction accumulating near the interface of the lower two layers (fraction 3 in Fig. 2.8) is collected and slowly diluted 1:4 with isolation buffer. This mixture is centrifuged at 16 700 g for 10 min in a JA 21 rotor, producing a loose pellet as in method A. The pellet is stored on ice. Suitable aliquots for respiratory measurements or enzyme assays are transfered to an Eppendorf conical microfuge tube (1.5 ml volume), 100 μl of fatty acid-free bovine serum albumin (10 mg/ml) are added, and finally the volume is adjusted to 1 ml with isolation buffer. After centrifugation for 10 min in an Eppendorf microfuge (Eppendorf, Hamburg, Germany) at 7300 g, the supernatant is removed and the pellet resuspended directly in the appropriate buffer for assay [119].

2.9.2.4 Separation of Various Cell-Organelles by Isopycnic Centrifugation

Liver tissue biopsies may be subfractionated by a single centrifugation run [112]. A liver homogenate is mixed with Percoll in 0.25 mol l^{-1} sucrose and cell-organelles are separated upon formation of an almost linear gradient of Percoll ranging from 1.035 to 1.15 g/ml (cf. Table 2.11). The recoveries of cell-organelles are 86 – 100 %. Gradients of Percoll generated by centrifugation in situ may also be used to separate plasma membranes from rat liver [120] or human platelets [121].

Rat liver lysosomes were fractionated in a zonal rotor into a light fraction of 1.06 g/ml and a heavy fraction of 1.09 g/ml [122]. It could be shown that the densities of the lysosomes in vivo increased with their age.

Rat liver peroxisomes were isolated on a Percoll gradient generated in situ (cf. Table 2.11) and used to study fatty acid oxidation [122].

Light granules of density 1.04 g/ml and heavy granules of density 1.08 – 1.10 g/ml were isolated from mast-cell tumor cells by different consecutive centrifugation steps [123]. The granules were isolated from the mast-cell tumor by centrifugation of the homogenate in 0.25 M sucrose at 40 000 g for 20 min. The pellet was then resuspended in 0.25 mol l^{-1} sucrose and layered on top of a Percoll solution ($p = 1.065$ g/ml in 0.25 mol l^{-1} sucrose). The two species of granules were separated while forming the Percoll gradient at 55 000 g for 30 min.

Separation of cell organelles in a linear gradient of Percoll

Lysosomes

Peroxisomes

2.9.3 Enzyme Composition of Mammalian Cell-Organelles

2.9.3.1 Enzymes Present in Cell Nuclei

The nucleus is the densest subcellular organelle ($p = 1.35$ g cm^{-3}) which results from the high concentrations of macromolecules within it (around 34 g protein, 9.5 g DNA and 1.5 g RNA per 100 g nuclei) [124]. The perinuclear membrane which envelops the nucleus contains all the enzymes that are also found in the endoplasmic reticulum except cytochrome c oxidase. Enzymes of the soluble compartment of the nucleus are readily washed out while those of the second group listed in Table 2.13 are not, but may be extracted with concentrated salt solutions (1 mol l^{-1} NaCl). Enzymes of the fourth group are only extracted by use of detergents.

Table 2.13. Enzymes present in the cell nucleus [117]

1. Enzymes in the soluble space of the nucleus
  Glycolytic enzymes
  Pentose phosphate pathway enzymes
  Arginase
  Isocitrate dehydrogenase
  Lactate dehydrogenase
  Malate dehydrogenase
	2. Enzymes bound to chromatin
	  RNA nucleotidyltransferase II
	  RNA nucleotidyltransferase III
	  Nucleoside triphosphatase
	  DNA nucleotidyltransferase
	  NMN adenylyltransferase

3. Enzymes concentrated in the nucleolus
  RNA nucleotidyltransferase I
  RNA methyltransferases
  Ribonuclease
	4. Enzymes bound to membranes
	  Glucose-6-phosphatase
	  Acid phosphatase

Table 2.14. Principal enzymes or groups of enzymes present in mitochondria [117]

1. Matrix
   Tricarboxylic acid cycle enzymes except succinate dehydrogenase
   Enzymes catalyzing the ß-oxidation of fatty acids
   Carbamoylphosphate synthetase
   Glutamate dehydrogenase
   Phosphoenolpyruvate carboxykinase (pigeon liver)
   Pyruvate carboxylase
   Ornithine carbamoyltransferase

2. Inner membrane
   Adenosinetriphosphatase (ATP synthase)
   α-Aminolaevulinate synthase
   Carnitine palmitoyltransferase
   Cytochrome c oxidase
   Glycerol 3-phosphate dehydrogenase (FAD enzyme)
   3-Hydroxybutyrate dehydrogenase
   NADH dehydrogenase
   Succinate dehydrogenase (+ associated) respiratory chain

3. Intermembrane space
   Adenylate kinase
   Nucleosidediphosphate kinase
   Nucleosidemonophosphate kinase
   L-xylulose reductase

4. Outer membrane
   Acyl-CoA synthetase
   Adenylate kinase
   Cholinephosphotransferase
   Cytochrome b_5 reductase
   Glycerophosphate acyltransferase
   Hexokinase
   Kynureninase
   Monoamine oxidase (flavin-containing)
   NADH dehydrogenase (rotenone insensitive)
   Phospholipase A_2

2.9.3.2 Principal Enzymes of Mitochondria

The density of mitochondria of animal cells is about $p = 1.1$ g cm^{-3}. On the inner surface of the inner membrane reside the mushroom-shaped particles containing mitochondrial ATP synthase. The mitochondrial matrix is granular in appearance and has a very high concentration of protein (500 mg cm^{-3}) [117]. In the liver about 70% of the mitochondrial protein is in the matrix, 20% is part of the inner mitochondrial membrane and only 4% is part of the outer membrane [117]. A list of enzymes present in mitochondria is presented in Table 2.14.

2.9.3.3 Enzymes of Lysosomes

Lysosomes are roughly spherical organelles or vacuoles. They are slightly smaller than mitochondria and of similar density (p is approximately 1.2 g cm^{-3}). Lysosomes

are bounded by a single membrane enclosing a dense granular matrix. They contain
about 60 hydrolytic enzymes of which most have an acid pH-optimum (Table 2.15).
About 35% of the total protein is in the membranes and the remainder is in the
matrix. The protein concentration in the matrix is about 200 mg cm^{-3} [117] and the
intralysosomal pH is about 1.5 units below that of the cytosol. Several of the proteins
of the membrane and matrix are glycoproteins and many are acidic. The proteins on
the inner surface of the lysosomal membrane contain about 16 μg sialic acid per mg
protein (and this is thought to maintain the acid intralysosomal pH) [125]. It has been
suggested [126, 127] that there are two types of lysosomes, primary and secondary.
Primary lysosomes have a full complement of hydrolytic enzymes but do not participate in intracellular digestion. Secondary lysosomes result when primary lysosomes
fuse with other membranous structures containing substrates to be degraded. The
primary lysosomes are believed to be formed by pinching off portions of the Golgi
apparatus [117].

**Primary
and secondary
lysosomes**

Table 2.15. Enzymes present in lysosomes [117]

Class/subclass of enzymes	Name (EC number)
Oxidoreductases	NADPH oxidase (1.6.2.*); Peroxidase (1.11.1.7)
Hydrolases acting on carboxylic esters	Arylesterase (3.1.1.2); Triacylglycerol lipase (triglyceride lipase) (3.1.1.3); Phospholipase A$_2$ (phospholipid 2-deacylase) (3.1.1.4); Cholesterol esterase (3.1.1.13); Phospholipase A$_1$ (phospholipid 1-deacylase) (3.1.1.32)
Hydrolases acting on phosphoric monoesters	Acid phosphatase (3.1.3.2); Phosphatidate phosphatase (3.1.3.4); Phosphoprotein phosphatase (3.1.3.16)
Hydrolases acting on phosphoric diesters	Deoxyribonuclease II (3.1.4.6); Sphingomyelin phospho-diesterase (sphingomyelinase) (3.1.4.12); Phosphodiesterase II (acid exonuclease, spleen exonuclease) (3.1.4.18); Ribonuclease II (3.1.4.23); Acyl di(glycerophosphoryl) glycerol phospho-diesterase (3.1.4.*)
Hydrolases acting on sulphuric esters	Sulphatase A (arylsulphatase A) (3.1.6.1); Cerebroside sulphatase (3.1.6.8); Sulphatase B (arylsulphatase B) (3.1.6.1); Chondroitin-4-sulphatase (3.1.6.4); Chondroitin-6-sulphatase (3.1.6.*); Iduronosulphatase (3.1.6.*)
Hydrolases acting on glycosides	Lysozyme (muramidase) (3.2.1.17); Neuraminidase (3.2.1.18); α-Glucosidase (acid maltase) (3.2.1.29); β-Glucosidase (3.2.1.21); Glucosylceramidase (glucocerebrosidase) (3.2.1.45); α-Galacto-sidase (3.2.1.22); Galactosylgalactosylglucoxylceramidase (ceramide trihexosidase) (3.2.1.47); β-Galactosidase (3.2.1.23); GM$_1$ Gangliosidase (3.2.1.*); Galactosylceramidase (3.2.1.46); α-Mannosidase (3.2.1.24); β-Mannosidase (3.2.1.25); β-N-Acetyl-glucosaminidase (3.2.1.30); (β-N-acetylhexosaminidase) (3.2.1.52); β-Glucuronidase (3.2.1.31); Hyaluronate endoglucos-aminidase (hyaluronoglucosaminidase) (3.2.1.35); α-N-Acetyl-galactosaminidase (3.2.1.49); α-N-Acetylglucosaminidase

Table 2.15 (continued)

Class/subclass of enzymes	Name (EC number)
Hydrolases acting on glycosides	(3.2.1.50); α-L-Fucosidase (3.2.1.51); α-L-Iduronidase (3.2.1.76); Heparin endoglucuronidase (3.2.1.*); Glycopeptide endoglucosaminidase (3.2.1.*); Heparan sulphate endoglycosidase (3.2.1.*); NAD(P)$^+$ nucleosidase (3.2.2.6)
Hydrolases cleaving peptide bonds near the ends of polypeptides: exopeptidases	Lysosomal aminopeptidase (amino acid naphthylamidase, cathepsin H) (3.4.11.*); Lysosomal carboxypeptidase A (cathepsin A, cathepsin I) (3.4.12A.1); Lysosomal carboxypeptidase B (cathepsin IV; catheptic carboxypeptidases A, B and G, cathepsin B2) (3.4.12A.1); Lysosomal carboxypeptidase C (proline acid carboxypeptidase, angiotensinase C, acid angiotensinase, catheptic carboxypeptidase C, prolylcarboxypeptidase) (3.4.12A.2); Tyrosine acid carboxypeptidase (3.4.12A.3); Lysosomal dipeptidase (Ser-Met dipeptidase, cysteinyltyrosinase, Cys-Tyr hydrolase) (3.4.13.*); Dipeptidyl-peptidase I (cathepsin C; dipeptidyl aminopeptidase I, dipeptidyl arylamidase I, glucagon-degrading enzyme, dipeptidyl transferase) (3.4.14.1); Dipeptidylpeptidase II (dipeptidyl aminopeptidase II, dipeptidyl arylamidase II, carboxytripeptidase) (3.4.14.2)
Hydrolases cleaving peptide bonds away from the ends of polypeptides: endopeptidases	Acrosin (3.4.21.10); Lysosomal elastase (leucocyte elastase) (3.4.21.11); Cathepsin G (chymotrypsin-like enzyme) (3.4.21.20); Cathepsin B (cathepsin B', cathepsin B1) (3.4.22.1); α-Glutamyl hydrolase (glutamate carboxypeptidase, folate conjugase, pteroyl polyglutamate hydrolase, pteroyl-α-oligoglutamyl endopeptidase) (3.4.22.11); Cathepsin L (3.4.22.*); Cathepsin D (3.4.23.5); Cathepsin E (3.4.23.*); Granulocyte collagenase (3.4.24.7)
Hydrolases acting on amide bonds other than peptides	Acylsphingosine deacylase (ceramidase) (3.5.1.23); Aspartylglucosylaminase (3.5.1.26)
Hydrolases acting on acid anhydrides	Nucleoside triphosphatase (acid pyrophosphatase) (3.6.1.15); Adenylylsulphatase (3.6.2.1); Phosphoadenylylsulphatase (3.6.2.2)
Hydrolases acting on nitrogen-sulphur bonds	Heparin sulphamatase (heparin sulphamidase or N-sulphatase) (3.10.1.*)

* not further specified

2.9.3.4 Enzymes of the Endoplasmic Reticulum

When a cell is disrupted the endoplasmic reticulum is broken and the membrane fragments form vesicles. These vesicles constitute the "microsome fraction" in subcellular fractionations. The principal type of liver cell, the hepatocyte, and the cells of the exocrine pancreas (the cells concerned producing enzymes for digestion) secrete a number of proteins and both have dense endoplasmic reticula, mainly with attached ribosomes. The principal enzymes of the endoplasmic reticulum are presented in Table 2.16. Cytochrome P 450 is the most abundant enzyme comprising 4 % of the total protein. Many proteins appear on the cytoplasmic surface of the endoplasmic reticulum and some in the lumen and the luminal surface [117].

Table 2.16. Principal enzymes of the endoplasmic reticulum [117]

Location	Enzyme or group of enzymes
smooth endoplasmic reticulum	Cholesterol biosynthetic enzymes Steroid hydroxylation enzymes Fatty acid elongating enzymes (C_{16} – C_{24}) Drug-metabolizing enzymes (aromatic ring hydroxylation, side chain oxidation, deamination, dealkylation, dehalogenation) Glycerolphosphate acyltransferase Carnitine acyltransferase
rough endoplasmic reticulum, cytoplasmic side	Protein synthesis
rough endoplasmic reticulum, luminal side	Glucose-6-phosphatase
cytoplasmic side	Adenosinetriphosphatase Cytochrome b_5 reductase NADPH-cytochrome reductase GDPmannose α-D-mannosyltransferase
luminal side	Nucleosidediphosphatase β-D-glucuronidase UDP-glucuronosyltransferase
cytoplasmic side	5′-Nucleotidase Cholesterol acyltransferase

2.9.3 Enzymes of the Cytosol

The cytosol is the soluble phase which remains as the supernatant after a tissue homogenate has been centrifuged for sufficient g × min to sediment the microsome fraction and free ribosomes (usually 100 000 g, 60 min). It is also named the "high-speed" supernatant. It comprises most of the soluble phase from both the cytoplasm and the luminal side of the endoplasmic reticulum. A large number of enzymes are present in the cytosol, particularly those responsible for glycolysis, gluconeogenesis, fatty acid synthesis, nucleotide biosynthesis, and amino acyl-tRNA synthesis (Table 2.17) [117].

2.9.4 Plant Subcellular Organelles

2.9.4.1 Prerequisites for the Isolation of Chloroplasts

Chloroplasts with an intact envelope and high photosynthetic activity can be obtained from spinach and pea by mechanical disintegration of leaves or from protoplasts

Table 2.17. Enzymes or groups of enzymes present in the cytosol [117]

1. Carbohydrate metabolism
   Glycolytic enzymes including phosphorylase, phosphorylase kinase, and protein kinase A
   Enzymes of the pentose phosphate pathway
   Citrate (*pro-3S*)-lyase
   Fructose-bisphosphatase
   Glucose-1-phosphate uridylyltransferase
   Glycogen synthase
   Isocitrate dehydrogenase (NADP-dependent)
   Lactate dehydrogenase
   Malate dehydrogenase (cytosolic isoenzyme)
   Malate dehydrogenase (oxaloacetate-decarboxylating) (NADP$^+$)
   Phosphoenolpyruvate carboxykinase (rat liver)

2. Lipid metabolism
   Acetyl-CoA carboxylase
   Fatty acid synthetase complex
   Glycerol-3-phosphate dehydrogenase (NAD$^+$)

3. Amino acid and protein metabolism
   Alanine aminotransferase
   Aminoacyl-*t*RNA synthetases
   Arginase
   Argininosuccinate lyase
   Argininosuccinate synthetase
   Aspartate aminotransferase

4. Nucleic acid synthesis
   Nucleoside kinase
   Nucleotide kinase

of these and a number of other plants. By mechanical disintegration the yield in intact chloroplasts is about 5 – 7 % while about 10 – 20 % of leaf mesophyll cells can be obtained as protoplasts and the yield of intact chloroplasts from these is 95 – 100 %. During mechanical disintegration of whole leaf tissues as well as during the disruption of protoplasts a number of deleterious compounds such as phenolics and terpenoids may be set free. Therefore chloroplast isolating media must meet certain requirements:

Composition of chloroplast media

1. *osmolarity,* the osmolarity of the medium must be of that of the chloroplasts to prevent their lysis or shrinking;
2. *buffer,* the medium must be buffered to a pH-value which stabilizes the chloroplasts and avoids clotting;
3. *phosphate,* to maintain photosynthesis,
4. *bovine serum albumin* (optional), to bind free fatty acids and phenolics which destroy the chloroplast envelope, and also avoid clotting of the chloroplasts;
5. *manganese,* a co-factor of the enzymes which are part of photosystem II, responsible for the hydrolysis of water;
6. *magnesium,* a co-factor of enzymes of the Calvin-cycle and part of the chlorophyll molecule; Mg activates the enzymes ribulose-bisphosphate carboxylase/oxygenase, fructose-1,6-bisphosphatase and seduheptulose-1,7-bisphosphatase [128],

7. *ethylenediaminetetraacetic acid* (EDTA), stabilizes the surface charge of chloroplasts and avoids clotting [128];
8. *HCO_3^-*, keeps the Calvin-cycle running;
9. *polyvinylpyrrolidone (PVP)* (optional), binds phenolics and avoids the oxidation of o-diphenols to quinones which disintegrate the chloroplast envelope;
10. *catalase* (optional), disintegrates H_2O_2 which may be formed during chloroplast isolation and which has a deleterious effect on the envelope.

Leaves of spinach (*Spinacia oleracea*) yield the best chloroplasts and their behaviour is very well documented. A number of other species are often called spinach, for example various cultures of beet (*Beta vulgaris*) such as "perpetual" or "summer" spinach, Swiss chard and "New Zealand spinach" (*Tetragonia expansa*) but these yield chloroplasts with indifferent CO_2-fixing rates [128].

2.9.4.2 Culture of Spinach and Peas

In regions away from the equator it becomes difficult to grow spinach. In Europe it produces copious leaves only in spring and autumn unless the days are artificially shortened to about 8.5 h. It grows well in water-culture. In low light it may be necessary to limit the temperature to 15 – 20 °C while in strong light conditions spinach is relatively indifferent to high temperatures. When chloroplasts are extracted by mechanical tissue disruption, chloroplasts rich in starch are to be avoided. Therefore spinach is best harvested in early morning and avoided entirely once flowering has started [128]. White rings of starch and calcium oxalate occurring in the chloroplast pellet upon centrifugation are indicative of poor spinach [128].

Peas are easy to grow especially if a water culture (Hoagland's nutrient solution) is used [42]. In order to obtain chloroplasts with an optimum CO_2-fixing rate, it is important to manipulate growth conditions to give healthy, uniform and actively growing seedlings of about 5 cm height after 9 – 12 days. The variety used is less important. Spinach and pea leaves are the plants from which intact chloroplasts can be isolated from crude extracts prepared by mechanical disruption of the tissues. If it is intended to isolate chloroplasts from other plants, mostly protoplasts must be prepared first.

Nutrient solution

2.9.4.3 Isolation of Type A1 Chloroplasts from Pea Leaves

Intact chloroplasts may be prepared from pea leaves by the following procedure [42]. Pea leaves are weighed, put in a beaker and stored submerged in distilled 4 °C water for at least 1 h. The water is then carefully decanted and seven parts of isolation medium are added to one part of fresh matter (v/w). The isolation medium consists of 103 g sucrose, 4.4 g $Na_4P_2O_7 \cdot 10\,H_2O$, 2 g bovine serum albumin, and 1 g $MgCl_2 \cdot 6H_2O$ in 1 l of distilled water. Before the solution is adjusted to 1 l, it is titrated to pH 7.8 using 1 mol $l^{-1}\,H_3PO_4$. The density of the medium is 1.041 g cm^{-3}. Then, 15 min after the addition of the isolation medium, the plant material is homogenized for 20 – 30 s in a Braunmixer (Braun, Frankfurt/M, Germany) at speed 1. The resulting slurry is filtered through two layers of a loosely woven cotton diaper and the filtrate centrifuged for two min at 300 g to remove cell debris. The resulting supernatant is then centrifuged for 2 min at 4000 g to precipitate chloroplasts. The sediment is resuspended in

Isolation medium

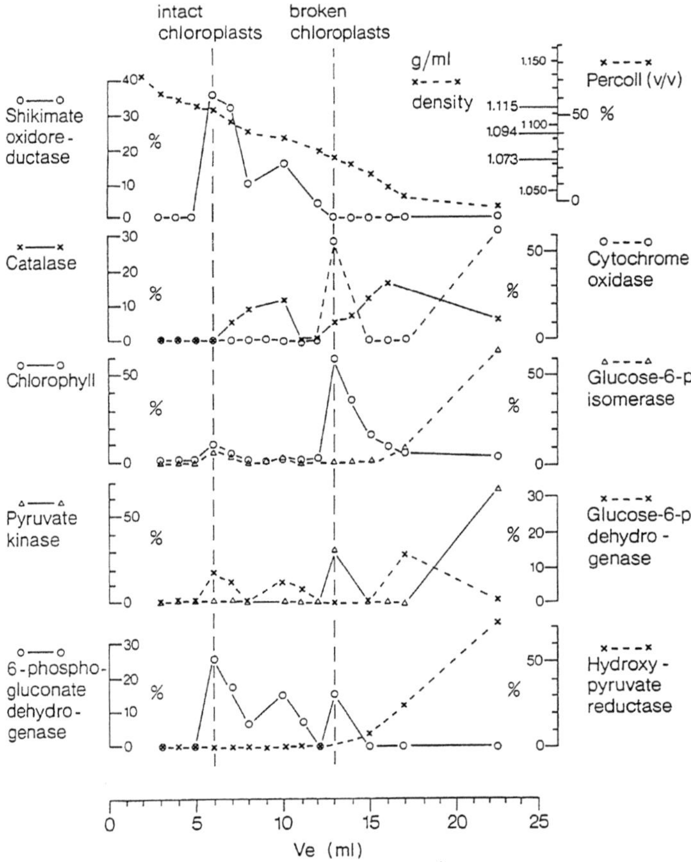

Fig. 2.9. Distribution of marker enzymes in a stepwise Percoll-gradient after the centrifugation of a crude particulate homogenate prepared from the leaves of pea plants which were grown for about 14 days under 7 W m^{-2} of white light and which were supplied with a modified Hoagland nutrient solution. Percentages are those % enzyme activity of total enzyme activity of gradient and supernatant. 100 % equals 0.6 nkat for glucose-6-phosphate isomerase, 0.74 nkat for glucose-6-phosphate dehydrogenase, 0.78 nkat for 6-phosphogluconate dehydrogenase, 37.3 nkat for hydroxypyruvate reductase, 3.7 nkat for pyruvate kinase, 95.7 units for catalase, and 0.5 units for cytochrome c oxidase. With respect to chlorophyll 100 % = 33.2 μg [42]

Composition
of a stepped
Percoll-gradient

a few milliliters of isolation medium and layered on top of a step gradient of Percoll containing the same amounts of substances as the isolation medium does. From bottom to top, the gradient consists of 2 ml 90 % v/v (density (p): 1.162 g cm^{-3}), 3 ml 60 %, 2 ml 55.4 %, 3 ml 40 %, 3 ml 24 %, and 2 ml of 10 % Percoll. After about 15 min centrifugation at around 4000 g, the layer containing intact chloroplasts (p = 1.115 g cm^{-3}) can be removed from the gradient. If the gradient is fractionated into 1-ml fractions the location of intact and broken chloroplasts can be analyzed by measuring chlorophyll contents and the activities of marker enzymes (Fig. 2.9). For routine assays linear Percoll gradients ranging from 72 % ($p = 1.138$ g cm^{-3}) to 0 % Percoll ($p = 1.041$ g cm^{-3}) can also be used [42, 129]. Aliquots of the fraction of intact chloroplasts are taken for the determination of chlorophyll, enzyme activities and to count the number of chloroplasts.

The main part of the suspension containing the chloroplasts is centrifuged for 15 min at 4000 g, the supernatant is decanted, and the intact chloroplasts are ruptured osmotically by adding a few milliliters 10 mmol l^{-1} Tris-HCl buffer of pH 9.0. The resulting suspension is centrifuged for 15 min at 4500 g, and the clear supernatant is used for enzyme assays. When fractionating the gradient into 1-ml portions, each fraction is diluted with an equal amount of distilled water, and before measuring enzyme activities the reaction solutions captured in plastic filter troughs are centrifuged for 1 min at 2000 g, except for catalase and cytochrome c-oxidase [42]. Marker enzymes of chloroplasts may be taken from Table 2.18.

When using the described procedure to isolate intact chloroplasts from two-week old pea plants three bands of chloroplasts may be obtained (Fig. 2.9). The densities of these bands were estimated to be 1.115, 1.094 and 1.073 g cm^{-3} [42]. The light band ($p = 1.073$) contains broken chloroplasts while the two heavier bands contain intact chloroplasts. Of these the most dense chloroplasts ($p = 1.115$) are free of catalase activity whereas the lighter band of intact chloroplasts which bands at a density of 1.094 contains catalase activity; catalase is the marker enzyme of peroxisomes. Under the phase contrast microscope, intact chloroplasts appear as double, lightbreaking particles. They are free of the mitochondrial marker enzyme cytochrome c oxidase and the cytosol marker enzyme pyruvate kinase (Fig. 2.9). Broken chloroplasts, however, having a density of 1.073 g cm^{-3} were contaminated by mitochondria and cytosol debris. For serial measurements only the most dense chloroplasts are used. The CO_2-dependent O_2-production of these chloroplasts was estimated to be larger than 100 μmol mg^{-1} chlorophyll h^{-1} [135], and they have been classified as type A$_1$ chloroplasts [136]. The intactness of their outer and inner membranes has been demonstrated [131]. The recovery of intact chloroplasts by the method described is 5 – 7 % [42]. Table 2.19 gives the data obtained when estimating the intracellular location of the enzyme shikimate dehydrogenase in pea leaves.

Range of linear Percoll-gradient

Rupture of intact chloroplasts

Types of chloroplasts

Type A1-chloroplasts

Table 2.18. Marker enzymes of plant cell organelles

Organell	Marker enzyme [Ref.]
Chloroplasts	ribulosebiphosphate carboxylase (4.1.1.39) [130], NADP-glyceraldehyde-phosphate dehydrogenase (1.2.1.12) [131], ferredoxine-NADP-reductase (1.18.1.2) [132], Mg-dependent ATPase (3.6.1.3) [133]
Mitochondria	cytochrome c oxidase (1.9.3.1) [42]
Peroxisomes	catalase (1.11.1.6) [42], hydroxypyruvate reductase (= glyoxylate reductase) (1.1.1.26) [42], glycollate oxidase (1.1.3.1) [134]
Glyoxisomes	citrate (si)-synthase (4.1.3.7)
Cytosol	pyruvate kinase (2.7.1.40) [42] cytochrome c reductase

Table 2.19. Data obtained when eluting chloroplasts from pea leaves [42]

Fresh weight (FW):
1.64 g FW (leaves of one plant)$^{-1}$

Dry weight (DW):
0.17 g DW (leaves of one plant)$^{-1}$

Chlorophyll (a + b) (Chl):
1.785×10^{-3} g Chl (g FW)$^{-1}$
0.307 g Chl (l of chloroplast suspension)$^{-1}$
2.84×10^{-12} g Chl (chloroplast)$^{-1}$

Chloroplasts:
1.08×10^{11} chloroplasts (l of chloroplast suspension)$^{-1}$
0.63×10^{9} chloroplasts (g FW)$^{-1}$
1.03×10^{9} chloroplasts (leaves of one plant)$^{-1}$

Shikimate dehydrogenase (SDH):
1.915×10^{-9} kat SDH (g FW)$^{-1}$ [70 % are chloroplast bound]
0.75×10^{-6} kat SDH (g Chl)$^{-1}$
0.11×10^{-6} kat SDH (l of chloroplast suspension)$^{-1}$
1.02×10^{-18} kat SDH (chloroplast)$^{-1}$
3.14×10^{-9} kat SDH (leaves of one plant)$^{-1}$

Further recipes to isolate intact chloroplasts by mechanical desruption of leaves are given in [128].

2.9.4.4 Isolation of Chloroplasts from Protoplasts

2.9.4.4.1 Preparation of Tissue

The time of protoplast preparation and chloroplast isolation should be kept as short as possible. Protoplasts are prepared from mesophyll cells by enzymatically digesting the pectin and cellulose material of their cell walls. Therefore, it is necessary to bring the cells into contact with the digestive enzymes which are cellulases and pectinases. Since these enzymes cannot easily penetrate through the epidermis cell layer leaves are either cut in small segments or the epidermis is stripped off or broken by other methods. In the case of monocots such as wheat, barley and maize, leaves are cut transversely into 0.5 – 1 mm segments with a razor blade [128]. Larger leaves of dicots such as tobacco, spinach and peas can be deprived of the epidermis by use of carborundum or gentle brushing (e. g. with a toothbrush) or by stripping it off with a pincette [128]. Gentle brushing of the epidermis is especially effective with sunflower. Upon slicing, the spaces between the cells are filled with air which hinders the digestive enzymes coming into contact with the cell walls. This difficulty can be avoided either by infiltrating the leaf segments under vacuum or by slicing the tissue under 0.5 M sorbitol, draining the strips and transferring them to the incubation medium. The latter method is less time consuming than stripping the epidermis and works particularly well with spinach [128].

2.9.4.4.2 Enzymic Digestion of Cell Walls

Active protoplasts are best prepared when the pH of the incubation medium is 5.0 – 5.8. The temperature is usually maintained between 25 and 30 °C. Digestion time is a compromise between yield, activity and convenience. Digestion of wheat leaves, leaves of *Zea maize*, *Digitaria sanguinalis* and *Helianthus annus* can be accomplished in 2 – 3 h at 28 °C (cf. 128). A mixture of cellulase and pectinase from *Aspergillus japonicus* (Pectolyase Y 23) releases active protoplasts from tobacco and grasses in less than 1 h [137]. The yield of pure protoplasts is 10 – 20 %. Shaking at any stage of the digestion procedure has a deleterious effect on the subsequent rate of photosynthesis [128]. During digestion the tissue is illuminated with a 150 W bulb mounted approximately 25 cm above the incubation medium [128]. The digestive enzymes move more rapidly into illuminated tissue and the rates of photosynthesis are improved in illuminated protoplasts [128]. The digestive enzymes become irreversibly bound to the cell walls and therefore it may be beneficial to renew the enzyme solution during incubation.

2.9.4.4.3 Purification of Protoplasts

After incubation, undigested tissue is removed by filtration through a nylon sieve of about 500 μm mesh (200 μm to retain vascular strands, 80 μm to retain bundle-sheath strands). Protoplasts are then purified by flotation in a step-gradient of sucrose and sorbitol. Mesophyll protoplasts float when centrifuged in a medium of high density (≥ 0.5 mol l^{-1} sucrose) while chloroplasts and debris collect in the pellet. For that purpose a crude protoplast preparatian in 0.5 mol l^{-1} sorbitol is overlayered with a buffered solution containing 0.1 mol l^{-1} sorbitol and 0.4 mol l^{-1} sucrose. On top of this solution a buffered 0.5 mol l^{-1} sorbitol solution is layered. Intact protoplasts will collect after 5 min centrifugation at the interface of the upper two solutions [138]. If the protoplasts are denser than a 0.5 mol l^{-1} sucrose solution, its concentration can be increased to 0.6 mol l^{-1}, or 5 – 10 % of dextran (MW 15 000 – 20 000) [139 – 142] or Ficoll [142] can be added. The protoplasts are removed from the step gradient, diluted with 0.5 mol l^{-1} sorbitol and centrifuged at 100 g for 1 min to precipitate them. They are generally stable in 0.5 mol l^{-1} sorbitol, 1 mmol l^{-1} CaCl$_2$, 20 mmol l^{-1} MES (2-[*N*-mopholino]-ethanesulfonic acid), pH 6.0 if stored at 4 °C for not longer than 8 – 10 h.

Separation
of protoplasts
and chloroplasts

2.9.4.4.4 Protocols to Prepare and Purify Protoplasts from C4-Species, Spinach and Peas

Isolation media [128]: 0.5 mol l^{-1} sorbitol, 1 mmol l^{-1} CaCl$_2$, 5 mmol l^{-1} MES-buffer; the medium is only lightly buffered and should be readjusted to pH 5.0 – 5.5 with dilute HCl after addition of the digestive enzymes cellulase and pectinase. Cellulase is used at a 2 – 3 wt/vol% concentration, pectinase concentrations are 0.3 – 0.5 wt/vol%. The following enzyme compositions have been found suitable:

Isolation media

wheat, maize, other C4-species [128]: 2 wt/vol% Onozuka R 10, 0.3 wt/vol% macerozyme R 10, pH 5.5;
spinach and peas: 3 wt/vol% Onozuka R 10, 0.5 wt/vol% macerozyme R 10, pH 5.5;
sunflower: 2 wt/vol% Onozuka R 10, 0.5 wt/vol% Rohament P, pH 5.5.

Digestive
enzymes

(Onozuka R 10 is supplied by Yakult Biochemical Co. Ltd, Nishinomiya, Japan and represents cellulase from *Trichoderma viride*; macerozyme R 10 is from the same producer and represents pectinase from Rhizopus spp.; Rohament P comes from Röhm, Darmstadt, Germany and represents pectinase from *Aspergillus* spp.)

Purification media

Purification media [128]:

Medium I:
0.5 mol l^{-1} sorbitol, 1 mmol l^{-1} $CaCl_2$, 5 mmol l^{-1} MES, pH 6.0;
Medium II:
0.4 mol l^{-1} sucrose, 0.1 mol l^{-1} sorbitol, 1 mmol l^{-1} $CaCl_2$, 5 mmol l^{-1} MES, pH 6.0;
Medium III:
0.5 mol l^{-1} sucrose, 1 mmol l^{-1} $CaCl_2$, 5 mmol l^{-1} MES, pH 6.0.

Purification in a step-gradient

Purification in a step-gradient [128]: all operations are done with chilled solutions. The incubation medium is carefully removed from the tissue which is then washed three times with 20 ml medium I. Each washing is passed through a 500-μm and a 200-μm mesh. The washings are combined and centrifuged (polycarbonate centrifuge tubes are unsuitable) at 100 g for 5 min in a swing out rotor. Then the tubes are inverted rapidly to discard the supernatant. The crude pellet is gently resuspended in a few drops of medium III. Then 5 ml of medium I are added and the tube is gently inverted to suspend the protoplasts. Then 2 ml of medium II are layered on top, which acts as a wash layer, followed by 1 ml sorbitol medium I. The pure protoplasts are collected by a 5 min centrifugation at 250 g at the interface between the sorbitol and the sucrose-sorbitol layer. They are taken off with a Pasteur pipette. The pellet is inspected microscopically; if it contains many protoplasts some dextran is added to the sucrose medium III and the purification is repeated.

A two-phase purification system

Purification in a polymer two-phases system: protoplasts may also be purified in an aqueous two-phase system generated by the mixture of the two polymers dextran and polyethylene glycol [144]. Then 0.6 ml of crude protoplast preparation are thoroughly mixed with 5.4 ml of separation medium and centrifuged at 300 g for 5 min at 4 °C. Protoplasts are enriched at the interface of both phases while the chloroplasts are found in the lower phase. The separation medium consists of 5.5 % polyethylene glycol 6000, 10 wt/vol% dextran ($M_r = 20\,000$), 10 mmol l^{-1} sodium phosphate, pH 7.5, and 0.46 mol l^{-1} sorbitol.

2.9.4.4.5 Photosynthetic Activity of Protoplasts

CO_2-dependent O_2 evolution is easily measured with an O_2 electrode. The chlorophyll concentration should be about 50 μg chlorophyll ml^{-1}. A suitable medium for wheat, barley and spinach is [128]: 0.4 mol l^{-1} sorbitol, 10 mmol l^{-1} $NaHCO_3$, 5 mmol l^{-1} $CaCl_2$, 50 mmol l^{-1} Tricine-KOH, pH 7.6. $CaCl_2$ is included to prevent clumping and to inhibit photosynthesis by chloroplasts which are set free upon rupture of protoplasts.

2.9.4.4.6 Chloroplast Isolation from Protoplasts

Chloroplasts can be isolated from protoplasts by passing them (in 0.5 mol l^{-1} sorbitol) through a fine nylon mesh of about 20 µm affixed to the cut end of a 1 ml disposable syringe. A ring cut from an automatic- pipette tip can be used to fix the mesh to the syringe [128]. The syringe lubricant is removed with alcohol as it inhibits chloroplast photosynthesis [128]. A suitable nylon mesh is Nybolt P-20 (16% open pore area, Schweizerische Seidengazefabrik, Zürich, Switzerland). The protoplasts are passed through the mesh for at least three times into an appropriate isolation medium. If chloroplasts are to be isolated from wheat, pea or sunflower the isolation medium should have an alkaline pH and contain a chelating agent such as EDTA since otherwise the properties of the chloroplast envelope and its translocators are altered and photosynthetic enzymes are partially inactivated. EDTA may be replaced by other chelators such as pyrophosphate, citrate, 150 – 200 mmol l^{-1} choline chloride, KCl or a Good's buffer [128]. The following isolation medium has been suggested for spinach and C4-plants [128]: 0.33 mol l^{-1} sorbitol, 2 mmol l^{-1} EDTA, 1 mmol l^{-1} MgCl$_2$, 1 mmol l^{-1} MnCl$_2$ and 50 mmol l^{-1} HEPES (N-[2-hydroxyethyl] piperazine-N′-[2-ethane-sulfonic acid), adjusted to pH 7.6. A medium for wheat is [128]: 0.4 mol l^{-1} sorbitol, 10 mmol l^{-1} EDTA, 25 mmol l^{-1} Tricine, adjusted to pH 8.4 (10 mmol l^{-1} NaHCO$_3$ in the medium results in a shortening of the induction period).

Composition of isolation medium

Isolated chloroplasts are centrifuged at 250 g for 45–90 s to yield a chloroplast pellet largely free of cytosol. Afterwards the chloroplasts can be resuspended by gentle agitation in fresh medium. Although chloroplasts may maintain maximum rates of photosynthesis for only a short time (30 min for wheat chloroplasts) this does not affect their intactness. Chloroplasts purified in silica sol gradients retain intactness for one to three days [128]. They are free from soluble vacuolar hydrolases which are thought to be responsible for the decline in chloroplast integrity.

2.9.4.5 Isolation of Plant Mitochondria

The preparation of intact, physiologically active mitochondria from etiolated plant tissues or green leaves is achievable by successive cycles of differential centrifugation. Such fractions, however, are generally contaminated with other subcellular structures. They can be reduced when mitochondria are isolated from etiolated tissues by use of continuous and discontinuous sucrose gradients, or by centrifugation through a sucrose cushion. These procedures cannot be used to separate broken chloroplasts from mitochondria that were isolated from green leaves. However, an effective separation of mitochondria from chloroplast material under iso-osmotic conditions is possible if polyvinyl pyrrolidone-coated silica sol (Percoll) gradients are used [145]. This technique is also extremely effective in purifying mitochondria from etiolated tissue.

A general method for isolating plant mitochondria is available [145], and it has been modified to suit the characteristics of different tissues, e.g., potato tuber [146], etiolated corn seedlings [147], green leaf tissues [148, 149], tomato fruit [150] and cotyledons of germinating seedlings [151].

Washed mitochondria from etiolated tissues such as mung bean hypocotyls or potato tubers can be isolated by the procedure shown in Fig. 2.10 [145]. NADH-cytochrome c oxidase activity increases if the outer mitochondrial membrane in damaged [144]. Leaf tissue mitochondria may also be isolated by the procedure shown in Fig. 2.10. Of particular importance is the high tissue-to-medium ratio (500 – 600 g laminar tissue to 2 l medium) and the provision of insoluble polyvinylpyrrolidone (PVPP) to encounter the inactivating effects of phenols released from the vacuoles of

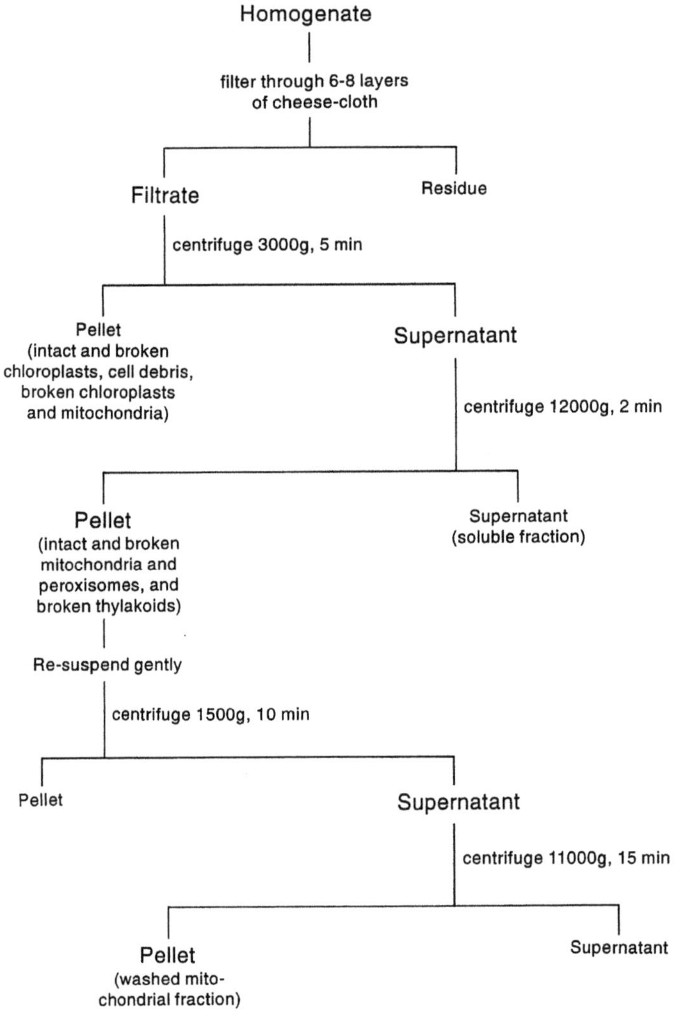

Fig. 2.10. Isolation of plant mitochondria from etiolated tissues [144]

Fig. 2.11. Typical distribution of markers following separation of *C. album* mitochondria on discontinuous Percoll-gradients. Similar results have been observed for spinach and cos lettuce. Glycolate oxidase activity is expressed as nmol O_2 consumed/min · mg protein and catalase activity as I.U./mg protein. Cytochrome c oxidase activity is expressed as nmol cytochrome c oxidized/min · mg protein. Histogram denotes % of the original activity recovered in the fraction, and represents specific activity. *Vertical dashed lines* indicate interfaces between various layers and the figures in the protein distribution are vol.% Percoll in these layers [134]

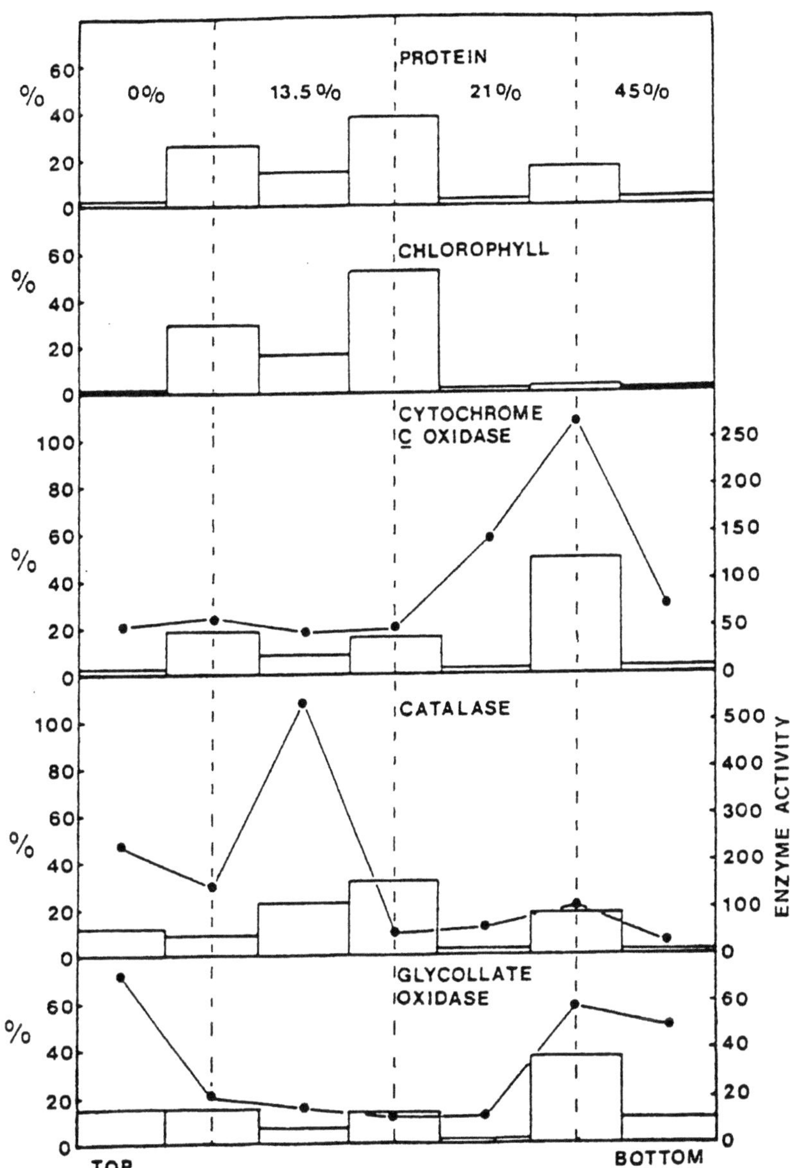
PROTEIN
0%
13.5%
21%
45%
%
60
40
20
0
CHLOROPHYLL
%
60
40
20
0
CYTOCHROME
C OXIDASE
%
100
80
60
40
20
0
250
200
150
100
50
0
CATALASE
%
100
80
60
40
20
0
500
400
300
200
100
0
ENZYME ACTIVITY
GLYCOLLATE
OXIDASE
%
60
40
20
0
60
40
20
0
TOP
BOTTOM

the cells upon disruption [144]. The low speed wash steps are extremely important in removing contaminants such as chloroplasts and peroxisomes.

A rapid purification of mitochondria from green leaves and etiolated plant tissues is possible under iso-osmotic and low-viscosity conditions if a three-step discontinuous system of Percoll solutions is used. The three-step gradient consists of 6 ml of 45%, 12 ml of 21% and 12 ml of 13.5 vol% Percoll mixtures containinq sucrose (0.25 mol l⁻¹), MOPS (3-[N-morpholino]propane-sulfonic acid) (10 mmol l⁻¹, pH 7.2) and bovine serum albumin (BSA) (0.2%).

Mitochondria from etiolated tissues and purified by the procedure shown in Fig. 2.10 are re-suspended in wash medium (around 4 ml) containing mannitol (0.3 mol l⁻¹), EDTA (1 mmol l⁻¹), BSA (0.2%) and MOPS (20 mmol l⁻¹, pH 7.2), at a protein concentration of 10 – 20 mg/ml, and layered onto the three-step gradient of Percoll. Fractionation is undertaken for 30 min at 7000 g in an angle rotor at 0 – 4 °C. Mitochondria assemble at the interface of the 45% and 21% Percoll layers. They are re-suspended in around 60 ml of wash medium (see above) and finally pelleted at 11000 g for 15 min [144]. Mitochondria purified in this way show improved rates of substrate oxidation, respiration coefficient and ADP : O_2 ratios.

This three-step procedure had to be modified in order to separate mitochondria from contaminants of green leaves. The inclusion of 50 mmol l⁻¹ propane-1,2-diol in the 13.5% Percoll layer proved successful to alter the distribution of chlorophyll. The fraction of washed mitochondria is re-suspended in a medium containing propane-1,2-diol (50 mmol l⁻¹), sucrose (0.239 mol l⁻¹), BSA (0.2%) and MOPS (10 mmol l⁻¹, pH 7.2), such that the chlorophyll concentration is not more than 0.5 mg/ml. The three-step gradient consists of 6 ml 45%, 8 ml 27% and 14 ml 21 vol% Percoll mixtures, each containing sucrose (0.25 mol l⁻¹), BSA (0.2%) and MOPS (10 mmol l⁻¹, pH 7.2). Mitochondria assemble at the interface of 45% and 27% Percoll layers. The bulk of chlorophyll is removed at the interface of the re-suspending medium and 21% Percoll layers. The goodness of separation results from the position and amount of the markers chlorophyll (thylakoids), glycollate oxidase and catalase (peroxisomes), and cytochrome c oxidase (mitochondria) (Fig. 2.11).

2.9.4.6 Isolation of Plant Golgi Apparatus

The following purification procedure uses green onions (*Allium cepa*) [152].

Roots and lignified stem regions are cut off from the base of an onion. Then a scalpel is used to remove a cone of tissue 0.5 – 1 cm in diameter from the base, and 0.5 – 1 cm high from the central portion of the onion bulb. Included in the explant are stem, meristematic region, and leaf bases. Scale leaves and the green portions of the onion are discarded.

Approximately 5 g of stem explants from 30 – 50 onions are washed and blotted. Then the tisssue is chopped with razor blades (ten double-edged razor blades spaced 5 mm apart, mounted on an onion chopper or a motor driven device). The chopping surface is a Plexiglass trough. The stem explants are combined with 5 ml of homogenization medium on the chopping surface. The tissue is chopped for about 5 min at a rate of 30 chops per second when using the motor driven device [118]. The homogenization medium consists of 0.5 mol l⁻¹ sucrose, containing 10 mmol l⁻¹ sodium phosphate, pH 6.8 and 1 wt/vol% dextran.

The homogenate is filtered through a single layer of Miracloth and squeezed by hand in order to remove cell walls and debris. The filtrate is centrifuged at 10 000 g for 30 min to remove nuclei, plastids and mitochondria. Then the entire 10 000 g supernatant is layered onto a sucrose cushion composed of a bottom layer of 0.25 ml of 1.8 mol l⁻¹ sucrose and a top layer of 0.5 ml of 1.6 mol l⁻¹ sucrose contained in a 5.4 ml lusteroid centrifuge tube. The Golgi apparatus are centrifuged for 30 min at 35 000 g to bring them on the sucrose cushion. After centrifugation, the supernatant is removed by suction and replaced by additional sucrose layers as follows: 1.0 ml 1.5 mol l⁻¹; 1.0 ml 1.25 mol l⁻¹, and 1.0 ml 0.5 sucrose. Purification is achieved by centrifugation in the sucrose gradient for 3 h at 100 000 g. The dictyosomes band at the interface of the 0.5 mol l⁻¹/1.25 mol l⁻¹ sucrose layers. They are removed with a Pasteur pipette, resuspended in distilled water and concentrated by a final centrifugation at 35 000 g for 30 min.

The density of the collected dictyosomes is about 1.125 g/ml. Plant Golgi apparatus isolated in the absence of coconut milk or glutaraldehyde exhibit a marked tendency to unstack. Purification beyond the first differential centrifugation step in the absence of coconut milk or glutaraldehyde results in preparations that consist largely of single, separated cisternae or aggregations of dictyosomes.

2.10 Catalytic Activities of Enzymes Forming a Common Metabolic Sequence

The catalytic activities of enzymes which form a common metabolic sequence may differ by several orders of magnitude as has been shown for the enzymes of the glycolytic pathway [31]. While the activity of liver hexokinase was estimated to be around 40 nkat (g fresh weight)⁻¹ that of the enzyme triose phosphate isomerase was found to be around 13 000 nkat (g fresh weight)⁻¹ [31]. Besides, the activities of the enzymes triose phosphate isomerase, glyceraldehyde-phosphate dehydrogenase, phosphoglycerate kinase, phosphoglyceromutase and enolase may deviate from each other by one order of magnitude (Fig. 2.12). It is of interest that despite the activity-differences of the latter enzymes they form a regulatory unit [32]. The activity ratios observed in cortex cerebri, liver, heart and musculus quadriceps were the following: triose phosphate isomerase/glyceraldehyde-phosphate dehydrogenase: (6 – 12/1), phosphoglycerate kinase/glyceraldehyde-phosphate dehydrogenase: (0.6 – 1.2/1), phosphoglyceromutase/glyceraldehyde-phosphate dehydrogenase (0.6 – 1.2/1), and enolase/glyceraldehyde-phosphate dehydrogenase (0.15 – 0.3/1) [32]. In rabbit muscle the observed rates have an equimolar basis [152].

Corresponding observations were made for enzymes of the citrate-cycle, the β-oxidation of fatty acids, and the amino acid metabolism [31]. The large differences in enzyme activity may be one reason why some enzymes can be stained after electrophoretic separation whereas others cannot. In order to exclude insufficient enzyme activity as a reason for a failure to visualize enzymes following electrophoresis, a photometric activity determination should be performed before starting electrophoresis. Enzymes should have a catalytic activity of about 10 to 60 μkat l⁻¹ if applied per cm² of sample area of a starch gel and double or five times of that when applied to the same area of a polyacrylamide gel.

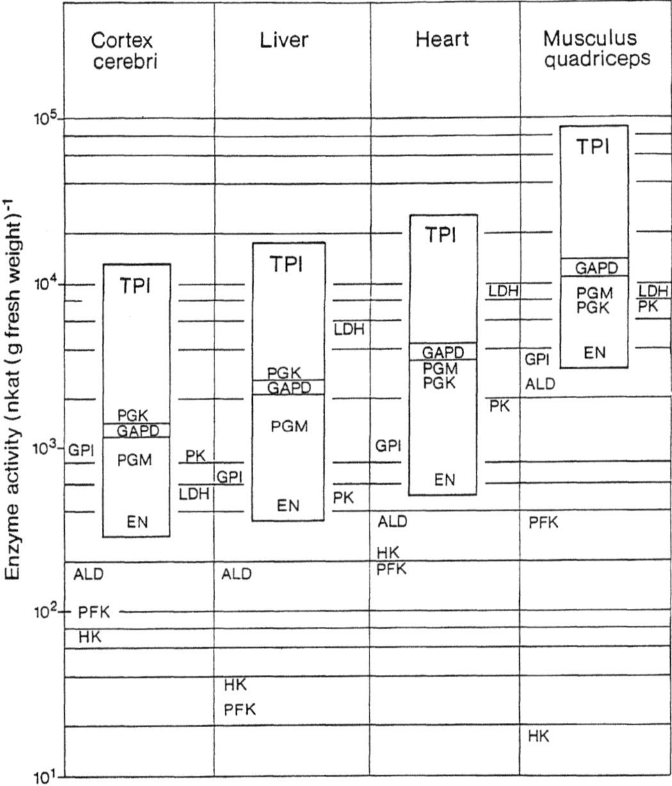

Fig. 2.12. Catalytic activities of glycolytic enzymes estimated in various tissues of the rat and plotted on a logarithmic scale. Data taken from [136, 153] and transformed to nkat (g fresh weight)⁻¹. Abbreviations: *ALD*, aldolase (4.1.2.13); *EN*, enolase (4.2.1.11); *GAPD*, glycerinaldehyde phosphate dehydrogenase (1.2.1.12); *GPI*, glucose phosphate isomerase (5.3.1.9); *HK*, hexokinase (2.7.1.1); *LDH*, lactate dehydrogenase (1.1.1.27); *PFK*, phosphofructokinase (2.7.1.11); *PGK*, phosphoglycerate kinase (2.7.2.3); *PGM*, phosphoglyceromutase (2.7.5.3); *PK*, pyruvate kinase (2.7.1.40); *TPI*, triosephosphate isomerase (5.1.1.1). The five glycolytic enzymes which occur in nearly constant activity ratios in the four tissues investigated are framed [32]

2.11 References

1. Cooper TG (1977) Tools in biochemistry. Wiley, New York, pp 355-405
2. Evans WH (1979) Preparation and characterization of mammalian plasma membranes. In: Work TS, Work E (eds) Laboratory techniques in biochemistry and molecular biology, vol 7. Elsevier-North Holland, Amsterdam, pp 11 – 44
3. Hetherington PJ, Follows M, Dunnill P, Lilly MD (1971) Trans Inst Chem Eng 49: 142 – 148
4. Kemmerer V, Griffin CC, Brand L (1975) Phosphofructokinase from *Escherichia coli*. In: Wood WA (ed) Methods in enzymol vol 42 Academic Press, New York San Francisco London, pp 91 – 98
5. South DJ, Reeves RE (1975) Pyruvate, orthophosphate dikinase from *Bacteroides symbiosus* (1975). In: Wood WA (ed) Methods in enzymol vol 42 Academic Press, New York San Francisco London, pp 187 – 191
6. Suelter CH (1985) A practical guide to enzymology, Biochemistry: A series of monographs. John Wiley & Sons, New York, pp 64 – 132
7. Follows M, Hetherington PJ, Dunnill P, Lilly MD (1971) Biotechnol Bioeng 13: 549 – 560
8. Durchschlag H, Biedermann G, Egger H (1981) Eur J Biochem 114: 255 – 262
9. Dethmers JK, Ferguson-Miller S, Margoliash EJ (1979) Biol Chem 254: (23) 11973 – 11981
10. Brewer GJ, Sing CF (1970) An introduction to isozyme techniques. Academic Press, New York, pp 53 – 61
11. Warburg O, Christian W (1941) Biochem Z 310: 384 – 421
12. Taylor BL, Utter MF (1974) Anal Biochem 62: 588 – 591
13. Thorner JW (1975) Glycerol kinase. In: Wood WA (ed) Methods in enzymol vol 42 Academic Press, New York San Francisco London, pp 148 – 156
14. Achstetter T, Ehmann C, Wolf DH (1981) Arch Biochem Biophys 207: 445-454
15. Barrett AJ (1977). Introduction to the history and classification of tissue proteinases. In: Barrett AJ (ed) Proteinases in mammalian cells and tissues, Elsevier-North Holland, Amsterdam, pp 1 – 55
16. Volkova AP, Grin NR (1967) Tr Tseuntr Nauchno-Issled Dezinfekp Inst, 16 – 21. In: Chem Abstr 70: 10, 606 m
17. Reuber MD (1981) Clin Toxicol 18: 47 – 84
18. Braid PE, Nix M (1969) Can J Biochem 47: 1 – 6
19. Knight CG, Barrett AJ (1976) Biochem J 155: 117 – 125
20. James GT (1978) Anal Biochem 86: 574 – 579
21. Sekar V, Hageman JH (1979) Biochem Biophys Res Commun 89: 474 – 478
22. Beaty NB, Lane MD (1982) J Biol Chem 25: 924 – 929
23. Laura R, Robison DJ, Bing DH (1980) Biochemistry 19: 4859 – 4864
24. Westkaemper RB, Abeles RH (1983) Biochemistry 22: 3256 – 3264
25. Umezawa H (1976) Structures and activities of protease inhibitors of microbial origin In: Methods in enzymol vol 45 Academic Press, New York San Francisco London pp 678 – 695
26. Birk Y (1976) Proteinase inhibitors from plant sources In: Methods in enzymol vol 45 Academic Press, New York San Francisco London, pp 695 – 697
27. Brand K, Hess B (1983) Cell and tissue disintegration. In: Bergmeyer HU (ed) Methods of enzymatic analysis 3rd edition vol II (samples, reagents, assessment of results) Verlag Chemie, Weinheim, Deerfield Beach, Florida, Basel, pp 26 – 49
28. Crabtree B, Leech AR, Newsholme EA (1979) Measurement of enzyme activities in crude extracts of tissues. In: Pogson C (ed) Techniques in metabolic research, B 211. Elsevier North-Holland, Amsterdam, pp 1 – 37
29. Bartels H (1960) Planta, 55: 573 – 597
30. Read G, Crabtree B, Smith GH (1977) Biochem J 164: 49-355
31. Pette D (1968) Aktivitätsmuster und Ortsmuster von Enzymen des energieliefernden Stoffwechsels. In : Schmidt FW (ed) Praktische Enzymologie. Verlag Hans Huber, Bern Stuttgart, pp 15 – 51
32. Pette D, Luh W, Bücher Th (1962) Biochem Biophys Res Comm 7: 419 – 424
33. Fishman L (1974) Biochem Medicine 10: 309 – 315

34. Esterbauer H, Grill D, Beck G (1975) Phyton (Austria), 17: 87 – 99
35. Pitel JA, Cheliak WM (1985) Physiol Plantarum 65: 129 – 134
36. Martin B, Bassham JA (1980) Physiol Plantarum 48: 213 – 220
37. Weimar M, Rothe GM (1986) Physiol Plantarum 69: 692 – 698
38. Bucher-Wallin IK, Bernhard L, Bucher JB (1979) Europ J For Pathol 9: 6 – 15
39. Imbert MP, Wilson LA (1972) Phytochemistry 11: 29 – 36
40. Hoyle MC, Routley DG (1974). Promotion of the IAA oxidase-peroxidase complex enzyme in yellow birch. In: Bieleski RL, Ferguson A, Cresswell R (eds) Mechanisms of regulation of plant growth. R Soc NZ Bull vol 12. Wellington, pp 743 – 750
41. Schnarrenberger C, Oeser A, Tolbert NE (1973) Arch Biochem Biophys 154: 438 – 448
42. Rothe GM, Hengst G, Mildenberger I, Scharer H, Utesch D (1983) Planta, 157: 358 – 366
43. Wendel JF, Parks CR (1982) J Heredity 73: 197 – 204
44. Rothe GM (1988) Electrophoresis 9: 307 – 316
45. Schneider V, Hallier UW (1970) Planta 94: 134 – 139
46. Andersen RA, Sowers JA (1968) Phytochemistry 7: 293 – 301
47. Anderson JW (1968) Phytochemistry 7: 1973 – 1988
48. Haard NF, Tobin CL (1971) J Food Sci 36: 854 – 857
49. Cheliak WM, Pitel JA (1984) J Heredity 75: 34 – 40
50. Hoyle MC (1978) Physiol Plantarum 42: 315 – 320
51. Kelley WA, Adams RP (1977) Phytochemistry 16: 513 – 516
52. King EE (1971) Phytochemistry 10: 2337 – 2341
53. Feret PP (1971) Silvae Genetica 20: 46 – 50
54. Pierpoint WS (1966) Biochem J 98: 567 – 580
55. Craker LE, Gusta LV, Weiser CJ (1969) Can J Plant Sci 49: 279 – 286
56. Drawert F, Lessing V, Leupold G (1977) Chem Mikrobiol Technol Lebensmittel 5: 65 – 70
57. Maresh GA, Dunbar BS (1988) Electrophoresis 9: 54 – 57
58. Husain SS, Lipinski B, Gurewich V (1981) Proc Natl Acad Sci USA 78: 4265 – 4269
59. Böhme H-J, Kopperschläger G, Schulz J, Hofmann E (1972) J Chromatogr 69: 209 – 214
60. Anonymus. Affinity Chromatography, Principles and Methods. Pharmacia, Laboratory Separation Division, Uppsala.
61. Stellwagen E, Baker B (1976) Nature 261: 719 – 720
62. Thompson ST, Cass KH, Stellwagen E (1975) Proc Nat Acad Sci USA 72: 669 – 672
63. Easterday RL, Easterday I (1974). In: Dunlap RB (ed) Immobilized biochemicals and affinity Chromatography. Plenum Press, New York, pp 123 – 134
64. Lamkin GE, King EE (1976) Biochem Biophys Res Commun 72: 560 – 565
65. Mol JG, Piskiewicz D (1976) FEBS Lett 72: 147 – 150
66. Yoshino M, Kawamura Y, Ogasawara N (1974) Biochem (Tokyo) 75: 1391 – 1393
67. Witt JJ, Roskoski R (1975) Biochemistry 14: 4503 – 4507
68. Marshall J (1970) J Chromatogr 53: 379 – 380
69. Ahmad A, Surolia A, Bachhawat BK (1977) Biochim Biophys Acta 481: 542 – 548
70. Baxter A, Curry LM, Durham JP (1978) Biochem J 173: 1005 – 1008
71. Staal GEJ, Visser J, Veeger C (1969) Biochim Biophys Acta 185: 39 – 48
72. Uotila L, Koivusalo M (1975) Eur J Biochem 52: 493 – 503
73. Tormanen CD, Redd WL, Srikantaiah MV, Scallen TJ (1976) Biochem Biophys Res Commun 68: 754 – 762
74. Heyns W, De Moor P (1974) Biochim Biophys Acta 358: 1 – 13
75. Seelig GF, Colman RF (1977) J Biol Chem 252: 3671 – 3678
76. Solomonson LP (1975) Plant Physiol 56: 853 – 855
77. Cheng Y-C, Domin B (1978) Anal Biochem 85: 425 – 429
78. Oka J, Ueda K, Haynaishi O (1978) Biochem Biophys Res Commun 80: 841 – 848
79. Nichols BP, Lindell TD, Stellwagen E (1978) Biochim Biophys Acta 526: 410 – 417
80. Pai S-H, Ponta H, Rahmsdorf H-J, Hirsch-Kauffmann Herrlich P, Schweiger M (1975) Eur J Biochem 55: 299 – 304
81. Baksi K, Rogerson DL, Rushinsky GW (1978) Biochemistry 17: 4136 – 4139
82. White H, Jencks WP (1976) J Biol Chem 251: 1708 – 1711
83. Adinolfi A, Hopkinson DA (1978) Ann Human Genet 41: 399 – 407

84. Grunbaum BW, Crim M (1981) Handbook for forensic individualization of human blood and bloodstains. Sartorius GmbH, Göttingen, Germany, Hayward California USA, pp 55 – 161
85. Ulmer AJ, Flad H-D (1979) J Immunol Methods 30: 1 – 10
86. Dulbecco R, Vogt M (1954) J Exp Med 99: 167 – 175
87. Earle WR (1943) J Nat Cancer Inst 4: 165 – 212
88. Hanks JH, Wallace RE (1949) Proc Soc Exp Biol Med 71: 196 – 203
89. Harris H, Hopkinson DA (1976) Handbook of enzyme electrophoresis in human genetics. Elsevier-North Holland, Amsterdam Oxford, pp 3 – 4
90. Harris R, Ukaejidfo EO (1970) Brit J Haemat 18: 229 – 235
91. Boyum A (1968) Scand J Clin Lab Invest 21: Suppl. 97, 77 – 89
92. Boyum A (1968) Scand J Clin Lab Invest 21: Suppl. 97, 31 – 50
93. Boyum, A (1976) Scand J Immunol 5: Suppl. 5, 9 – 15
94. Chi DS, Harris NS (1978) J Immunol Methods 19: 167 – 172
95. Ho CK, Babiuk LA (1978) Immunol 35: 733 – 740
96. Taylor DW, Marchette NJ, Siddiqui WA (1978) Develop Comp Immunol 2: 539 – 546
97. Pearson TW, Roelants GE, Ludin LB (1979) J Immunol Methods 26: 271 – 282
98. Yang TJ, Jantzen PA, Williams LF (1979) J Immunol 38: 85 – 93
99. Musson RA, Henson PM (1979) J Immunol 122: 2026 – 2031
100. Huesco P, Rocha M (1978) Rev Esp Fisiol 34: 339 – 344
101. Wittman G (1980) Zbl Vet Med B 27: 253 – 256
102. Blaxhall PC (1981) J Fish Biol 18: 177 – 181
103. Gutierrez C, Bernabe RR, Vega J, Kreisler M (1979) J Immunol Methods 29: 57 – 63
104. Kurnick JT, Östberg L, Stegagno M, Kimura AK, Örn A, Sjöberg O (1979) Scand J Immunol 10: 563 – 573
105. Morgan JF, Morton HJ, Parker RC (1950) Proc Soc Exp Biol Med 73: 1 – 8
106. Morton HJ (1970) In vitro 6: 89 – 108
107. Hutchins D, Steel CM (1979) Separation of human lymphocytes on the basis of volume and density. In: Peeters H (ed) Separation of cells and subcellular elements. Pergamon Press, Oxford New York, pp 77 – 82
108. Saksela E, Timonen T, Ranki, A (1979) Immunological Review 44: 71 – 123
109. Anonymous. Percoll, methodology and applications, Density marker beads for calibration of gradients of Percoll. Pharmacia, Uppsala
110. Rennie C, Thompson S, Parker A (1979) Clin Chim Acta 98: 119 – 125
111. Buitrago A, Gylfe E, Henriksson Ch, Pertoft H (1977) Biophys Res Comm 79: 823 – 828
112. Jenkins WJ, Clarkson E, Milsom J, Sherlock S (1979) Clin Sci 57: 29 P
113. Osmundsen H, Neat CE (1979) FEBS Lett 107: 81 – 85
114. Pertoft H, Laurent TC (1977) Isopycnic separation of cells and cell organelles by centrifugation in modified colloidal silicia gradients. In: Catsimpoolas N (ed) Methods of cell separation vol 1. Plenum Press, New York, pp 25 – 65
115. Perret B, Chap H-J, Douste-Blazy L. (1979) Biochim Biophys Acta 556: 434 – 446
116. Terland O, Flatmark T, Kryvi H (1979) Biochim Biophys Acta 553: 460 – 468
117. Price N, Stevens L (1982) Fundamentals of enzymology. Oxford University Press Oxford, New York, pp 309 – 362
118. Moré DJ (1971) Isolation of golgi apparatus. In: Jakoby WB (ed) Method in Enzymology, XXII. Academic Press, New York, pp 130 – 144
119. Sims NR (1990) J Neurochem 55: 689 – 707
120. Öbrink B, Wärmegård B, Pertoft H (1977) Biochem Biophys Res Commun 77: 665 – 670
121. Perret B, Chap HJ, Douste-Blazy L (1979) Biochem Biophys Acta 556: 434 – 446
122. Pertoft H, Wärmegård B, Höök M (1978) Biochem J 174: 309 – 317
123. Lindahl U, Pertoft H, Seljelid R (1979) Biochem J 182: 189 – 193
124. Siebert G (1968) Nucleus. In: Florkin M, Stotz EH (eds) Comprehensive biochemistry vol 23 Elsevier, Amsterdam, pp 1 – 32
125. Reijnagoud D-J, Tager JM (1977) Biochim Biophys Acta 472: 419 – 449
126. Dean RT, Barrett AJ (1976) Essays Biochem 12: 1 – 40
127. De Duve C, Wattiaux R (1966) A Rev Physiol 28: 435 – 492

128. Leegood RC, Walker DA (1983) Chloroplasts. In: Hall DO, Moore AC (eds) Isolation of membranes and organelles from plant cells. Academic Press, London, pp 185 – 210
129. Morgenthaler J-J, Marsden MPF, Price CA (1975) Arch Biochem Biophys 168: 289 – 301
130. Lilley RMc, Walker DA (1974) Biochim Biophys Acta 358: 226 – 229
131. Heldt HW, Flügge UI, Stitt M (1986) Biologie in unserer Zeit 4: 97 – 105
132. Avron M, Jagendorf A (1956) Arch Biochem Biophys 65: 475 – 490
133. Joyard J, Douce R (1975) FEBS Letter 51: 35 – 340
134. Jackson Ch, Dench JE, Hall DO, Moore AL (1979) Plant Physiol 64: 150 – 153
135. Mills WR, Joy KW (1980) Planta 148: 75 – 83
136. Nagata T, Ishii S (1979) Can J Bot 57: 1820 – 1823
137. Edwards GE, Lilley R, McCraig S, Hatch MD (1979) Plant Physiol 63: 821 – 927
138. Edwards GE, Robinson SP, Tyler NJC, Walker DA (1978) Plant Physiol 62: 313 – 317
139. Edwards GE, Robinson SP, Tyler NJC, Walker DA (1978) Arch Biochem Biophys 190: 421 – 433
140. Spalding MH, Schmitt MR, Ku SB, Edwards GE (1979) Plant Physiol 63: 738 – 743
141. Shahak Y, Crowther D, Hind D (1980) FEBS Letter 114: 73 – 78
142. Spalding MH, Edwards GE (1980) Plant Physiol 65: 1044 – 1048
143. Kanai R, Edwards GE (1973) Plant Physiol 52: 484 – 490
144. Jackson C, Moore AL (1979) Isolation of intact higher-plant mitochondria. In: Reid E (ed) Plant organelles. Ellis Horwood, New York Chichester Brisbane Toronto, pp A1 1 – 12
145. Bonner JWD (1967) A general method for the preparation of plant mitochondria. In: Estabrook RW, Pullman ME (eds) Methods in enzymology X, pp 126 – 133
146. Laties G (1974) Isolation of mitochondria from plant material. In: Fleischer S, Packer L (eds) Methods in enzymology vol 31A. Academic Press, New York San Francisco London, pp 589 – 600
147. Day DA, Hanson JB (1977) Plant Sci Lett 11: 99 – 104
148. Douce R, Moore AL, Neuburger M (1977) Plant Physiol 60: 625 – 628
149. Gardeström P, Ericson I, Larsson C (1978) Plant Sci Lett 13: 231 – 239
150. Ku H-S, Pratt HK, Spurr AR, Harris WM (1968) Plant Physiol 43: 883 – 887
151. Stinson RA, Spencer M (1969) Canad J Biochem 48: 541 – 546
152. Milr PD, Cotton DWK (1966) Nature 209: 1022 – 1023
153. Dallner G (1974) Isolation of rough and smooth microsomes - General. In: Fleischer S, Packer L (eds) Methods in Enzymol vol 31 A. Academic Press, New York San Francisco London, pp 191 – 201

3 Methods for Separating Native Enzymes

3.1 General Considerations

In the course of electrophoresis the stability of an enzyme depends on such conditions as (a) pH-value, (b) ion strength and ion species, (c) effector molecules, (d) temperature and (e) properties of the separation matrix. These parameters were empirically optimized for starch gel electrophoresis [1 – 3] and cellulose acetate electrophoresis [4, 5] when analyzing predominantly animal and human specimen. A major advantage of these types of separation media is that practically every buffer system can be used to separate enzymes whereas in disc-gel electrophoresis [6 – 8] the number of applicable buffer systems is limited. When using isoelectric focusing to separate native enzymes no buffer choice at all is possible [9 – 10]. On the other hand, starch gel electrophoresis is more time consuming than cellulose acetate electrophoresis. Cellulose acetate membranes are commercially available and easy to handle, and separation and enzyme visualization together do not take longer than 1 – 1.5 h. In

Table 3.1. Comparison of electrophoretic methods that have been used in combination with enzyme visualization techniques

Matrix	Type of electrophoresis			
	Cellulose acetate	Starch gel	Polyacryl-amide	IEF
Concentration	as given	10 – 15 %	5 – 10 %	5 % PAA
Length (mm)	70	300 (150)	85	120
Width (mm)	140	300 (200)	85	120
Thickness (mm)	0.11	10 (5)	0.2 – 1 (ø 6)	1
Sample volume (μl per mm gel width)	10 – 50	3000 – 4000	10 – 30	100 – 1000
Separation-time (h)	0.5 – 2	5 – 18	1.5 – 4	2
Separation-distance (mm)	70	250 (130)	80 (110)	110
Field strength (V/cm)	10 – 30	2 – 13.5	20 (40)	60 – 130
Separation basing on	charge difference	charge and size difference	charge and size difference	charge difference

Table 3.2. Comparison of visualization techniques which have been used in combination with different electrophoretic separation procedures

Parameter	Type of electrophoresis			
	Cellulose acetate	Starch gel	Polyacrylamide	IEF
Time of procedure (h)	1 – 3	8 – 24	2.5 – 4	2 – 3
Type of staining	overlay	dip or overlay	dip or overlay	dip or overlay
Volume of stain per gel volume (ml/ml)	0.2 – 0.4	6 – 50	1 – 2 per round gel	1 – 2 per round gel
Duration of staining (min)	10 – 60	60 – 240	10 – 120	10 – 60
Approximate number of enzymes visualized	10	80	70	20
Propagation	small	very wide	widest	small

clinical diagnosis therefore, both methods are favoured. Disc-gel electrophoresis, gradient gel electrophoresis and isoelectric focusing are used if the separation capacities of cellulose acetate membranes are insufficient or when the isoelectric point or the molecular mass of an enzyme is to be estimated. The data compiled in Tables 3.1 and 3.2 can be used to decide between methods for a particular case.

3.2 Cellulose Acetate Electrophoresis

Cellulose acetate membranes are sold dried by several manufactures such as Sartorius (Membranfiltergesellschaft, Göttingen, Germany). Cellulose acetate membranes are quite brittle when dry, but become pliable and easy to handle when wet. With proper treatment the strips may be rendered transparent. They may also be dissolved in organic solvents, facilitating quantification of enzymes. The membranes sold by Sartorius are sized 70 × 145 mm or 57 × 145 mm. They are perforated parallel to the shorter sides so that the central 70 mm long part can be used as the separation area while the upper and lower parts are used as electrode wicks (Fig. 3.1). While the strips are still dry, a line may be drawn with an unsharpened pencil approximately 1/3 apart from that end which will face the cathode, allowing about 5 mm margin on each side of it. Any other markings to be made on strips should also be done before wetting them. Later on the samples are applied along the pencil line. It has to be pointed out, however, that for optimum resolution the sample position has to be carefully evaluated. After wetting the membrane with buffer it is mounted on a tensioning membrane bridge. The electrophoresis apparatus is filled with buffer and the bridge, holding the membrane, is put into the apparatus. The samples are applied and electrophoresis is performed under the conditions given in Table 3.3. Later on, enzymes can be located directly in the membrane or on an overlay. The detailed procedure with one of the commercially available devices "Microzone Zell" (Beckmann Instru-

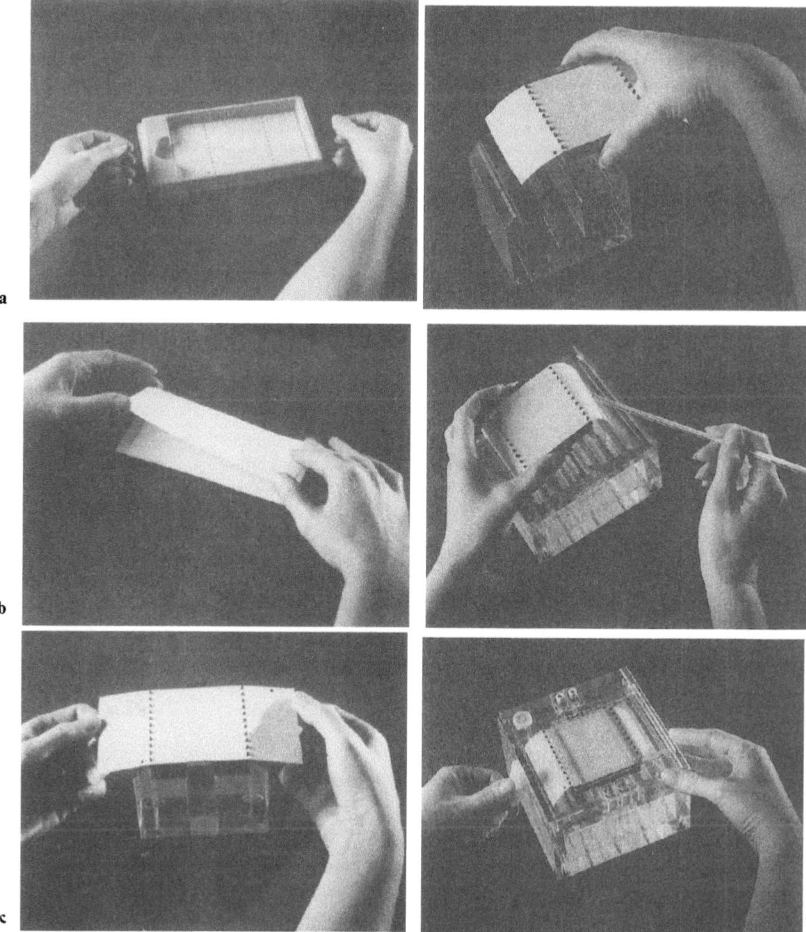

Fig. 3.1 a – f. Preparing and mounting a cellulose acetate membrane into the electrophoretic apparatus Sartophor (Sartorius Göttingen, Germany): **a** the membrane is soaked in buffer; **b** it is blotted to remove excessive liquid; **c** the membrane is mounted into two rows of teeth and stretched by the membrane holder, a hole outside the separating part of the membrane marking the start of sample application; **d** the membrane holder together with the membrane is taken from mounting support; now a spring stretches the membrane; **e** the bridge is inserted into the electrophoresis apparatus; the surmounting membrane parts serve as wicks and dip into the electrode buffer solutions; **f** the apparatus is covered with a specially designed cover through which the samples are applied; its perforations allow one to find the best position of sample application and to use the optimum position in future experiments. (By courtesy of Sartorius, Göttingen, Germany)

Table 3.3. Buffer systems used to separate multiple molecular forms of enzymes from man and other enzyme sources by cellulose acetate electrophoresis

Enzyme; EC No; Source; Run (h); V (mA), T °C; p.a.; [Ref.]	Stock solution	Buffer system	
		Electrode buffer	Membrane buffer
Acid phosphatase; 3.1.3.2; human blood; 0.75 – 1.0; 250 V (1 – 3 mA), 25 °C; A; [11, 12]	24 mmol l⁻¹ NaH₂PO₄ (2.88 g/l) 15 mmol l⁻¹ Citrate-Na₃-2H₂O (4.41 g/l), pH 5.9	same as stock solution (chill to 4 °C before use)	1:5 diluted stock solution
Adenosine deaminase; 3.5.4.4; human blood; 0.5; 250 V (0.5 – 2 mA), 25 °C, A; [13]	68.5 mmol l⁻¹ KH₂PO₄ (9.32 g/l) 31.5 mmol l⁻¹ K₂HPO₄-3H₂O (7.19 g/l), pH 6.5	same as stock solution (chill to 4 °C before use)	1:15 diluted stock solution [a]
Adenylate kinase; 2.7.4.3; human blood; 0.75; 300 V (2 – 3.5 mA), 25 °C, B; [13]	11 mmol l⁻¹ KH₂PO₄ (1.50 g/l) 2.9 mmol l⁻¹ Na₂HPO₄ (0.41 g/l), pH 6.25	same as stock solution (chill to 4 °C before use)	same as stock solution [a]
Amylase; 3.2.1.1; man; 3; 300 V (2 – 3.5 mA), 2 °C; G; [14]	20 mmol l⁻¹ K₂HPO₄-3H₂O (4.57 g/l) adjust pH to 7.4 with 0.1 N NaOH	stock solution with 0.02 % NaCl and 0.1 % bovine serum albumin	same as electrode buffer [d]
Carbonic anhydrase II; 4.2.1.1; man; 0.5; 500 V (1 – 1.6 mA), 25 °C; C; [13]	900 mmol l⁻¹ Tris (109 g/l) 500 mmol l⁻¹ Boric acid (31 g/l) 20 mmol l⁻¹ EDTA (free acid) (5.8 g/l), pH 8.6	1:14 diluted stock solution	1:14 diluted stock solution [a]
Chymotrypsin; 3.4.21.1; bovine; 0.5, 200 V; [15]	24 mmol l⁻¹ Diethyl-barbiturate-Na (4.95 g/l) adjust to pH 8.6 with 1 N HCl	same as stock solution	same as stock solution [c]
Endo-β-N-acetylgluco-saminidase; 3.2.1.96; man; 1/3; 250 V, 22 C°; [16]	50 mmol l⁻¹ Diethyl-barbiturate-Na (10.31 g/l), adjust to pH 7.9 with 1 N HCl	same as stock solution	same as stock solution [e]
Enolase; 4.2.1.11; rabbit; 5/6; (8 mA), 25 °C; F; [17]	Stock A: 9.33 mmol l⁻¹ Diethylbarbituric acid (1.83 g/l), 49.3 mmol l⁻¹ Diethyl-barbiturate-Na (10.3 g/l), pH 8.6 Stock B: 200 mmol l⁻¹ Acetate-Na (16.4 g/l)	Add sufficient Stock B to Stock A to obtain a pH of 8.4, then make up to 1 l with distilled water	same as electrode buffer

Table 3.3 (continued)

Enzyme; EC No; Source; Run (h); V (mA), T °C; p. a.; [Ref.]	Stock solution	Buffer system	
		Electrode buffer	Membrane buffer
Esterase D; 3.1.1.1; human blood; 1; 200 V (1 – 2.5 mA), 25 °C; D; [13]	100 mmol l⁻¹ Tris (12.11 g/l) 100 mmol l⁻¹ Maleic acid (11.62 g/l) 8.7 mmol l⁻¹ EDTA, free acid (2.92 g/l) 10 mmol l⁻¹ MgCl₂ · 6H₂O (2.03 g/l), adjust to pH 7.4 with 0.1 N NaOH	same as stock buffer (chill to 4 °C before use)	1:10 diluted stock buffer [a]
Glucose-6-phosphate dehydrogenase; 1.1.1.49; human blood; 1/3; 350 V (1 mA), 25 °C; D; [13]	900 mmol l⁻¹ Tris (109 g/l) 500 mmol l⁻¹ Boric acid (31 g/l) 200 mmol l⁻¹ EDTA-Na₂ (7.4 g/l) pH 8.7	cathode buffer: 1:7 diluted stock solution; anode buffer: 1:5 diluted stock solution	1:20 diluted stock buffer plus 20 mg NADP per 40 ml membrane buffer [a]
Glucose-6-phosphate isomerase; 5.3.1.9; human blood; 0.5; 350 V (1 mA), 25 °C; A; [1]	900 mmol l⁻¹ Tris (109 g/l) 500 mmol l⁻¹ Boric acid 30.9 g/l) (20 mmol l⁻¹ EDTA-Na₂ (7.4 g/l), pH 8.7	cathode buffer: 1:7 diluted stock solution; anode buffer: 1:5 diluted stock solution	1:20 diluted stock buffer
Glutamate pyruvate transaminase; 2.6.1.2; human blood; 0.75; 270 V (1 – 4 mA), 25 °C; E; [13]	100 mmol l⁻¹ Tris (12.1 g/l), 100 mmol l⁻¹ Maleic acid (11.6 g/l), 8.7 mmol l⁻¹ EDTA, free acid (2.9 g/l), 10 mmol l⁻¹ MgCl₂ · H₂O (2.0 g/l), adjust to pH 7.4	same as stock solution (chill to 4 °C before use)	1:22 diluted stock solution [a]
Glutathione reductase; 1.6.4.2; man; 0.75; 250 V (1.5 – 3.5 mA), 25 °C; E; [18]	100 mmol l⁻¹ Tris (12.1 g/l), 100 mmol l⁻¹ Maleic acid (11.6 g/l), 8.7 mmol l⁻¹ EDTA, free acid (2.9 g/l), 10 mmol l⁻¹ MgCl₂ · 6H₂O (2.0 g/l), adjust to pH 7.4 with 1 N NaOH	same as stock solution (chill to 4 °C before use)	1:20 diluted stock solution

Table 3.3 (continued)

Enzyme; EC No; Source; Run (h); V (mA), T °C; p.a.; [Ref.]	Stock solution	Buffer system	
		Electrode buffer	Membrane buffer
Glyoxalase I; 4.4.15; human blood; 0.5; 300 V (1.5 – 2 mA), 25 °C; F; [19, 20]	85 mmol l^{-1} KH_2PO_4 (11.5 g/l), 65 mmol l^{-1} $K_2HPO_4 - 3H_2O$ (14.8 g/l), pH 6.7	1:15 diluted stock buffer	1:15 diluted stock buffer
Lactate dehydrogenase; 1.1.1.27; human blood; 0.5; 250 V (3.2 – 4.0 mA), 25 °C; G; [21, 22]	Stock A: 14 mmol l^{-1} Diethylbarbituric acid (2.76 g/l), 74 mmol l^{-1} Diethyl-barbiturate-Na (15.4 g/l), pH 8.4 Stock B: 200 mmol l^{-1} Tris (24.2 g/l), adjust pH to 8.3 with 1 N HCl	mix equal volumes of Stock A and Stock B. The pH should read 8.4. (chill to 4 °C before use)	same as electrode buffer
Peptidase A; 3.4.11.*; man; 1; 200 V (0.7 – 2.2 mA), 25 °C; F; [13, 23, 24]	100 mmol l^{-1} Tris (12.1 g/l) 100 mmol l^{-1} Maleic acid (11.6 g/l) 10 mmol l^{-1} $MgCl_2 \cdot 6H_2O$ (2.0 g/l), adjust to pH 7.4 with 1 N NaOH	same as stock solution (chill to 4 °C before use)	1:20 diluted stock solution [a]
6-Phosphogluconate dehydrogenase; 1.1.1.44; human blood; 5/6; 200 V (1 – 2 mA); 25 °C; G; [1]	100 mmol l^{-1} Tris (12.1 g/l), 100 mmol l^{-1} Maleic acid (11.6 g/l), 8.7 mmol l^{-1} EDTA, free acid (2.9 g/l), 10 mmol l^{-1} $MgCl_2 \cdot 6H_2O$ (2.0 g/l), adjust pH to 7.4 with 1 N NaOH	same as stock solution (chill to 4 °C before use)	1:15 diluted stock solution
Phosphoglucomutase; 2.7.5.1; human blood; 1; 240 V (0.5 – 2.0 mA), 25 °C; G; [13]	100 mmol l^{-1} Tris (12.1 g/l), 100 mmol l^{-1} Maleic acid (11.6 g/l), 8.7 mmol l^{-1} EDTA, free acid (2.9 g/l), 10 mmol l^{-1} $MgCl_2 \cdot 6H_2O$ (2.0 g/l), adjust to 7.4 with 1 N NaOH	same as stock solution (chill to 4 °C before use)	1:22 diluted stock solution [a]

Table 3.3 (continued)

Enzyme; EC No; Source; Run (h); V (mA), T °C; p.a.; [Ref.]	Stock solution	Buffer system	
		Electrode buffer	Membrane buffer
Tetrahydrofolate dehydrogenase; 1.5.1.3; *E. coli*; 1–2, 30 V/cm, E; [25]	18.7 mmol l^{-1} Diethylbarbituric acid (3.69 g/l), 99 mmol l^{-1} Diethylbarbiturate – Na (20.60 g/l), pH 8.6	same as stock solution	same as stock solution [b]
Trypsin; 3.4.21.4; bee; 0.5, 200 V; [15]	24 mmol l^{-1} Diethylbarbituric acid (4.95 g/l), adjust to pH 8.6 with 1 N HCl	same as stock solution	same as stock solution [c]

p.a. : point of application: A = 20, C = 30, D = 15, E = 25, F = 10, and E = 50 mm apart from the cathodic perforation, B = 30 mm apart from the anodic perforation; G = center of membrane; the length of the membrane was 350 mm.

* not further specified

[a] Size of cellulose acetate membrane: 70 × 55 mm (separation distance).
Sample solution: 200 µl distilled water are added to 75 µl whole blood cells (which have been separated from plasma by allowing to stand for several minutes). Cap tubes and freeze. Thaw in warm water and freeze. Thaw and centrifuge. Sample volume: 1 µl.

[b] Size of cellulose acetate memebrane: 39 × 315 mm.
Sample buffer: 50 mmol l^{-1} Tris-HCl, pH 7.5, containing 100 mmol l^{-1} KCl. Sample volume: 2 to 5 µl corresponding to an enzyme activity of 0.26 to 0.52 mU.

[c] Size of cellulose acetate membrane: 25.5 x 145 mm.
Sample volume: 1 µl, corresponding to 150 µg Protein.

[d] Size of cellulose acetate membrane: 25 x 170 mm.
Sample buffer: 20 mmol l^{-1} phosphate, pH 7.4, containing 1% bovine serum albumine. Sample volume: 3 µl.

[e] Sample buffer: 150 mmol l^{-1} NaCl.

ments GmbH, München 40, Germany) or "Sartophor" (Sartorius, Göttingen, Germany [13]) is to be performed as follows (Fig. 3.1, page 73).

The pre-cut cellulose acetate membrane is floated in electrode buffer in a shallow dish; within seconds the liquid completely penetrates the membrane. Excess liquid is removed by carefully blotting the membrane between two sheets of filter paper avoiding the entrappment of any air bubbles. The blotted membrane is carefully lifted and positioned on a membrane holder; the membrane has a reference hole which is always positioned in the lower left-hand corner of the membrane so that sample numbers are always properly identifyable. The square electrophoresis tank, separated into two equal sized chambers, is filled with buffer, the membrane holder is positioned on the buffer tank (Fig. 3.1), the tank cover is put on the tank and the power is put on a few minutes before samples are added to the membrane. The power is turned off and the samples are applied along a pencil line using a microliter syringe or samples

Loading of membranes

are applied by taking a special sample applicator as sold by Sartorius. The slots in the tank cover are closed with a protective lid and electrophoresis is performed maintaining the conditions listed in Table 3.3. When visualizing enzymes occurring in freshly taken human blood, only 1 to 3 µl samples are needed. Enzyme separation can be achieved in 30–60 min applying a current of 16–20 V cm⁻¹. Recipes to visualize enzymes on cellulose acetate membranes are to be found in Chap. 6.

If densitometric scanning is desired, cellulose acetate strips can be cleared after staining, depending upon the solubility characteristics of the stain. If possible the membrane can be rendered transparent by washing it in 5% acetic acid, then treating it for 1 min with 95% ethanol and finally immersing it in 10% acetic acid and 90% ethanol for 5 min. Afterwards the transparent membrane is dried and heated on a glass plate at 60–70 °C for 20 min so that it can finally be scanned [3]. The property of being completely soluble in 10% ethanol and 90% chloroform allows, in some applications, the membrane to be cut into strips, dissolved, and the dye in them to be measured with a spectrophotometer.

Cellulose acetate membranes are partially charged which causes electroendosmosis. Electroendosmosis helps to separate the proteins. Therefore, the placement of the sample on the supporting medium relative to the cathode and anode will affect and determine the degree of separation and resolution of enzymes and their multiple molecular forms [13]. For the reproduction of separation patterns, e.g., in population genetics and forensic medicine, it is a pre-requisite to apply the samples to the very same horizon on the membrane. This is easily accomplished with the electrophoretic equipment shown in Fig. 3.1 [13]. Once a favourable position is established for a given buffer and enzyme system, it should be used in subsequent determinations. If the electrophoretic conditions are changed the proper sample position placement must be reestablished.

3.2.1 Enzyme Visualization

In order to visualize multiple molecular forms of enzymes occurring in human blood (Fig. 3.2) cellulose acetate membranes may be soaked in staining solution [13] or the following overlay technique may be used. The reagents needed are dissolved in 15 ml of buffer. Then 250 mg agar are dissolved in 10 ml of the same buffer and heated in a water bath to 90–95 °C until the agar powder is dissolved. Afterwards the solution is cooled to 50–55 °C and mixed with the staining solution. (To obtain firmer gels the concentration of the agar may be increased to up to 500 mg.) Finally the mixture of agar solution and staining solution is poured into a disposable square plastic petri dish until the dish is slightly more than half full. As quickly as possible, the contents of the dish are emptied into a second petri dish. The contents of the second dish are emptied into a third dish and so on until the bottom of four to ten petri dishes, depending on the concentration of the agar solution, are covered with a smooth complete layer of agar. When handling light-sensitive chemicals the procedure should be performed in semi-darkness. The petri dishes that receive the mixture should rest on a flat and level surface. As soon as the gel solidifies, the petri dishes that will not be used immediately are sealed in individual plastic bags and aluminum foil and stored

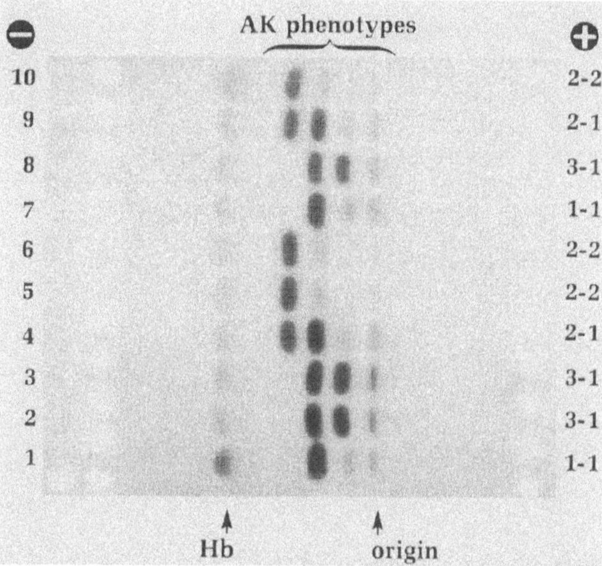

Fig. 3.2. Enzyme polymorphisms of adenylate kinase as occurring in human blood samples and made visible by cellulose acetate electrophoresis: membrane and electrode buffer: 0.014 mol l⁻¹ phosphate buffer, pH 6.25, around 22 °C; 200 V (2–3.5 mA per membrane); running time 45 min; staining with agar overlay; composition of overlay: to 15 ml 0.075 mol l⁻¹ Tris-HCl, pH 7.9, containing 0.025 mol l⁻¹ MgCl$_2$ · 6H$_2$O are added: 38 mg adenosine-5′-diphosphate, 190 mg dextrose, 12 mg NADP, 25 µl (8.75 U) glucose-6-phosphat dehydrogenase, 25 µl (7 U) hexokinase, 2 mg MTT and 2 mg PMS; 250 mg of agar in 10 ml of 0.075 mol l⁻¹ Tris-HCl, pH 7.9; preparation of overlay as described in the text. (Figure kindly provided by Sartorius, Göttingen, Germany.) Adenylate kinase phenotype as described in [13]. The variant 1-1 occurs in 93 % of humans, the other variants are rare [13]

in a completely dark refrigerator. They may be kept for several days or weeks without loss of reactivity.

Immediately after electrophoresis, the ends of the cellulose acetate membrane which extend on either side of the perforation are cut off just outside the perforations with scissors or a razor blade leaving the membrane still under tension in the membrane holder. The perforation indicating the order of the applied samples must remain intact. The membrane is then lifted and rolled down with its upper side onto the agar gel containing the staining reagents without trapping any air bubbles. The placement of the membrane should be done in semi-darkness if light-sensitive chemicals are involved. The cover is then placed over the gel and the petri dish is placed in an incubator in an inverted position, which allows for a maximum interaction between the enzyme molecules trapped in the cellulose acetate membrane and the staining components in the gel.

3.3 Starch Gel Electrophoresis

A large number of enzyme proteins from almost any organism has been separated by starch gel electrophoresis which uses partially hydrolyzed starch as separation medium [26]. Starch gels are non toxic, they have no denaturing effects on enzymes, a good resolving power and nearly any buffer system between pH 2 and 11 can be used [27]. The method is preferentially taken to screen for multiple molecular forms of enzymes. On the other hand starch gels are somewhat friable and not easy to handle, they require a longer electrophoretic run than cellulose acetate membranes do, and they cannot be scanned densitometrically unless especially processed.

Properties
of starch gels

Starch gels are prepared from commercially available partially hydrolyzed potato starch. The process of gel preparation consists of making a "solution" of starch, cooking it, and pouring it into a mould (starch gel tray). As the starch solution cools, it solidifies forming an opalescent gel which can be used as separation medium. The following descriptions are limited to the horizontal mode of starch gel electrophoresis since the vertical version is seldom used.

3.3.1 Gel Tray

The solubilized starch is cast in a gel tray, the set- up of which can be very simple, consisting of a thick glass plate on which a frame of Plexiglas^R (Perspex^R) of variable hight and size is attached with a little vacuum grease [28]. A more solid set-up is demonstrated in Fig. 3.3. It consists of a bridge-like tray. The bottom ends of each side of the tray are covered with a strip of Perspex being attached with self-adhesive tape. Then the tray is placed on a horizontal level and filled with gel until the liquid stays about 1 mm above the tray. Afterwards it is covered with a frame into which a slot former (comb) can be inserted while the starch is still liquid (Fig. 3.3). After the starch solution has solidified into a gel, removal of the template will leave trays in the gel in which the samples can be placed. The teeth of the comb must be long enough to penetrate about 1 mm of the gel tray bottom. The dimensions of the teeth of the comb determine the volume of the solution which can be filled into the sample slots. The sample volume in a slot can be varied to make banding patterns either darker or lighter using combs of 1 to 2 mm thickness.

Sample
application

There are other sample application methods in addition to the use of slot formers: cuts may be made in the gel with a razor blade or knife, and small filter paper strips, into which sample solution has been soaked, may be inserted into the cuts, or else the whole starch block may be cut perpendicularly to the migration direction of proteins into a 1/5 and 4/5 sized part. The sample-containing wicks are then placed on the cut surface of the larger gel block and the smaller one is afterwards firmly attached back to the larger gel part without trapping air bubbles [28]. Using a comb minimizes the risk of getting the two gel blocks separated at the edge during electrophoresis which spoils separation of proteins. When using the gel-tray method it is good practise to enrich the samples with sucrose to a concentration of 10 %.

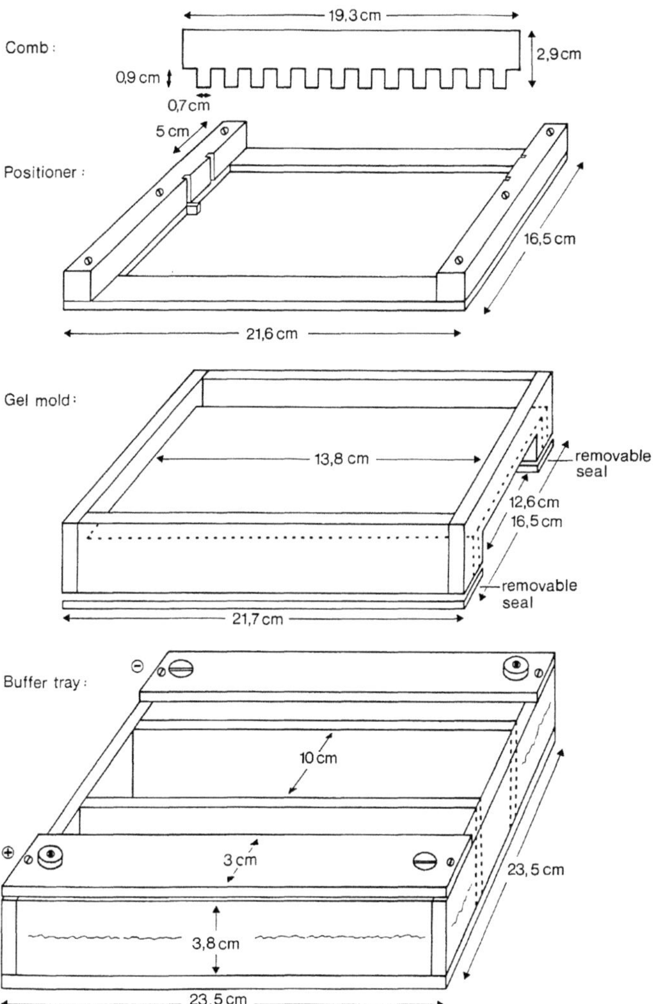

Fig. 3.3. The three main parts of a horizontal starch gel electrophoresis apparatus (all parts made of Perspex). *Top*-frame (positioner) to position the comb used to produce a number of trays into which the samples can be applied. The assembled positioner is mounted on the gel mould immediately after the starch gel has been poured into the mould; *middle*-gel mould; *bottom*-buffer tray

3.3.2 Preparation of Starch Gels

Starch gels are prepared from commercially available starch which has been partially hydrolyzed by the manufacturer according to the method of Smithies [26] (Conaught Medical Research Laboratories, Toronto, Canada; Electrostarch Corp., Madison, WI., U.S.A.; Sigma, St. Louis, MO, U.S.A.; Serva, Heidelberg, Germany). To prepare a gel to be cast in the gel tray shown in Fig. 3.3 (150 × 120 × 10 mm; 12 – 20 slots) 500 ml of a 11 – 12 % starch gel solution are needed. There are a number of different ways to heat a starch suspension until it cooks; bunsen burners, heating plates or hot water baths were used for that purpose [3, 28]. However, with these methods, starch particles may agglomerate and form lumps with insufficiently hydrated starch or starch particles may sediment on the bottom of the flask forming a tough cake which dissolves so slowly that it might burn. These problems can be circumvented by use of a powerful microwave device such as that supplied by Sharp (Sharp Co., type R-10R50, Japan) equipped with a turntable and operating variably in 100 watt-steps from 100 to 1000 W [29] as follows.

First, 54 g of hydrolyzed (Toronto) starch are weight on a dish, and 500 ml of gel buffer are subdivided into three fractions, placing 150 ml into a 500-ml beaker, 30 ml into a 200-ml beaker and 320 ml into a 750-ml Erlenmeyer flask. The latter solution is degassed by ultrasonication. The 750 ml flask is placed on the turntable of the microwave, covered with a watch glass and heated at 1000 W for 2 min while the 500-ml beaker is placed on a magnetic stirrer and a stirring bar added. The 54 g of dry starch are quickly added to the 150 ml of buffer, pressing the starch through a (tea) strainer with the aid of a porcelain pestle to obtain a fine powder. Afterwards the resulting suspension is degassed by ultrasonication. The flask containing the heated 320 ml of gel buffer is removed from the microwave using heat-resistant safety gloves, a stirring bar is added and the flask is put on a magnetic stirrer. The starch suspension is briefly stirred in the 500-ml beaker using a glass rod and poured along this rod into the heated gel buffer while vigorously stirring and avoiding any lump formation. The remaining starch is rinsed out from the 500-ml beaker with the 30 ml of buffer contained in the 200-ml beaker. The washings are poured into the 750 ml flask to obtain the final volume of 500 ml starch suspension and the flask is put on the turntable of the microwave and covered with a watch glass. It is heated first at 1000 W for 4 min and then at 500 W for 2 min, then heating is continued at 500 W for 3 × 1 min with intermissions of about 10 s to avoid any surging. The boiling process is completed only when large bubbles form at the bottom of the flask and ascend in the mixture. If necessary heating is continued for 1 min, after which the flask is carefully removed from the microwave using heat resistant safety gloves. It is slightly slanted to allow any air bubbles to escape without splashing. Finally the very hot starch gel solution is poured into the gel tray standing on a sheet of paper so that the excess of starch which overflows can be subsequently discarded. Then a Perspex plate with a frame made to take up the slot former is put on top of the gel without trapping any air bubbles and the comb is inserted. The starch solution is allowed to cool at room temperature. After approximately 1 h the gel will have settled. It is stored overnight (in a refrigerator) before use. If desired, it can also be stored refrigerated in the gel tray for several days.

3.3.3 Buffer Systems

In starch gel electrophoresis, continuous and discontinuous buffer systems are used. For most enzymes the buffer systems are continuous which means that the same buffer ions in the gel and the electrode vessels are used, though the gel buffer is usually ten times less concentrated than the electrode buffer. Buffer systems are named discontinuous if gel and electrode buffer are different in their composition. A discontinuous system has been used, e.g., to separate nucleoside phosphorylase isozymes of human erythrocytes. Electrode buffer solutions should be fresh if small buffer reservoirs are used (Fig. 3.3). If larger buffer volumes are used the buffer may be taken two to four times. But in this case it is essential to mix anodal and cathodal buffer and check the pH-value before starting a new experiment.

The main function of the buffer is to keep the pH within the separation medium constant. The extent to which the pH is altered is directly related to the voltage gradient, the current and the duration of electrophoresis. Buffers function at optimum rate efficiency when the desired pH is close to the pK-value of the buffering substance, and when they have an appropriate ion strength and stability. In practice, however, such ideal buffers may not give optimum resolution or allow sensitive staining of enzymes, and therefore compromises are necessary. The composition of most electrophoresis buffers is determined by trial and error. Since the stability properties of enzymes vary greatly, a large number of buffer systems have been suggested to separate the various enzyme and isozyme systems [1, 28]. In practice, however, it is more desirable to use only a limited number of gel and electrode buffers, and it is indeed possible to separate a number of human isozyme systems by only two different buffer systems [2] (cf. Chap. 6). When starting a study on an enzyme of which the stability properties are unknown, it is good practise to commence with buffers of a pH-value close to the pH-optimum of the enzyme studied and to carry out electrophoresis for a relatively short period (2–3 h) at a low voltage (2–3 V · cm⁻¹). Then, based on the information obtained, the separation conditions may gradually be altered, for example by increasing the voltage and duration of the run and changing the pH and buffer until optimum separation is achieved. It is of advantage to incorporate known activators or cofactors of an enzyme into the gel or corresponding to its charge into the anodic or cathodic electrode vessel in order to improve enzyme stability and resolution of bands. Phosphoglucomutase (EC 2.7.5.1), for example, is activated by Mg^{2+} and EDTA [1] while several dehydrogenases (e.g., alcohol dehydrogenase, EC 1.1.1.1; glucose-6-phosphate dehydrogenase, EC 1.1.1.49 and 6-phosphogluconate dehydrogenase, EC 1.1.1.44) show improved stability, separation and staining intensity in the presence of NAD(P). Pure buffer systems often have lower separating and resolving properties than mixtures of cationic (e.g. Tris) and anionic (e.g. citric acid) buffers. A compilation of buffer systems and visualization recipes for the location of enzymes following starch gel electrophoresis is given in Chap. 6.

Chosing an appropriate buffer system

Improving enzyme stability

3.3.4 Sample Application

When using starch gels with preformed sample slots, samples can be placed directly into the trays. The slot volumes may be varied by using combs of various sizes. It is not necessary to put samples in all slots, but the empty slots must be filled with sample buffer. All slots should be filled to approximately 90 % to avoid pattern distortion. Another method is to soak sample solutions into small pieces of filter paper (Whatman No. 3 MM, or, if heavier loading is wanted, Whatman No. 17 filter paper) of the size 2 × 8 mm and insert these into the gel with the aid of tweezers and a small piece of razor blade. The sample inserts should be well-spaced and not allowed to protrude above the surface of the gel. The samples are usually applied to the prospective cathodal end since most enzymes behave chemically as anions at the pH-values of commonly used buffer systems (Chap. 6) [1 – 3, 26, 28]. When using the gel tray shown in Fig. 3.3 the slot former is taken out of the gel and the covering Perspex plate removed from the gel. Then the samples are applied.

3.3.5 Electrophoresis

Before the gel tray containing the applied samples is placed into the electrophoretical apparatus, approximately 180 ml of buffer are filled in each electrode vessel. Except for a few enzymes, both buffer trays contain the same buffer solutions (Chap. 6). When using the gel tray shown in Fig. 3.3 no wicks are needed to link the starch gel with the electrode buffer because the ends of the gel dip directly into the cathodal and anodal buffer. Electrophoresis is carried out in a cold room or a refrigerator at 4 – 8 °C but leaving the power supply outside the cooling device. The voltage gradients and the times of electrophoresis depend on the buffer systems used and some of the most often applied conditions have been compiled in Chap. 6. The voltage is adjusted to the appropriate level by taking measurements directly across the ends of the gel at the start of the experiment using a voltmeter. Current is applied for approximately half an hour. Then the current is switched off again and the entire gel is covered with Saran foil to prevent undue drying. Afterwards a plastic tray having the basal surface of the gel and filled with ice is placed on the covered gel to reinforce cooling. To start the experiment the current is turned on again by applying the voltage evaluated before.

The power (wattage) put into the gel is proportional to the product of voltage (V) and current (A). It is also proportional to the resistance multiplied by the square of the current, since voltage is equal to current multiplied by resistance. The heating of a gel is more markedly affected by increasing the current. Therefore, buffer systems are designed to attain a relatively high voltage at the expense of the current. It is good practice to check the voltage by reading the power supply after 30 – 60 min to determine whether it is holding constant. The temperature in the coldroom or refrigerator can also be checked at this time. If the gel cracks and does not conduct current well it must be discarded and the experiment must be started again. The duration of electrophoresis is usually determined empirically by the length of time required to obtain full separation of the enzymes under investigation. Separation is influenced by the current flow, the molecular size of the enzymes, their isoelectric points, and the pH and ionic strength of the buffer. The higher the ionic strength of the buffer, the slower the rate of migration, the sharper the bands, and the greater the heating effect. In practice, a compromise must usually be reached between these factors since, at a certain buffer strength, several enzymes will lose their solubility while, on the other hand, enzymatic activity is a function of time and temperature. In practice the duration of starch gel electrophoresis is usually either 3-5 h (one day procedures) or about 18 h (overnight procedures). The shorter running times makes it possible to run and stain the gels within one work day (for details see Chap. 6).

3.3.6 Gel Slicing

After electrophoresis has been terminated the power supply is turned off and the gel tray is removed from the electrophoretic apparatus. The plastic foil is taken away from the gel and its cathodal and anodal ends are cut off with a kitchen knife and removed. Then the gel is loosened from the smaller sides of the gel tray by running a spatula along each side. A piece of thin Perspex as wide as the smaller internal side of the gel tray and double the length of the gel is pushed under the gel to remove it from the tray. Afterwards the gel is transferred to a gel slicing table (Fig. 3.4). Often the gel will break at the site of the slots but this is no reason to discard it, because only in a few cases enzymes migrate into the cathodal end. It is advisable to mark the gel, e. g., by punching a hole with a 4-6-mm-wide piece of glass tubing into the right hand side of its anodal end so that orientation can be maintained regarding the order of the sample slots. Prior to enzyme staining a starch gel must be sliced at least once. There are several types of gel slicers. In the one shown in Fig. 3.4, five thin stainless wires are stretched under tension in a U-shaped device. The slicer is moved from the free end

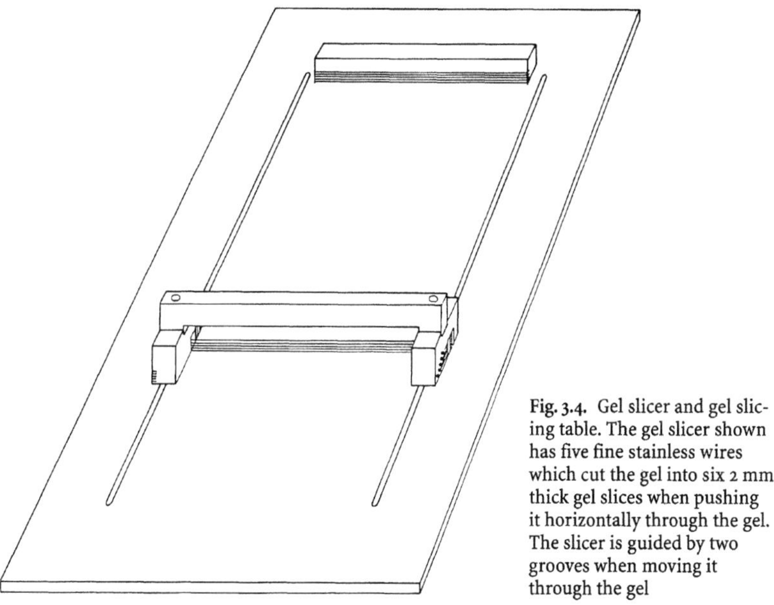

Fig. 3.4. Gel slicer and gel slicing table. The gel slicer shown has five fine stainless wires which cut the gel into six 2 mm thick gel slices when pushing it horizontally through the gel. The slicer is guided by two grooves when moving it through the gel

of the slicing table to the supporting end while slightly pressing a thin Perspex plate on the top of the gel. By doing so the gel is sliced into six equal layers. The two upper and one bottom gel slices are discarded because they do not show good banding patterns. The staining solution is applied to the cut surfaces of the three remaining slices from the centre part of the gel. Slicing provides the opportunity to stain more than one enzyme on the same gel. Slices cannot be made thinner than 2 mm because of the friability of starch gels.

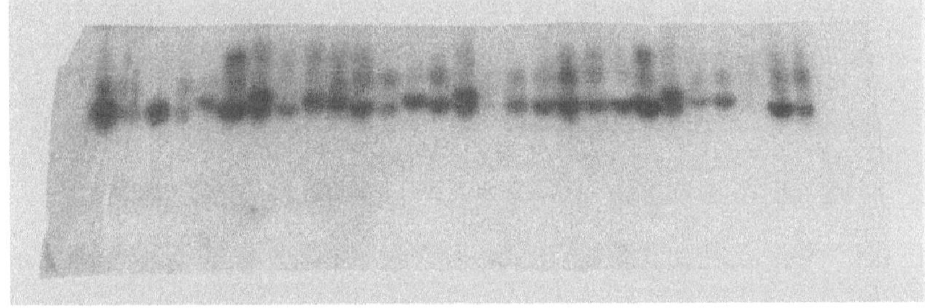

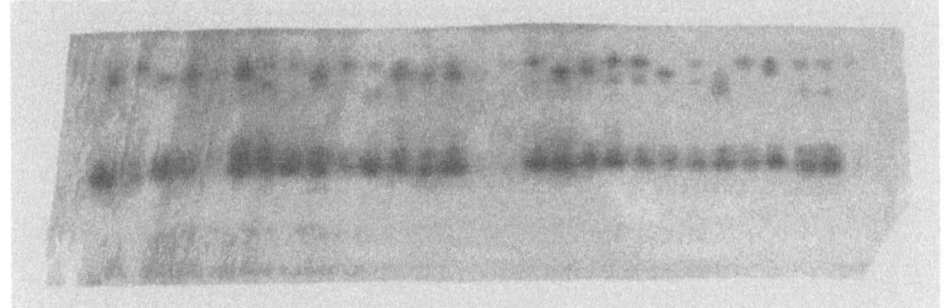

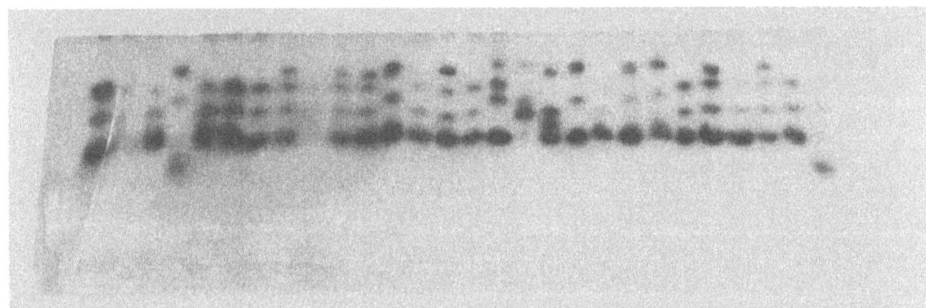

Fig. 3.5 Enzyme polymorphism as found in some tree species by starch gel electrophoresis and enzyme visualization. Top: acid phosphatase of Scots pine (*Pinus sylvestris*) endosperm. Electrode buffer: 90 mmol l⁻¹ Tris, 260 mmol l⁻¹ boric acid, 4 mmol l⁻¹ EDTA-Na$_2$, pH 7.5; gel-buffer: one in ten diluted electrode buffer; running conditions: 200 V for 5 h [30] (staining protocol: Chap. 7); middle: phosphoglucomutase of lime (*Tilia cordata*) buds. Electrode buffer: 150 mmol l⁻¹ Tris-citrate, pH 7.5; gel buffer: 20 mmol l⁻¹ Tris-citrate, pH 7.5; running conditions: 80 mA for 5 h [31] (staining protocol: Chap. 7); bottom: malate dehydrogenase of Scots pine (*Pinus sylvestris*) endosperm. Electrode buffer: 135 mmol l⁻¹ Tris plus 43 mmol l⁻¹ citric acid, pH 7.0; gel buffer: a 15-fold dilution of the electrode buffer; running conditions: 50 mmol l⁻¹ for 6 h [30] (staining protocol: Chap. 7)

3.3.7 Enzyme Visualization

After the gel has been sliced in at least two equal layers it is ready for staining. Each slice is placed into a flat dish with its cut surface facing the viewer. Ordinary household plastic dishes can be used but they should not be much larger than the base of the gel in order to save expensive staining reagents. There are three staining methods: (a) the dish is filled with staining solution until the gel is just covered with it; (b) staining solution is applied dropwise with a Pasteur pipette onto the gel surface so that the gel is able to soak up the reagents and (c) the staining reagents are dissolved in liquid agar and poured on the starch gel so that a so-called agar overlay is formed. Several enzyme systems visualized with method a) or b) are shown in Fig. 3.5. An agar overlay can be prepared as follows.

    The required amount of agar to obtain a final concentration of 0.8 to 1.0 wt/vol.% is weighed and dissolved in half the volume of buffer needed to dissolve the staining reagents. The agar solution is then heated until boiling, cooled down to around 55 °C and mixed with the staining reagents being dissolved in the remaining part of buffer. Then the whole is poured on the starch gel and when the agar solidifies a gel overlay is formed. At the interface of starch and agar, enzyme and reagents diffuse forming coloured bands. Bands are more discrete when using this method and little staining reagent is needed. The agar may be replaced by a cellulose acetate or a cellulose nitrate membrane moistened with staining solution. Upon incubation the bands appear primarily on the overlay. The membrane can be taken off, dried and stored as a permanent record. Staining is usually carried out for 1 – 3 h in a warm-air incubator at 37 °C. Incubation at this temperature increases the rate of the enzyme reaction causing the stain to form in a shorter period of time. If staining is too slow, there may be considerable diffusion of the enzyme molecules resulting in less sharpness to the banding patterns. When studying a new enzyme system it is advisable to perform a number of control reactions in which the substrate, the cosubstrate or any auxiliary substance is omitted from the test system. This procedure, of course, requires analytical grade reagents. Some enzymes, catalyzing presumably specific reactions, may on occasion catalyze other reactions which are also thought to be specific. The detection of such cross reactivities usually depends upon staining with multiple substrates (cf. Chaps. 5 and 6).

3.3.8 Documentation and Gel Preservation

Gels stained with staining reagents containing the stain nitro blue tetrazolium (nitro BT) can be soaked in 50 % ethanol for 30 – 60 min. This causes the gels to become considerably tougher. They may then be sealed in a transparent plastic bag and preserved in a refrigerator almost indefinitely. Many stains, however, including some of the other tetrazolium salts, are soluble in alcohol and cannot be preserved in that way. In these cases a membrane overlay may help.

    Since only in a few cases can gels be conserved in an alcoholic solution and, since cellulose acetate or nitrate membranes are costly, it is advisable to photograph the stained gels. Starch gels are usually not used for densitometric scans. But if one intends to quantify separated enzymes, gels may be rendered transparent by placing them in glycerol heated to 70 – 80 °C for 5 min [3].

Photographic methods involve combining black and white or color films with filters of various types. Kodak color print film (ASA 100), in combination with an 80A color correction filter, improves band visibility. Fluorescent stains can be photographed in color using a Wratten 2E filter, or in black-and-white using a red R2 filter (TechPan 2415, Polaroid 667 or 55 films). Yellow or red filters may be used when photographing blue-purple stains with black-and-white film (TechPan-25, Panatomic X-32, Tri-X, Plus-X-125). Green filters are of advantage when photographing blue bands on a red background. Gels should be transferred to glass or Perspex plates for scoring and photography on a light table. Backlighting and photofloods mounted on a copy stand are helpful. The enzyme assayed, buffer system, date, gel number and sample codes are best inscribed below the gel on the plate or on a transparent foil before it is photographed.

Gels may be stored in 2% formaldehyde or dried (gel slab dryers, Bio Rad, München, Germany; Hoefer, San Francisco, CA, USA) between layers of cellophane film soaked briefly in water. Gels may be dried for 3 h with heat and vacuum, followed by overnight drying under vacuum only to keep a permanent record.

3.4 References

1. Harris H, Hopkinson, DA (1976). Handbook of enzyme electophoresis in human genetics, Elsevier, New York
2. Siciliano MJ, Shaw CR (1976). Separation and visualization of enzymes on gels. In: Smith I (ed) Chromatographic and electrophoretic techniques, vol 2. Heinemann, London pp 185 – 209
3. Brewer GJ, Sing CF (1970). An introduction to isozyme techniques. Academic Press, New York
4. Meera Khan P (1971). Arch Biochem Biophys 145: 470 – 483
5. Van Someren H, Van Henegouwen HB, Los W, Wurzer-Figurelli E, Doppert B, Veroloet M, Meera Khan P (1974). Humangenetik 25: 189 – 201
6. Williams DE, Reisfeld RA (1964). Ann New York Acad Sci 121: 373 – 381
7. Davis BJ (1964). Ann New York Acad Sci 121: 404 – 427
8. Lasky M (1978). Protein molecular weight determination using polyacrylamide gradient gels in the presence and absence of sodium dodecyl sulfate. In: Catsimpoolas N (ed) Electrophoresis '78, Elsevier-North Holland, Amsterdam, pp 195 – 210
9. Hayes MB, Wellner D (1969). J Biol Chem 244: 6636 – 6644
10. Smith AE, Yamada EW (1971). J Biol Chem 246: 3610 – 3617
11. Grunbaum BW, Zajac PL (1978). J Forens Sci 23: 84 – 88
12. Zajac PL, Grunbaum BW (1978). J Forens Sci 23: 615 – 618
13. Grunbaum BW, Crim M (1981). Handbook for forensic individualization of human blood and bloodstains, Sartorius GmbH, Göttingen, Germany
14. Takeuchi T, Matsushima T, Sugimura T, Kozu T, Takeuchi T, Takemoto, T (1974). Clin Chim Acta 54: 137 – 144
15. Dahlmann B, Jany Kl-D (1975). J Chromatogr 110: 174 – 177
16. Herd JK, Tschida J, Motycka L (1974). Anal Biochem 61: 133 – 143
17. Hullin DA, Thompson RJ (1977). Anal Biochem 82: 240 – 242
18. Long WK (1967). Science 155: 713 – 714
19. Kömpf J, Bissbort S, Gussmann S, Ritter H (1975). Humangenetik 27: 141 – 143
20. Parr CW, Bagster IA, Welch SG (1977). Biochem Genet 15: 109 – 113
21. Myers RC, Van Remortel H (1968). Clin Chem 14: 1131 – 1134
22. Grunbaum BW (1975). Microchem J 20: 495 – 510

23. Lewis WHP, Harris H (1967). Nature 215: 351 – 355
24. Lewis WHP (1973). Ann Hum Genet Lond 36: 267 – 271
25. Schalhorn A, Wilmanns W (1977). Res exp Med 169: 213 – 219
26. Smithies O (1955). Biochem J 61: 629 – 641
27. Ferguson A (1980). Biochemical Systematics and Evolution, Blackie and Son Ltd., Glasgow, Great Britain
28. Conkle MT, Hodgskiss PD, Nunnally LB, Hunter SC (1982). Starch gel electrophoresis of conifer seeds: a laboratory manual, United States Department of Agriculture, Forest Service, Pacific Southwest Forest and Range Experiment Station, P.O.Box 245, Berkely, California 94701
29. Maurer WD. Forstliche Versuchsanstalt Rheinland-Pfalz, Trippstadt, Germany, personal communication
30. El-Kassaby YA, Rudin D, Yadzani R (1989). Scand J For Res 4: 41 – 49
31. Cheliak WM, Pitel JA (1984). Techniques for starch gel electrophoresis of enzymes from forest tree species. Information Report PI-X-42, Petawa National Forest Institute, Canada

3.5 Analytical Polyacrylamide Gel Electrophoresis for Separating Native Enzymes

Many buffer systems have been suggested to separate native enzymes by polyacrylamide gel electrophoresis [1 – 5] but in practice only a few are frequently used. These are:

(a) the discontinuous system of Davis [6] and Ornstein [7];
(b) the Ornstein-Davis system without sample gel by Clarke [8];
(c) the discontinuous system of Williams and Reisfeld [9]; and
(d) the continuous system of Aronsson and Grönwall [10].

Systems (a), (b) and (d) separate enzymes at alkaline conditions (see also [11 – 17]) while others were designed to perform protein separation at neutral pH [18, 19] or at acidic conditions such as system (c) or those listed in [20, 21].

3.5.1 Homogenous Buffer Systems

The most famous homogeneous buffer system used in vertical polyacrylamide gel electrophoresis is the one suggested by Aronsson and Grönwall [10]. Gel and electrode buffer consist of 90 mmol l^{-1} Tris, 45 mmol l^{-1} boric acid and 2.5 mmol l^{-1} EDTA-Na$_2$, pH 8.4. Good results may also be obtained in vertical electrophoresis (e.g., with 110 m long and 6 mm thick round gels) when using as gel buffer 0.625 mmol l^{-1} glycine adjusted with Tris to pH 8.3 and as electrode buffer 5 mmol l^{-1} glycine, adjusted with Tris to pH 8.3 [22, 23]. The gels are submitted for 15 min to pre-electrophoresis at 1.5 mA/gel to remove unpolymerized constituents and catalysts. Then a small amount of sample solution enriched in 10 % sucrose is layered between the top of the gel and the electrode buffer and electrophoresis is performed for 2.5 h [24].

3.5.2 Disc-Electrophoresis

3.5.2.1 General Aspects and Buffer Systems

Disc-electrophoresis [24] employs two different gels: a large pore or stacking gel in which the proteins are concentrated to a very small band and a small pore or separation gel in which the proteins are separated according to size and charge. Ornstein and Davis used another large pore gel on top of the concentration gel in which they polymerized the sample. Nowadays the sample gel is omitted and protein samples containing 10 – 20 % sucrose or 10 – 20 % glycerol in a diluted phosphate buffer in-

**Small pore gel
large pore gel**

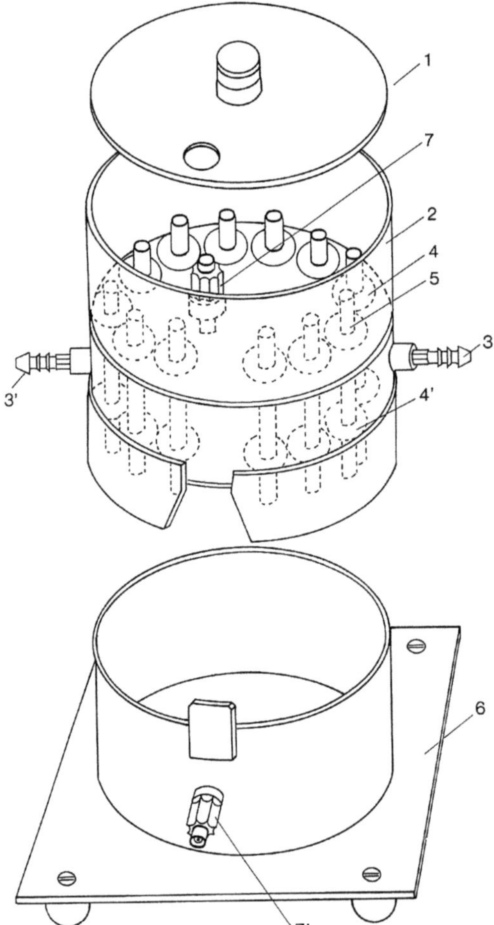

Fig. 3.6. Disc-electrophoresis apparatus. 1: cover, 2: gel tube holding device and upper electrode vessel, 3,3′: cooling water inlet and outlet, 4,4′: rubber gasket, 5: glass tube with rod gel, 6: lower electrode compartment, 7.7′: electrode plug

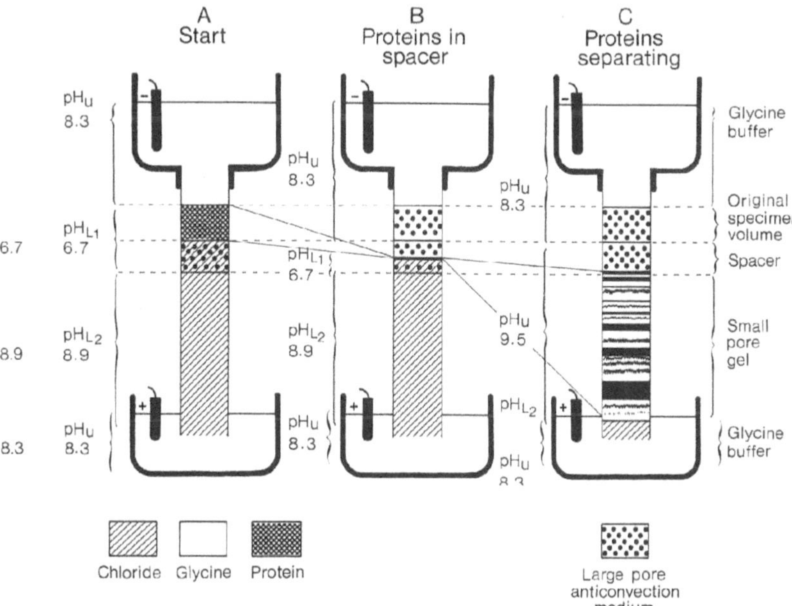

Fig. 3.7. Separation of proteins in disc-electrophoresis when using a neutral stacking gel buffer (pH_{L_1} 6.7), an alkaline separation buffer (pH_{L_2} 8.9) and an alkaline electrode buffer (pH_u 8.3) [7]

clusive of 0.003 % of a tracking dye (Bromophenol blue or Bromophenol red) are applied directly on top of the concentration gel. The total polymer concentration of the large pore gel is 3.125 (g (100 ml)$^{-1}$; acrylamide : BIS = 4:1), while the total polymer concentration of the separation gel is 7.7 wt/vol.% and the ratio of acrylamide to BIS is 30:1. The length of the large pore gel is approximately 5–10 % of the length of the separation gel. Both types of gel are usually polymerized in glass tubes of an internal diameter of 6 mm and a length of 100 mm (Fig. 3.6). In the large pore gel, proteins are concentrated into a very narrow zone between faster migrating buffer ions ("leading ions") and slower migrating buffer ions ("trailing ions"). At the beginning of electrophoresis the leading ions (Cl$^-$) are in the total gel column while the trailing ions (glycine) are exclusively in the electrode buffer. Under the influence of the electric field at alkaline conditions both types of ions start to migrate to the anode. However, the leading ions migrate faster than the proteins and the trailing ions thus leaving a zone of reduced conductivity behind. Since the specific conductivity is inversely proportional to the electric field strength, the migration of the negatively charged proteins and the trailing ions is accelerated until all the ions "migrate" at the same velocity thereby concentrating the proteins into a sharp zone (Fig. 3.7). Consequently a moving boundary is formed between leading and trailing ions which traverses the large pore gel towards the separation gel. When using a 100 mmol l^{-1} sodium phosphate buffer of pH 6.8 or the original large pore buffer suggested by Davis [24] without acrylamide and BIS but enriched with 5 – 10 % sucrose or glycerol to dissolve

**Leading ions
tailing ions**

the sample proteins, a large pore gel can be omitted without renouncing the concentration effect.

Upon entering the separation gel the migrating protein zone strikes discontinuities in gel concentration, buffer system, buffer concentration and pH-value. The pH-value of the separating gel is chosen so that the dissociation of the trailing ions increases manyfold. An increase in dissociation means an increase in electrophoretic mobility, and as a consequence the trailing ions overtake the proteins and start to migrate closely behind the leading ions (Fig. 3.7). The proteins now migrate in a homogeneous electric field and at a constant pH-value. A 7.7% T gel is used to separate human serum proteins at optimum according to their net negative charge and their Stokes' radius. This PAA-concentration is also used to separate most animal and plant enzymes although 3% and 10% gels were also used (cf. Table 3.4). In case the mobility of the trailing ion in the unstacking step remains lower than that of some sample proteins these will remain stacked and migrate with the (new) ionic boundary, i.e., these will not be subjected to separation [27].

Both concentration of diluted sample proteins to an extremely thin protein layer before entering the separation gel and resolution of proteins according to net charge and size differences cause the high resolution of disc-electrophoresis. The most popular disc-electrophoretic systems to separate enzymes are summarized in Table 3.4. Several enzymes which were visualized in situ following disc-electrophoresis are listed in Chap. 6.

Table 3.4. The most popular disc gel systems used to separate catalytical active enzymes

Electrode buffer [Ref.]	Large pore gel (acrylamide: BIS)	Separating gel (acrylamide: BIS)
5 mmol l^{-1} Tris, 38.4 mmol l^{-1} glycine, pH 8.3; [24]	60 mmol l^{-1} HCl 6.2 mmol l^{-1} Tris, pH 6.7; acrylamide + BIS = 3.125 g/100 ml; (10:2.5)	60 mmol l^{-1} HCl, 38 mmol l^{-1} Tris, pH 8.9; acrylamide + BIS = 7.7 g/100 ml; (30:0.8)

A sample gel may be used. It may be prepared from 3 µl sample (human serum) and 0.15 ml large pore gel solution. Nowadays a sample gel is not used. The migrating albumin front is marked by the addition of 1 ml of 0.001% Bromophenol blue to the upper buffer reservoir (1 l). Electrophoresis should be started within 1 h after the spacer gel has been set. To separate native enzymes it is recommended to perform electrophoresis in the cold (4 – 6 °C) at a current of 1 mA/gel until Bromophenol blue has reached the lower end of the gel (2 h).
The separating gel is prepared from 1 vol A, 2 vol C, 1 vol H_2O, and 4 vol F, with: A = 48 ml 1 N HCl, 36.6 g Tris, 0.46 ml Temed and H_2O to 100 ml, pH 8.9; C = 30.0 g acrylamide, 0.8 g BIS, 15 mg $K_3[Fe(CN)_6]$ and H_2O to 100 ml; F = 140 mg ammonium persulphate in 100 ml H_2O (freshly prepared). The large pore gel is prepared from 1 vol B, 2 vol D, 1 vol E and 4 vol H_2O, with B = 48 ml 1 N HCl, 5.98 g Tris and H_2O to 100 ml, pH 6.7; D = 10 g acrylamide, 2.5 g BIS and H_2O to 100 ml; E = 4 mg riboflavin in 100 ml H_2O. The large pore gel is photopolymerized under a 15 W fluorescent lamp placed 8 cm behind the tubes. Photopolymerization may be replaced by chemical polymerization using Temed and ammonium persulphate as catalysts. Gels are usually prepared in glass tubes of an inner diameter of 6 mm and a length of 100 mm with the separating gel being 75 mm long and the large pore gel being 5 mm in hight. The electrode buffer is prepared from a stock solution of 6 g Tris and 28.8 g glycine in 1 l of distilled water. It is diluted 1:10 before use.

Table 3.4 (continued)

Electrode buffer [Ref.]	Large pore gel (acrylamide: BIS)	Separating gel (acrylamide: BIS)
387 mmol l⁻¹ glycine, 50 mmol l⁻¹ Tris, 5 mmol l⁻¹ HCl, pH 8.1; [8]	Replaced by sample solution containing 2 to 5 % urea or 5 – 10 % sucrose or glycerol	6.2 mmol l⁻¹ Tris, 48.3 mmol l⁻¹ Glycine, pH 8.15; acrylamide + BIS = 7.625 g/100ml; (30:1.0)

Separation gels are prepared by mixing 2 vol A, 1 vol B, 4 vol C and 1 vol D, with: A = 30.0 g acrylamide, 1.0 g BIS and 123 ml H_2O; B = 0.28 vol.% Temed; C = 0.14 wt/vol.% ammonium persulphate; D = 29 g glycine, 6 g Tris and 980 ml H_2O. Stacking and sample gels are not used but replaced by 0.2 ml of a 5 – 10 % sucrose solution containing 200 – 400 µg of total (serum) protein (5 – 50 µg of a given protein). Dialysis prior to electrophoresis is recommended only if foreign electrolyte concentrations are high, as in urine. Samples may be dissolved in a 100 mmol l⁻¹ phosphate buffer of pH 6.5 – 7.0. The electrode buffer is prepared from a stock of 29 g glycine, 6.0 g Tris, 5 ml N HCl and 975 ml H_2O (final pH 8.1) and diluted 1:10 prior to use. Gels may be prepared in glass tubes of an inner diameter of 6 mm and a length of 100 mm with the separating gels being 75 mm long and the large pore gel being 5 mm high. Electrophoresis is performed for 30 min at 4 mA per gel.

140 mmol l⁻¹ β-alanine, 350 mmol l⁻¹ acetic acid, pH 4.5; [25]	60 mmol l⁻¹ KOH, 63 mmol l⁻¹ acetic acid, pH 6.8; acrylamide + BIS = 3.125 g/100 ml; (10:2.5)	60 mmol l⁻¹ KOH, 376 mmol l⁻¹ acetic acid, pH 4.3; acrylamide + BIS = 7.7 g/100ml; (30:0.8)

This system is used to separate basic proteins and peptides at an acid pH-value. A maximum of 50 µg protein can be separated per gel. 5 – 10 ml of sample solution are mixed with 0.15 ml large pore solution and 0.15 ml of this mixture are applied per gel. The separation gel has a volume of 0.85 ml and is prepared from a mixture of 1 vol A, 2 vol C, 1 vol H_2O and 4 vol F, with: A = 48 ml N KOH, 17.2 ml acetic acid (glacial), 4 ml Temed and water to make 100 ml; C = 30 g acrylamide, 0.8 g BIS and water to make 100 ml; F = 0.28 g ammonium persulphate per 100 ml of water. The large pore gel has a volume of 0.15 ml and is prepared by mixing: 1 vol B, 2 vol D, 1 vol E, and 4 vol H_2O, with B = 48 ml N KOH, 2.87 ml acetic acid (glacial), 0.46 ml Temed and water to make 100 ml; D = 10 g acrylamide, 2.5 g BIS and water to make 100 ml; E = 4 mg riboflavine dissolved in 100 ml H_2O. Polymerisation of the large pore and sample gel is initiated by a 15 W fluorescent lamp placed about 8 cm behind the tubes. The remaining part in the tubes is filled with electrode buffer (31.2 g β-alanine, 8.0 ml acetic acid (glacial) and water to make 1 l, pH 4.5). Gels may be prepared in glass tubes of 5 mm in diameter and 70 mm length. Electrophoresis is performed for 20 min at a current of 6 – 8 mA per tube.

35 mmol l⁻¹ diethylbarbituric acid, 8.26 mmol l⁻¹ Tris, pH 7.0; [18]	48.75 mmol l⁻¹ H_3PO_4, 51.09 mmol l⁻¹ Tris, pH 5.8; acrylamide + BIS = 3.125 g/100 ml; (4:1)	70.71 mmol l⁻¹ Tris, 60.00 mmol l⁻¹ HCl, pH 7.5; acrylamide + BIS = 7.70 g/100 ml; (30:1)

The electrode buffer contains 5.53 g 5,5-diethylbarbituric acid and 1 g Tris, pH 7.0, in 1 liter H_2O.
The buffer solution needed for the separation gel (solution A in the Davis system) is prepared from 48 ml 1 N HCl, 6.85 g Tris and 0.46 ml Temed, pH 7.5 and distilled water to make 100 ml. All other solutions needed to prepare the separation gels are the same as those used in the Davis system [24].
The buffer solution of the large pore gel (solution B in the Davis system) consists of 39 ml 1 mol l⁻¹ H_3PO_4, 4.95 g Tris, 0.46 ml Temed, and distilled water to make 100 ml; the pH is 5.5. All other solutions needed to prepare the gel are the same as those used in the first system of this table [24].

Table 3.4 (continued)

Electrode buffer [Ref.]	Large pore gel (acrylamide: BIS)	Separating gel (acrylamide: BIS)
5 mmol l⁻¹ Tris, 38.4 mmol l⁻¹ glycine, pH 8.3; [26]	125 mmol l⁻¹ H_3PO_4, 4.71 mmol l⁻¹ Tris, pH 6.9; acrylamide + BIS = 3.125 g/100 ml; (4:1)	378 mmol l⁻¹ Tris, 125 mmol l⁻¹ HCl, pH 8.9; acrylamide + BIS = 3.5 g/100 ml; (30:0.78 constant)

The electrode buffer is the same as in the system of Davis [24]. The buffer solution needed for the separation gel (solution A in the Davis system) is prepared from 36.62 g Tris, 10 ml of 32 wt/vol.% HCl (1.16 kg · l⁻¹), 0.46 ml Temed, and distilled water to make 100 ml, pH 8.9. All other solutions needed to prepare the separation gel are the same as those used in the Davis [24] system which is the first of this table.

The buffer solution of the large pore gel (solution B in the Davis system [24]) is made of 10 ml 10 mol l⁻¹ H_3PO_4 (57.30 ml H_3PO_4 85 wt/vol.% (1.71 kg l⁻¹) in 1 l of H_2O), 0.456 g Tris, and 0.46 ml Temed and distilled water to 100 ml, pH 6.9. All other solutions needed to prepare the separation gel are the same as those used in the Davis [24] system.

3.5.2.2 Preparation of Gels

Rod gels are mainly used in disc-electrophoresis when the purity of enzyme preparations is tested. Flat gels are preferably used to score the presence of proteins in many samples. Methods to prepare gel slabs are described in Sect. 3.5.3.1. Round gels are preferably prepared in glass tubes of a length of 70 to 150 mm with an inner diameter of 5 to 6 mm. Micromethods use capillaries of a volume of 2, 5 or 10 µl and a length of 32.5, 41.5 or 54.5 mm [28]. The glass tubes used in disc-electrophoresis should be made from the same length of tubing, should have plain and smooth ends and their outer diameter must be of a size that will fit tightly into the rubber gaskets of the upper electrode vessel (Fig. 3.6). Before use the tubes are cleaned by treating them with 10 wt/vol.% dichromate sulphuric acid and by rinsing them with distilled water. When completely dried one end is capped with a plastic cap surrounding the tube and the capped tubes are mounted in a rack. Before pouring the gels the various stock solutions needed (cf. Table 3.4) are warmed to room temperature while the electrode buffer is cooled down to a temperature of 4 – 8 °C. The stock solutions described in Table 3.4 can be kept at low temperatures for several weeks while the ammonium persulphate solution has to be freshly prepared. The separating gel solution must be used immediately after the two solutions containing the catalysts Temed and ammonium persulphate have been mixed together. During stirring the mixing of too much air into the solution must be avoided. By means of a Pasteur pipette the tubes are filled three-quarters full with separating gel solution. The gel solution is then carefully overlayered with 0.2 to 0.4 ml of distilled water. After approximately 1 h the polymerization is completed and the water is carefully removed from the gel surface without damaging it. Then the surface is rinsed twice with large pore gel solution and the large pore gel solution is layered on top to a height of approximately 5 mm (separation gel length 75 mm). Finally it is overlayered with 0.3 ml of distilled water and photopolymerized or chemically polymerized (cf. Table 3.4). Disc gels should be used on the day of preparation.

Before taking the gels the sealing cap is carefully removed without letting air penetrate between the gel and the glass wall. The outer side of the tubes is wetted and they are inserted, with the stacking gel uppermost, into the silicone gaskets of the disc gel apparatus where they are left for 30 min to cool down to the temperature of the circulating water. The water layer on top of the sample gel is taken off by means of a syringe carrying a piece of tygon tube. Lower and upper electrode vessels are filled with pre-cooled buffer. Finally the sucrose enriched sample solution (preferably in a 100 mmol l⁻¹ phosphate buffer of 6.5 to 7.0 when using the Davis system [24]) is layered between the top of the gel and the electrode buffer by means of a precision syringe. The sample volume should not exceed the volume of the large pore gel. The lower end of the gels dipping into the electrode buffer must be free of any adhering air bubbles.

Following electrophoresis the glass tubes are taken out of the apparatus and the gels removed using a syringe filled with water and fitted with a long hypodermic needle. The needle is inserted a few millimeters into the bottom end of the tube between the separation gel and the glass wall while rotating the tube and squirting water out of the needle. The gel will finally slip out of the tube and can be placed in the staining solution.

3.5.2.3 Determination of the Molecular Size of Native Proteins

Two different methods can be used to estimate the size of enzymes from their migration distances in polyacrylamide gel electrophoresis. The older method uses a series of homogeneous separation gels of different concentrations while the more recent method takes a flat gel with a linearly increasing polyacrylamide concentration from top to bottom. It is one of the advantages of both methods that crude tissue or cell extracts can be used to estimate the molecular size of enzymes (and their various multiple molecular forms), provided they can be located in the gel by a specific staining method. Various mathematical procedures to compute the molecular size of native proteins have been published. Here, only one reliable method using homogeneous PAA gels [26] and one taking gradient gels [29] will be described in detail.

The method with homogeneous gels uses a number of gels of different PAA concentrations in the range of 3.5 to 35 % T. A two-step mathematical procedure is used to estimate the molecular mass of native proteins: first the exclusion limit (the gel concentration at which protein mobility becomes zero) is calculated for a number of marker proteins and the sample proteins, then, in a second step, the exclusion limits of the calibration proteins are plotted versus their Stokes' radii. The constants of the resulting calibration line are calculated by regression analysis. Into the equation obtained the exclusion limit of a sample protein is inserted which then allows calculation of its Stokes' radius [26]. Using disc-electrophoresis with separating gels of 3.5 to 35 % T, Felgenhauer [26] found that most of the 25 proteins which he investigated, slowed to zero mobility on prolonged electrophoresis, while smaller proteins continued migrating.

It has often been suggested that electrophoretic mobilities be calculated by use of the so-called Ferguson equation. However, it is not recommended to use this equation in PAA gel electrophoresis because (a) the Ferguson plot has been developed for starch gels, (b) it cannot be used to calculate the exclusion limit of proteins and (c) progressive deviations from its predicitions occur when PAA gels exceeding 15 % T are used [26].

Exclusion limit

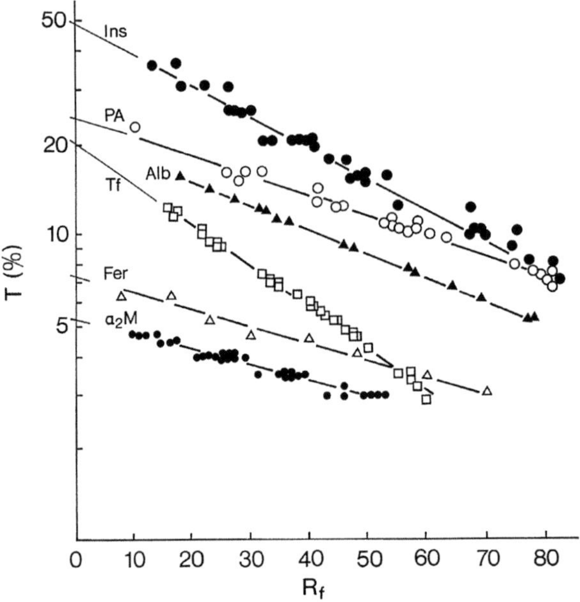

Fig. 3.8. Plot of log PAA concentration (%T) versus the relative migration (R_f) of six different proteins. The exclusion limit (T_{lim}) of each protein results from the intersection of the regression line with the %T-axis ($R_f = 0$). *Ins* = insulin; *PA* = pre-albumin; *Alb* = albumin; *Tf* = transerrin; *Fer* = ferritin; $\alpha_2 M$ = α_2-macroglobulin. Figure taken from [26] with permission of the author and publisher

The exclusion limit of proteins migrating in homogeneous PAA gels of different concentrations can be obtained by plotting relative protein mobilities referred to Bromophenol red (called R_f-values) against the logarithm of the concentrations of the PAA gels (log T) for which the R_f-values were measured (Fig. 3.8). When doing so a straight line with the underlying function is obtained:

R_f-value

$$\log T = -k\, R_f + \log T_{lim} \tag{1}$$

with: T, total concentration of acrylamide + BIS; R_f, relative protein mobility as referred to the tracking dye Bromophenol red (100 × protein mobility/Bromphenol red mobility); k, slope of the regression line; T_{lim}, PAA concentration at which zero mobility is given [26].

From the log T_{lim}-values the T_{lim}-values are calculated. To obtain a calibration line the reciprocal values of the T_{lim}-values (10^2/exclusion limit) of marker proteins are plotted vs their Stokes' radii (Fig. 3.9). T_{lim} and Stokes' radii (R_s) of globular proteins may therefore be correlated by the equation [26]:

Calibration line

$$R_s = a\, (T_{lim})^{-1} + b \tag{2}$$

with R_s (Stokes' radius); T_{lim}, exclusion limit; a and b, regression constants.

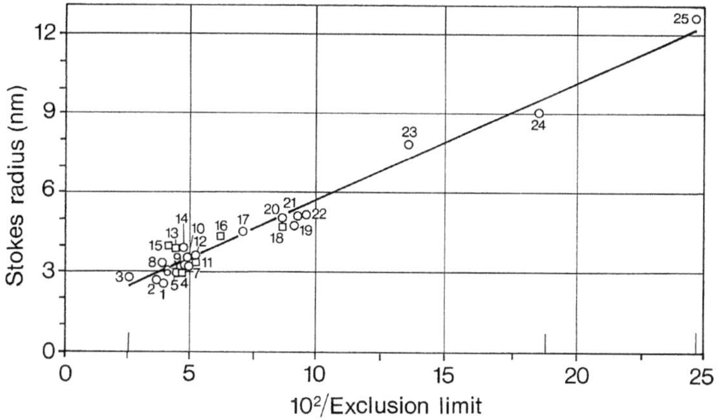

Fig. 3.9. Plot of the Stokes'radius (nm) of 25 standard proteins versus their exclusion limits. Exclusion limits (T_{lim}) are definded as those gel concentrations (%T) where protein mobility becomes zero (cf. Fig. 3.8). The isoelectric points of the proteins are below 7.0; glycoproteins with contents of carbohydrate residues above 18 % are marked by squares. Taken from [26] with permission of the author and publisher.

Index to proteins:

1. α-Amylase (from *Bacterium subtilis*, R_s = 2.67 nm, M = 48580, f/f_0 = 1.11)
2. Ovalbumin (from chicken, R_s = 2.74 nm, M = 43500, f/f_0 = 1.18);
3. β-Lactoglobulin (from cow, R_s = 2.90 nm, M = 35830, f/f_0 = 1.33);
4. Peroxidase (from horse radish, R_s = 3.08 nm, M = 41 280, f/f_0 = 1.34);
5. Hydroxynitril lyase (from *Bacterium subtilis*, R_s = 3.10 nm, M = 73 000, f/f_0 = 1.12);
6. Malate dehydrogenase (from heart mitochondria, R_s = 3.32 nm, M = 62000, f/f_0 = 1.27);
7. D-Amino-acid oxidase (from hog kidney, R_s = 3.34 nm, M = 114 250, f/f_0 = 1.04);
8. Hemoglobin (from ox, R_s = 3.39 nm, M = 67350, f/f_0 = 1.26);
9. Albumin (from human, R_s = 3.45 nm, M = 69000, f/f_0 = 1.27);
10. Albumin (from ox, R_s = 3.55 nm, M = 66200, f/f_0 = 1.34);
11. Follicle stimulating hormone (from swine, R_s = 3.56 nm, M = 67360, f/f_0 = 1.32):
12. Transferrin (from human, R_s = 3.72 nm, M = 81000, f/f_0 = 1.30);
13. Fetuin (from calf, R_s = 4.00 nm, M = 48470, f/f_0 = 1.66);
14. Conalbumin (from chicken, R_s = 4.04 nm, M = 86180, f/f_0 = 1.38);
15. α_1-Glycoprotein (from human, R_s = 4.06 nm, M = 44100, f/f_0 = 1.74);
16. Haptoglobin (from human, R_s = 4.51 nm, M = 80000, f/f_0 = 1.59);
17. Ceruloplasmin (from human, R_s = 4.72 nm, M = 152200, f/f_0 = 1.34);
18. Choriogonadotropin (from human, R_s = 48.6);
19. Leucine aminopeptidase (from lens, R_s = 4.90 nm, M = 326 000, f/f_0 = 1.08);
20. Phosphorylase (from muscle, R_s = 4.93 nm, M = 177000, f/f_0 = 1.46)
21. Catalase (from bovine liver, R_s = 5.22 nm, M = 241170, f/f_0 = 1.27);
22. Fumarate hydratase (from pig heart, R_s = 5.27 nm, M = 206000, f/f_0 = 1.27);
23. Apoferritin (from horse, R_s = 7.9 nm, M = 473450, f/f_0 = 1.53);
24. α_2-Macroglobulin (from human serum, R_s = 9.1 nm, M = 797 750, f/f_0 = 1.49);
25. β-Lipoprotein (from human, R_s = 12.6 nm, M = 2239000, f/f_0 = 1.46)

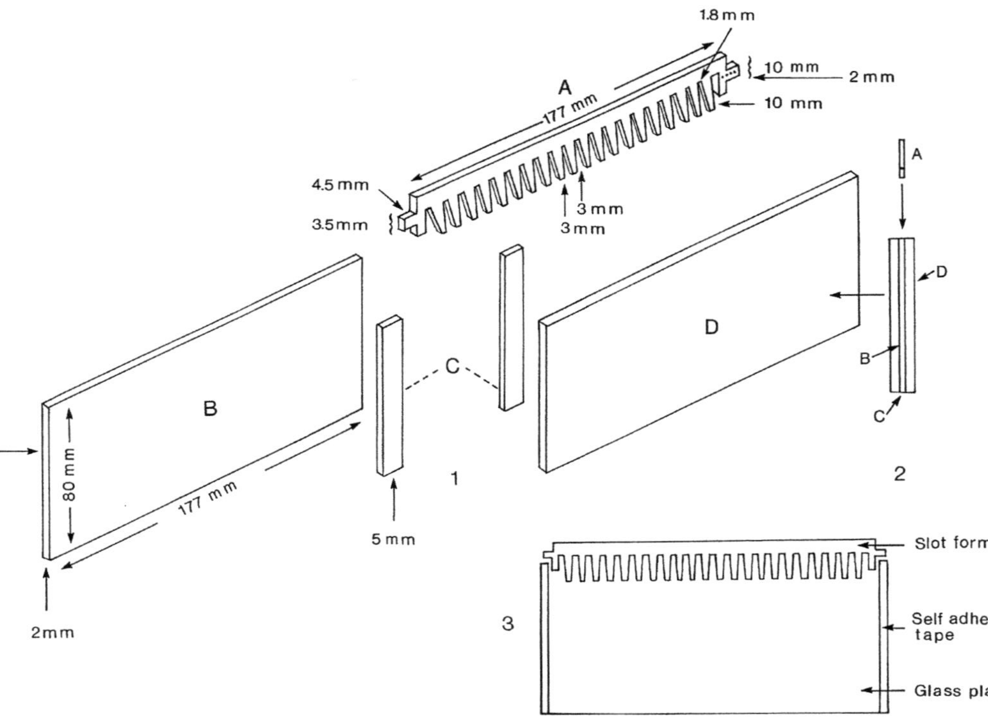

Fig. 3.10. Assembly of glass cassettes to cast PAA (gradient) gel slabs. A: slot former, B, D: front and rear glass plate, C: left and right distance bar. 1: explosive view of a glass cassette, 2: side-view, 3: front-view. Procedure according to Pharmacia, Uppsala, Sweden

Within the limits given, the experimental data Felgenhauer obtained fitted that function well. On the other hand, the straight line in Fig. 3.9 does not originate at zero so that the underlying function might be curvilinear over the total measuring range. For practical purposes, however, it is sufficient to use Felgenhauer's data fit when estimating the Stokes' radius of a globular protein in the range of 2.9 nm (50 000 mol mass) to 12 nm (2.000 000 mol mass). Felgenhauer's method has an interesting subsidary effect: proteins with larger frictional ratios ($f/f_0 = 1.48$) are situated above the regression line in a $(T_{lim})^{-1}$ vs R_s-plot, while proteins with smaller frictional ratios ($f/f_0 = 1.22$) are situated beneath (cf. Fig. 3.9). Electrophoretic alignment can be used to explain this phenomenon since, in the absence of electrophoretic alignment, the exclusion limits should scatter irrespective of molecular eccentricity. Three elongated proteins (tropomyosin: $f/f = 3.10$; L-meromyosin: $f/f_0 = 2.90$ and fibrinogen: $f/f_0 = 2.33$) do indeed deviate strongly from the regression line shown in Fig. 3.9. Overall globular non-denatured proteins exhibit similar hydrodynamic properties in partition chromatography and gel electrophoresis, and their hydrodynamic radius can be reliably determined with either method. Since elongated proteins are aligned parallel to the electric field in PAA electrophoresis [7] these proteins give spuriously small-sized values [26]. As a consequence, two proteins with identical Stokes' radii and isoelectric points may be separable in gel electrophoresis if they differ in their axial ratio.

Frictional coefficient

Electrophoretic alignment

From a range of globular proteins cited by Felgenhauer it can be taken that the average frictional ratio (f/f_0) of globular proteins in the range of 45 000 g/mol to 100 000 g/mol is $f/f_0 = 1.23$, while $f/f_0 = 1.28$ for proteins with molecular masses in the range of 100 000 to 500 000, and $f/f_0 = 1.43$ for proteins with molecular masses between 500 000 and 1.000 000. Knowing the hydrodynamic radius (Stokes' radius) of a globular protein and taking the average frictional coefficients given above, the molecular mass of a protein can easily be determined by applying eq. (3),

$$M_r = (f_0/f)^3 \times 3463 \times R_s^3, \tag{3}$$

with: M_r = molecular mass [g/mol], f/f_0 = frictional coefficient, R_s = Stokes' radius [nm].

This can be exemplified on mammalian liver alcohol dehydrogenase (EC 1.1.1.1), which has a molecular mass of $M_r = 80 000$ g/mol and a Stokes' radius of $R_s = 3.5$ nm; the average frictional coefficient of globular proteins in that range is $f/f_0 = 1.23$. By inserting these values into eq. (3) one obtains:

$$Mr \text{ [g/mol]} = (1.23)^3 \cdot 3463 \cdot 3.5^3 = 79 800 \tag{4}$$

3.5.3 Gradient Gel Electrophoresis

The high resolution of homogeneous PAA gel electrophoresis [6, 7, 24] can be improved by using gels of increasing concentration, so-called pore gradient gels [30 – 35]. The average pore radius of these gels decreases with increasing gel concentration, i.e. in the direction of the migrating proteins. This results in a sharpening of migrating protein bands and a separation of proteins over a larger molecular mass range than in homogeneous gels. Another advantage of using polyacrylamide gradient gels is that the range of PAA concentration (% T) can be adapted to the separation task. Large or slowly migrating proteins are better separated in a shallow gradient (e. g., from 4 to 10 % or 4 to 20 % T) while smaller or faster migrating proteins are preferably separated in a steeper gradient (e. g., from 4 to 30 % T). Besides, the pH-value of the electrophoretic buffer system can be choosen freely so that sufficient migration velocity of the sample proteins may be achieved. On the other hand it must be taken into consideration that marker proteins migrating at alkaline pH-values may not migrate at acid pH-values, which means that different marker protein sets have to be used at various pH ranges.

Several different devices for the preparation of gradient gels are now commercially available so that the preparation of such gels is no longer restricted to a few highly specialized laboratories. There are also ready-to-use pore gradient gels on the market so that pore gradient electrophoresis will probably become a routine method for characterizing proteins by their molecular size [29,31].

The migration of any *native* protein in a pore gradient gel depends on its charge and size. Therefore, it is likely that a sample protein migrates faster or slower than a marker protein of equal size [29]. Therefore, as with homogeneous gels, the exclusion limits of marker and sample proteins must be calculated before the molecular mass of a sample protein can be obtained [29]. The maximum migration distance is equivalent to the exclusion limit. The maximum migration distance (D_{max}) is the distance which a protein migrates in a pore gradient gel until its migration velocity becomes zero [29]. Application of a mathematical approximation procedure allows calculation of the maximum migration distance simply from a limited number of time dependent migration distances [29]. Once the maximum migration distances of a set of marker proteins has been achieved, these values can be correlated to their molecular mass or Stokes' radius. The resulting calibration lines can then be taken, together with the D_{max}-value of the sample protein, to calculate its relative mass (M_r) or Stokes' radius (R_s). The mass equivalent sphere radius (R_m) of a sample protein can be calculated from its molecular mass, and the quotient of its Stokes' radius and its spherical radius (R_s / R_m) equals its frictional coefficient (f/f_o).

Maximum migration distance

Time dependent migration of a protein in a linear gradient of polyacrylamide can also be used to estimate its free electrophoretic mobility. This quantity then provides the basis for calculating its charge at the pH of electrophoresis [29].

3.5.3.1 Preparation of Gradient Gels

Gradient gels are prepared from high and low concentration gel solutions usually by using the technique which was first described by Martin and Ames [36] for the preparation of linear sucrose gradients. Gradient gels are cast in glass cassettes of various dimensions. Gels may be made in a glass cassette without any further support, or they may be polymerized onto a specially prepared polyester film [37, 38] or a silanized glass plate [39]. The latter two methods are preferentially applied when the gels are taken off the casting glass cassettes and used in a horizontal arrangement. Support bound gels are preferably used in two-dimensional electrophoresis because their dimensions do not alter upon electrophoresis. Flat gels usually have dimensions of 82 × 82 (140) mm or 125 × 250 mm and a thickness of 1.0, 0.8 or 0.5 mm [31, 37, 38].

To prepare 6 or 12 "non supported" gradient gels simultaneously, each having the dimensions of 172 × 82 × 1.0 mm the following procedure may be used [31].

The glass cassettes are fitted with slot formers (Fig. 3.10) and inserted in a gel casting apparatus (Fig. 3.11). Linear PAA gradients with a constant ratio of acrylamide: BIS (24:1), ranging from 4 to 30 % (acrylamide plus BIS) per 100 ml of buffer, are prepared from a T_{min} solution containing 4.205 g acrylamide and 0.175 g BIS, and a T_{max} solution containing 31.54 g acrylamide and 1.314 g BIS per 100 ml of buffer. The buffer may be a TBE-buffer, i.e., a mixture of 45 mmol l⁻¹ Tris, 40 mmol l⁻¹ boric acid, and 1.25 mmol l⁻¹ EDTA-Na₂ of pH 8.4 or it may consist of the separation gel buffer of Ornstein and Davis [24] (Table 3.4). Then 22.5 µl Temed are given to 57 ml of the T_{min} and T_{max} solution, respectively. Both solutions are filled into the gradient mixer. A separate reservoir is filled with 35 ml of TBE-buffer and 50 mg ammonium persulphate are added. Prior to use, all solutions are degassed. The gradient is prepared by using a two-chamber gradient mixer, one reservoir for the catalyst peroxodisulphate, a two-channel pump, a 1-ml mixer and the gel casting apparatus (Fig. 3.11). The inner chamber of the gradient mixer (4) is filled with 57 ml of the T_{min} solution and 22.5 µl Temed, while the stop cock to the outer chamber is closed. Then 57 ml of the T_{max} solution and 22.5 µl Temed are pipetted into the outer chamber of the gradient mixer, and the reservoir (7) is filled with 35 ml of ammonium persulphate solution. The stop cock is opened, the two channel pump (6) is switched on and all lines are filled. Immediately after pumping the PAA gradient into the gel casting apparatus it is underlayered by a solution containing 50 wt/vol.% sucrose lifting the whole gradient into the cassettes.

To prepare thin gradient gels fixed to a clear and flexible polyester film (as manufactured by Gel Bond, Marine Colloids, Rockland, MN, USA) the following procedure may be applied.

A sheet of the specially prepared polyester film (125 × 250 mm) is placed with its hydrophilic side up on a 4 mm thick glass plate on which a few drops of water had been given. (The hydrophobic side of the polyester film is the side where a drop of water does not spread but stays as a drop.) A rubber roller may be used to roll the film firmly onto the glass plate. The casting cassette is formed of this plate, a U-shaped gasket and a glass plate with a firmly attached trough template. The template may be produced from one or more layers of a self adhesive tape. The U-shaped rubber gasket is about 5 mm broad and 0.5 mm thick. The two glass plates are clamped together (Fig. 3.12), the entire cassette is mounted upright and filled with gel solution. If the gradient mixer shown in Fig. 3.12 is used, then the light solution is filled in the distal tube of the gradient former while the dense solution is filled in the proximal tube. It is also possible to use a glass plate with an inlet on its base (Desaga, Heidelberg, Germany), so that the cassette can be filled from bottom to top. When the cassettes are assembled the slot formers must not touch the opposite glass wall but leave a space of 0.1 mm in between.

To prepare a flat gel of the dimensions 120 × 250 × 0.5 mm to be polymerized to a derivatized polyester foil (e.g. Gel Bond) the following solutions may be used together with the gel-forming devices shown in Fig. 3.12: (1) gel buffer: 90 mmol l⁻¹ (10.9 g) Tris 80 mmol l⁻¹ (4.95 g) boric acid, 2.5 mmol l⁻¹ (0.93 g) EDTA-Na₂-H₂O in 1 l double distilled water, pH 8.4; (2) electrode buffer: 1 in 2 diluted gel buffer; (3) stock acrylamide solution (30% T): add 28.8 g acrylamide plus 1.2 g BIS to 50 ml gel buffer, dissolve and make to 100 ml with distilled water; (4) dense acrylamide solution: shortly before use add to 6.5 ml stock solution 20 µl Temed (1 in 10 with H₂O diluted solution) and 5 µl ammonium persulphate solution (40 wt/vol.% in distilled water); (5) light acrylamide

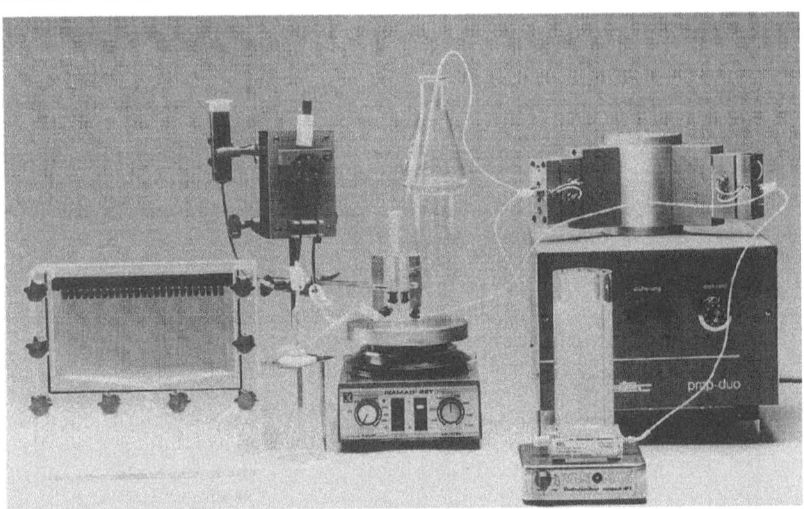

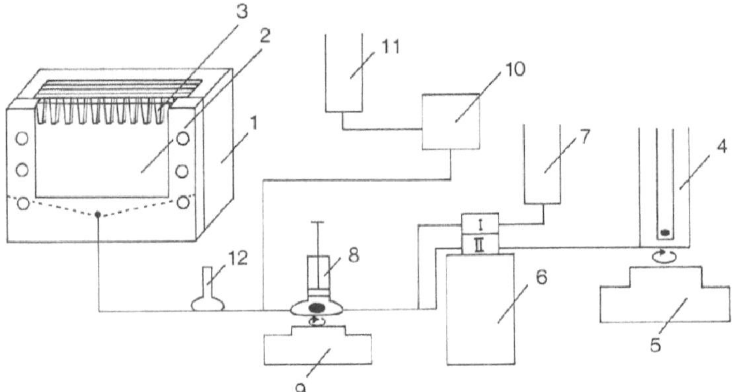

Fig. 3.11. Scheme of the equipment used to simultaneously prepare several gradient gel slabs. 1: gel casting apparatus made of Perspex, 2: front plate, 3: slot former inserted into glass cassette, 4: two-chamber gradient mixer (outer diameter = 55 mm, height = 150 mm) 9: magnetic stirrer, 7: reservoir for persulphate solution, 6: two channel pump I: 1.4 ml/min, II: 5.5 ml/min, 8: modified disposable 5 ml syringe, used as small chamber mixer, 10: one channel pump, 11: reservoir for 50 wt/vol.% sucrose solution to lift the PAA gradient into the glass cassettes, 12: air trap

solution (3% T): dilute 1 vol of dense acrylamide solution with 4.5 vol distilled water and 4.5 vol of gel buffer and, shortly before use, add to 9.5 ml light acrylamide solution 40 µl Temed (1 in 10 diluted with distilled water) and 10 µl ammonium persulphate solution (40 wt/vol.% in distilled water). Use 6.5 ml of dense and 6.5 ml of light acrylamide solution to prepare the gradient. Then overlayer the gradient with 3 ml of light acrylamide solution; the slots must be situated in the midst of the 3% T range.

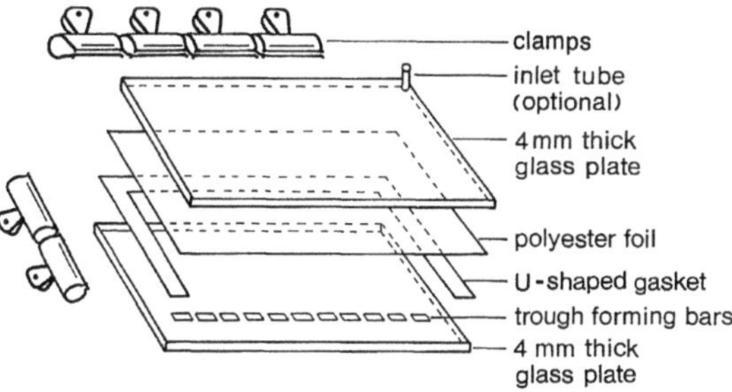

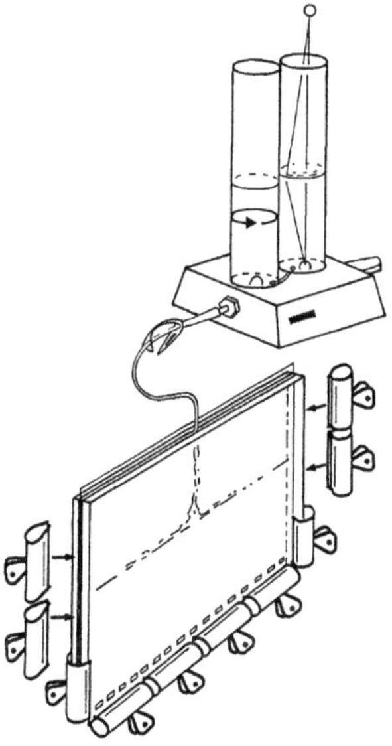

Fig. 3.12. Assembling a glass cassette to cast a PAA (gradient) gel for polymerization on a polyester foil (e.g., Gel Bond). The inlet tube is connected to a two-chamber gradient mixer. Taken from [41] with permission of the authors and the publisher

Before electrophoresis, the gels are taken out of the cassette. A few drops of kerosene are put on the cooling plate of the opened electrophoretic apparatus and the gel, firmly adhering to the polyester foil, is placed on it, carefully avoiding the inclusion of air bubbles. Both ends of the gel are connected with the buffer vessels by aid of paper wicks or a sponge-like material. A 15 – 30 min pre-electrophoresis is performed at 1000 V (50 V/cm). Then the slots are filled with protein solution and the power is turned on again at a voltage of 1000 V for approximately 2 h. Afterwards the gels, fixed on the polyester foil, are stained for enzymes by the dipping or overlay-method.

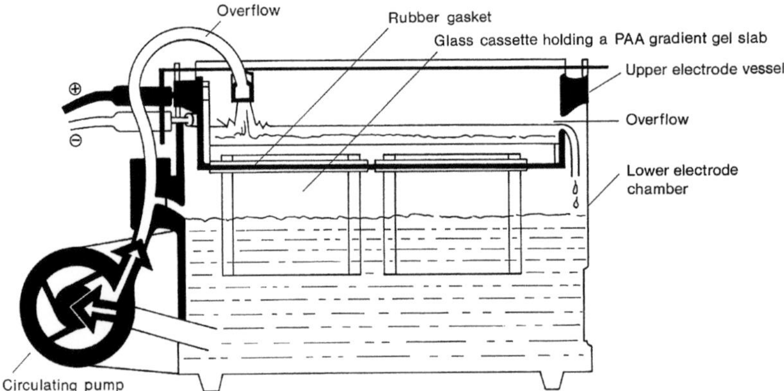

Fig. 3.13. Electrophoretic apparatus to run gradient gel slabs in a vertical position. Taken from an instruction leaflet of Pharmacia, Uppsala, Sweden

3.5.3.2 Conditions of Electrophoresis

Glass-cassette cast gradient gels are inserted vertically into an electrophoretic apparatus consisting of an upper and lower electrode vessel (Fig. 3.13). The upper buffer tank has four rubber gaskets which can hold up to four cassettes. For enzyme studies the buffer system suggested by Ornstein and Davis [24] is often used (Table 3.4). Samples are enriched with 10 % sucrose and filled into the sample slots with a 5 µl Hamilton syringe with the cathodal buffer already in the upper buffer reservoir. Then the current is switched on for 30 min and the proteins are allowed to penetrate into the gel slabs at a voltage of 300 V (40 V cm^{-1}) without circulating the buffer. Finally, the buffer is circulated from buffer tank to buffer tank at the same voltage. (The voltage per gel plate remains constant no matter whatever number of plates are run because of the construction of the apparatus.)

3.5.3.3 Determination of Molecular Size Properties of Native Proteins

The mathematical procedures presented in the following chapters are bound to several preconditions. These are:

(a) The concentration of polyacrylamide (%T) must increase linearly (at a constant ratio of acrylamide to N',N'-methylenebisacrylamide (BIS)). The gradient range can be chosen freely. Gradients may, for example, range from approximately 4 to 20 % T or from 4 to 30 % T.

(b) The charge of the marker (and sample) proteins must fit to the pH-value of the electrophoretic buffer system and the size of the marker and sample proteins must fit to the range of the polyacrylamide gradient.

(c) Times of migration must be adapted to the migration velocity of the marker and sample proteins.

(d) Sample and marker proteins must have been electrophoresed for at least five different durations of time.

(e) If the charge of a protein shall be calculated, the buffer system of the electrode compartment and the gel must be the same.

(f) Marker and sample proteins must have migrated on the very same gel plate.

(g) The voltage gradient must not extend 41 V cm⁻¹.

(h) To visualize marker and sample proteins in a gradient gel, it may be cut into several pieces and submitted to different visualization procedures (e.g., protein staining or enzyme visualization). Staining, however, should not have different influences on the length of the separated gel parts; otherwise each part must be equilibrated to the same length before measuring protein (enzyme) migration distances.

(i) Migration distances of proteins must have been measured with an accuracy of ± 0.5 mm (e.g., directly on the gel plate).

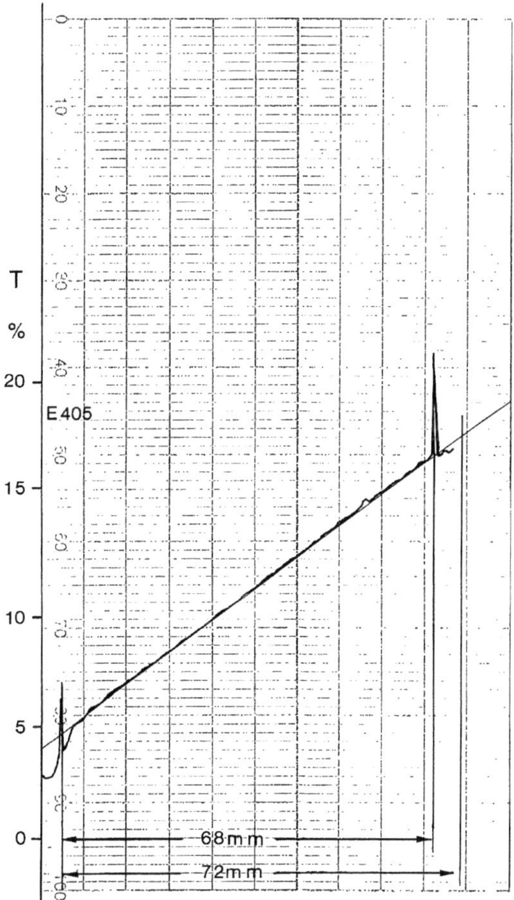

Fig. 3.14. Course of a linear PAA gradient produced as demonstrated in Fig. 3.11. E_{405}: absorbance at 405 nm of p-nitrophenol to mark the course of the gradient: the polyacrylamide concentration (% T = g acrylamide + g BIS/100 ml) was estimated by applying Eq. (5) given in the text

The course and % T range of self-made polyacrylamide gradient gels can be controlled by densitometry. A coloured dye may be added to the more dense acrylamide solution upon gradient formation. After polymerization the increase in colour intensity from bottom to top of the gel can be used to measure the course of gradient and its precise concentration in polyacrylamide [29]. The dye *p*-nitrophenol may be used for that purpose because it can be migrated (at alkaline conditions) out of the gel before protein application. For 1 mm thick gels 15 mg of *p*-nitrophenol may be added to 100 ml of the stock (20 or 30 % T) acrylamide solution, with all other steps of gradient preparation remaining the same as described above. Afterwards it can be seen that the gradient gels are intensely yellow coloured at their lower end with the colour intensity decreasing towards their upper end. The intensity of the colour is quantified by densitometry at 405 nm. Whilst the course of the gradient can be seen directly on the densitogram (Fig. 3.14) the % T-range of the gradient must be calculated with the aid of the following formula [29]:

$$T = \frac{T_s \, (E_{405} - E_p)}{c \, M_r^{-1} \, d \, E} \; [\%], \qquad (5)$$

where T_s [%] is the concentration of the stock acrylamide solution (e. g., $T_s = 30$); E_{405} is the absorbance of *p*-nitrophenol at 405 nm; E_p is absorbance of the glass walls of the cassette (or polyester foil to which the gel is backed) at 405 nm, c [g/l] is the *p*-nitrophenol concentration in stock acrylamide solution (c = 0.150); M_r [g/l] is the mol mass of *p*-nitrophenol ($M_r = 139.1$); d [mm] is thickness of gel (d = 0.5), and E [l/mol mm] is the molar extinction coefficient of *p*-nitrophenol at 405 nm (E = 1728).

3.5.3.3.1 Estimation of the Maximum Migration Distance of a Protein in a Linear Pore Gradient Gel

The maximum migration distance (D_{max} [mm]) for any native protein is obtainable if at least five time-dependent migration distances (D [mm]) have been registered for a protein [29]. The migration distances are transformed to the corresponding ln (ln D)-values and the times of electrophoresis (t [h]) are transformed to the $t^{-1/2}$-values and plotted vs each other.

A plot of ln (ln D) vs $t^{-1/2}$ can also be used to distinguish size isomers from charge isomers. Equally sized forms of an enzyme system are recognized by the fact that the straight lines obtained for each isozyme intersect at a common point on the ln (ln D)-axis as is for example the case with mammalian carbonic anhydrase [29] and mammalian lactate dehydrogenase (Fig. 3.15). On the other hand, migration of charge isomers should result in lines of equal slopes. Proteins differing in charge and size, however, give straight lines with different slopes and intercepts.

The transformed migration values (ln (ln D)) and times of electrophoresis ($t^{-1/2}$) are interrelated by the following equation [29]:

$$\ln(\ln D) = - a \, t^{-1/2} + b \qquad (6)$$

where "a" and "b" are the slope and the intercept of the corresponding straight line, respectively. The equation predicts that at very high values of "t" $t^{-1/2}$ reaches zero. This means that the maximum migration of a protein (D_{max} [mm]) can be taken from the intercept of the straight line with the ordinate in a plot of ln(ln D) versus $t^{-1/2}$ pro-

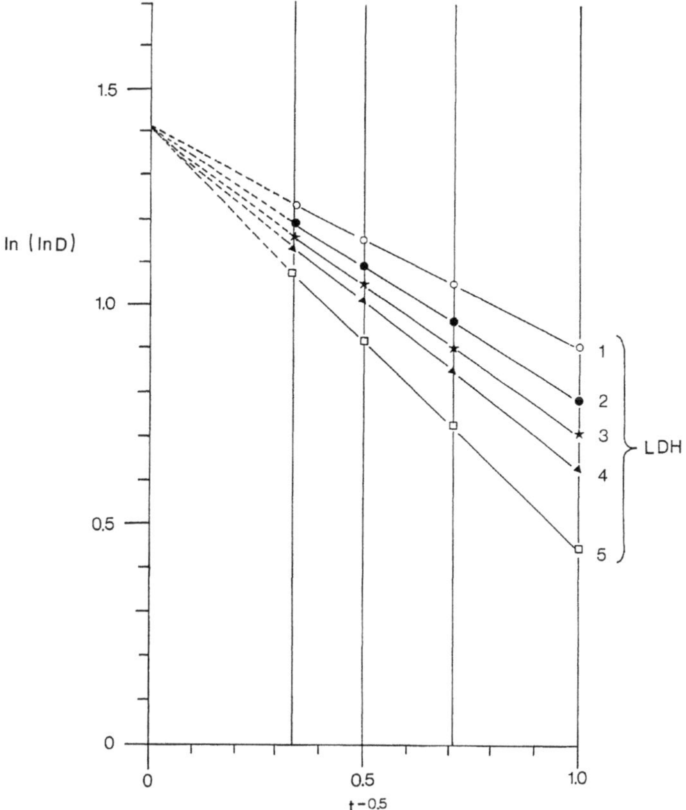

Fig. 3.15. Plot of ln (lnD) vs $t^{-0.5}$ of the five L-lactate dehydrogenase isoenzymes in human blood (Sigma, Isotrol). Electrophoretic conditions: linear PAA gradient gel from 4 to 27% T (acrylamide: BIS = 24:1); buffer system: 45 mmol l⁻¹ Tris, 40 mmol l⁻¹ boric acid, 1.25 mmol l⁻¹ EDTA-Na$_2$, pH 8.4; field strength: 41 V/cm (total gel length 73 mm); 5 °C; composition of LDH-visualization solution: 60 ml 0.05 mol l⁻¹ Tris HCl, pH 8.0, 450 mg Li-lactate, 45 mg NAD, 900 µl (0.5 mol l⁻¹) MgCl$_2$ · 6H$_2$O, 900 µl (10 mg/ml) PMS, 900 µl (50 mg/10 ml) MTT. The common point of intersection of the five straight lines on the ln (lnD)-axis indicates the size isomeric nature of the five isozymes

vided the protein has migrated at least 2 mm into a gradient gel and five different migration distances were registered. Letting t approximate infinity means that eq. (6) becomes

$$\ln(\ln D) = \ln(\ln D_{max}) = b \tag{7}$$

and

$$D = D_{max} = \exp(e^b). \tag{8}$$

3.5.3.3.2 Set Up of Calibration Lines to Calculate the Stokes' Radius and Molecular Mass of Sample Proteins

The maximum migration distance of globular proteins is related to the maximum gel pore radius at the respective gel concentration [29]. Therefore maximum migration distances (D_{max}) of proteins can be correlated to their known Stokes' radii (R_s). A linear relationship is obtained if the logarithm of the maximum migration distance ($\ln D_{max}$) of marker proteins is plotted versus the logarithm of their Stokes' radius ($\ln R_s$) [29]:

Calibration line to estimate R_S

$$\ln D_{max} = - m \ln R_s + b, \tag{9}$$

where $\ln D_{max}$ equals e^b of eq. (7), and m and b represent the slope and intercept of the straight line, respectively.

It has been shown [29] that a similar equation correlates the logarithm of the maximum migration distance ($\ln D_{max}$) to the logarithm of the molecular mass ($\ln M_r$):

Calibration line to estimate M_r

$$\ln D_{max} = - z \ln M_r + c \tag{10}$$

where $\ln D_{max}$ equals e^b of eq. (7), and z and c represent the slope and intercept of the straight line, respectively.

Knowing the maximum migration distance of any native globular protein or enzyme, the calibration line can be used to calculate the molecular mass of these proteins or enzymes by inserting the calculated $\ln D_{max}$-value and the values of the slope (z) and the intercept (c) of the calibration line into the equation $\ln D_{max} = - z \ln M_r + c$.

Using polyacrylamide (PAA) gradients of 4–20% T and a 45 mmol l⁻¹ Tris, 40 mmol l⁻¹ boric acid, 1.25 mmol l⁻¹ EDTA-Na₂ buffer of pH 8.4, the following marker proteins may be used (M_r g/mol; R_s nm):

Fig. 3.16. Photo of time-dependent migration patterns of aspartate aminotransferase isoenzymes (*left lanes 1 to 20*) and six different marker proteins (*right lanes 21 to 25*). Electrophoretic conditions: linear PAA gradient from 5 to 30% T (acrylamide : BIS = 24:1); buffer system: 45 mmol l⁻¹ Tris, 40 mmol l⁻¹ boric acid, 1.25 mmol l⁻¹ EDTA-Na₂, pH 8.4; field strength: 41 V/cm (73 mm total gel length); 4 °C. Aspartat aminotransferases were extracted from two individual seeds (I, II) of *Pinus strobus* which had been separated into the diploid embryo (2n) and the haploid endosperm (1n). The tissue of one embryo was homogenized in 125 µl extraction buffer, the endosperm of one seed was extracted into 250 µl extraction buffer (0.1 mol l⁻¹ sodium phosphate, pH 7.5, containing 0.5% Triton X-100), 10 µl of centrifuged (6000g × 15 min) extract were applied per gel trough. The marker proteins (high molecular weight marker proteins of one vial in 100 µl; Pharmacia, Uppsala, Sweden and 100 µg/100 µl ovalbumine, Boehringer Mannheim, Tutzing, Germany) consisted of thyroglobulin (*THY*), ferritine (*FER*), catalase (CAT), lactate dehydrogenase (*LDH*), bovine serum albumine (*BSA*) and ovalbumine (*OVA*). Marker proteins were stained as follows: gels were incubated for 10 min in 0.5 wt/vol.% Serva Violett 17 dissolved in 15 wt/vol.% phosphoric acid, destained in 3 wt/vol.% phosphoric acid for 30 min, fixed in 15 wt/vol.% glycerol for 10 min and stored in 3 wt/vol.% phosphoric acid. Aspartate aminotransferase isoenzymes were visualized by immersing the gel into a solution of 60 ml 0.1 mol l⁻¹ Tris-HCl, pH 8.0, 60 mg L-aspartate, 60 mg 2-oxoglutarate, 3 mg pyridoxal-5'-phosphate and 60 mg Fast Blue BB salt. Enzymes and marker proteins were electrophoresed for 2, 4, 6, 8 and 16 h. The 8 isozymes of AAT can be grouped to 3 different enzyme loci (A, B, C). Locus AAT-A is monomorphic (2 bands) while the loci B and C are polymorphic (1 or 3 bands). The existence of 3 AAT bands in (diploid) embryo tissue indicates that the enzyme is a dimer. Photo taken from [30] by permission of the publisher

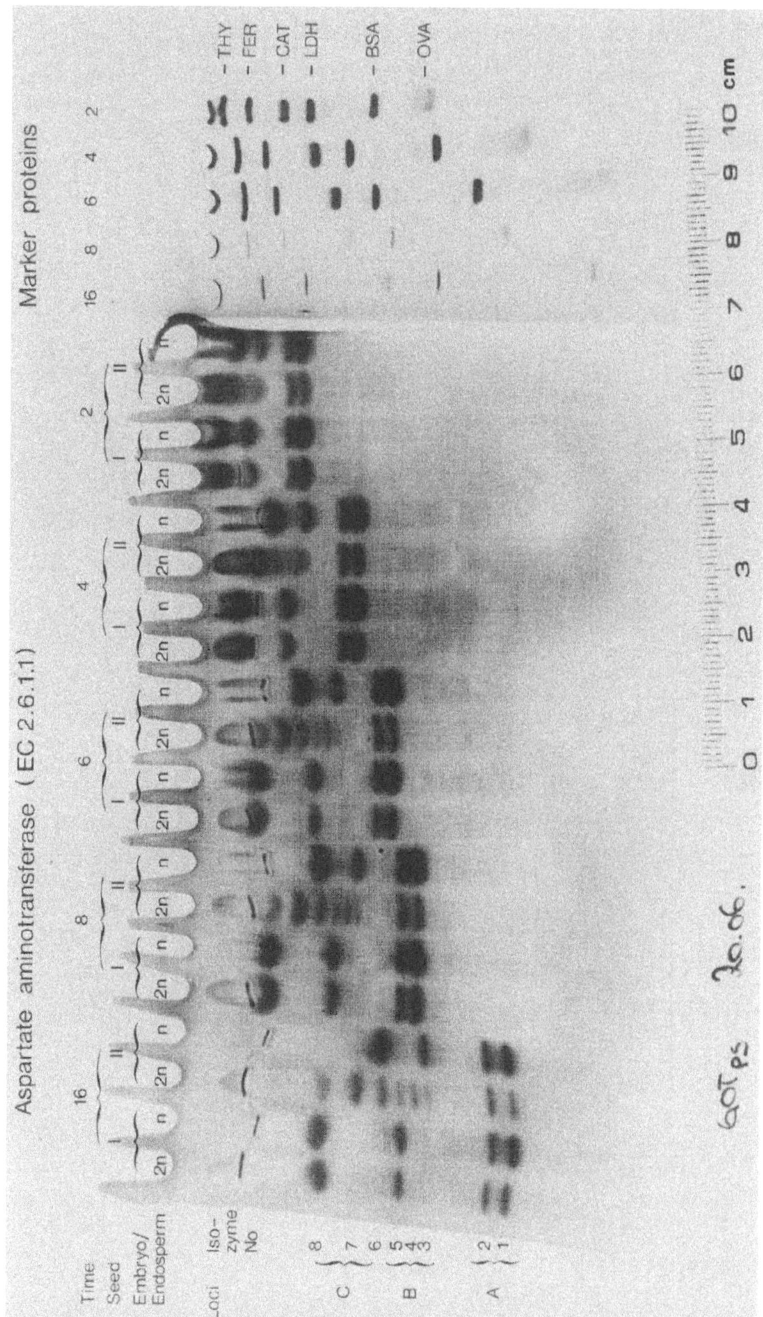
Marker proteins
Aspartate aminotransferase (EC 2.6.1.1)
THY
FER
CAT
LDH
BSA
OVA
2 4 6 8 16
0 1 2 3 4 5 6 7 8 9 10 cm
Time
Seed
Embryo/Endosperm
Loci
Iso-zyme No
Time
16 8 6 4 2
Marker proteins
C
B
A
8 7 6 5 4 3 2 1
GOT es 20.06.

OVA
BSA
LDH
CAT
FER
THY
ln(ln D)
(D[mm])
1.5
1.0
0.5
0
0.1 0.2 0.3 0.4 0.5 0.6
1/√t (t[h])

a

ln(ln D)
(D[mm])
1.5
1.0
0.5
0
AAT Ia, seed B
AAT I, seed A
AAT I, seed B
AAT 4, seed A
AAT 4, seed B
AAT 5, seed A
AAT 7, seed B
0.1 0.2 0.3 0.4 0.5 0.6 0.7
1/√t (t[h])

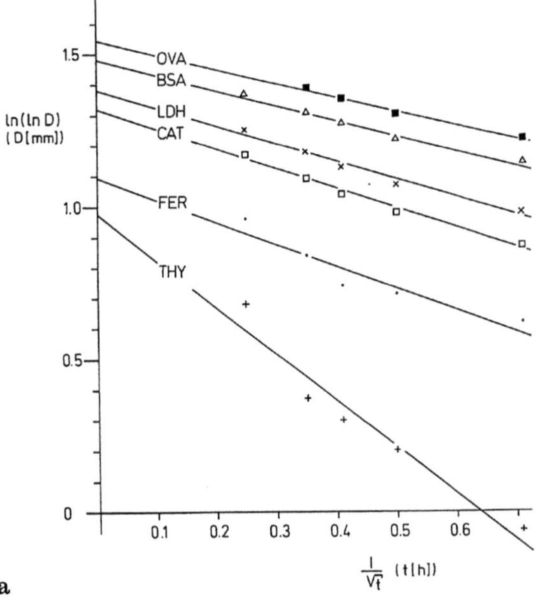

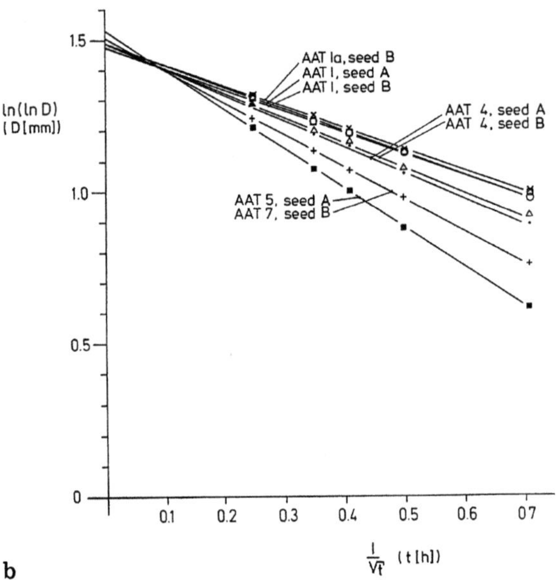

b

Fig. 3.17 a, b

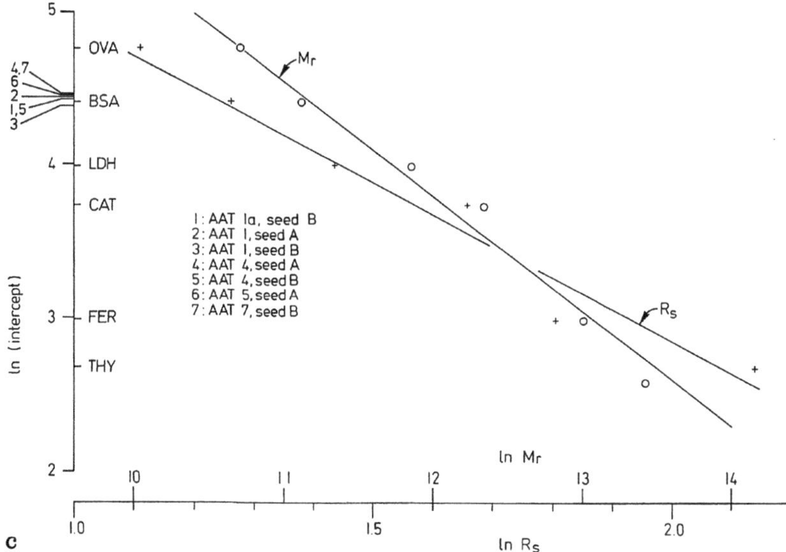

Fig. 3.17 a–c, a: Plot of ln (lnD) vs $t^{-0.5}$ for the marker proteins thyroglobuline (*THY*), ferritine (*FER*), catalase (*CAT*), lactate dehydrogenase (*LDH*), bovine serum albumine (*BSA*) and ovalbumine (*OVA*). **b:** Plot of ln (lnD) vs $t^{-0.5}$ for seven aspartate aminotransferase isoenzymes appearing in seeds of *Pinus strobus*. **c:** Plot of ln D_{max} (ln(intercept)) vs ln M_r (ln R_s) to calculate the molecular mass of seven aspartate aminotransferases. For further details see legends to Fig. 3.16. Graphs taken from [30] by permission of the publisher

(a) carbonic anhydrase (M_r: 30 000, R_s: 2.43),
(b) β-galactosidase (M_r: 116 000, R_s: 4.23),
(c) lactate dehydrogenase (M_r: 140 000, R_s: 4.20),
(d) catalase (M_r: 232 000, R_s: 5.25),
(e) ferritin (M_r: 440 000, R_s: 6.10) and
(f) thyroglobulin (M_r: 669 000, R_s: 8.50).

Beta-galactosidase and carbonic anhydratase (Sigma, St. Louis MO, USA) are run in separate lanes while the remaining marker proteins (Pharmacia, Uppsala, Sweden) can be loaded into the same gel trough. Separation times may be 0.5, 1, 2, 4 and 6 h (or longer).

With a PAA gradient of 4 – 30 % T and a 90 mmol l⁻¹ Tris, 80 mmol l⁻¹ boric acid, 1.25 mmol l⁻¹ EDTA-Na$_2$ buffer of pH 8.4, suitable marker proteins are are (M_r g/mol; R_s nm):

(a) ovalbumin (M_r: 45 000, R_s: 3.05),
(b) bovine serum albumin (M_r: 67 000, R_s: 3.55),
(c) lactate dehydrogenase (M_r: 140 000, R_s: 4.20),
(d) catalase (M_r: 232 000, R_s: 5.25),
(e) ferritin (M_r: 440 000, R_s: 6.10) and
(f) thyroglobulin (M_r: 669 000, R_s: 8.50).

Table 3.5. Time dependent migration distances of marker proteins and aspartate aminotransferase isozymes from *Pinus strobus* in a linear 4 – 27 % T PAA-gradient at pH 8.4

Protein/enzymes	Electrophoretic duration [h] and migration distances [mm] of marker proteins and aspartate aminotransferase isozymes				
	2	4	6	8	16
OVA	29.8	39.2	47.7	55.4	void
BSA	23.1	29.8	35.8	40.1	52.0
LDH	14.5	18.8	22.2	25.6	33.2
CAT	11.1	14.5	17.1	19.6	25.6
FER	6.4	7.7	8.2	10.2	13.6
THY	2.6	3.4	3.8	4.2	7.2
AAT 2, II	15.3	23.0	28.1	32.4	42.2
AAT 1, I	14.7	22.2	27.3	30.7	42.2
AAT 1, II	14.9	22.2	27.3	30.7	40.9
AAT 4, I	11.5	18.1	23.9	26.4	37.5
AAT 4, II	12.4	9.0	24.3	27.7	37.9
AAT 5, I	8.5	14.5	18.8	22.2	32.2
AAT 7, II	6.4	11.3	15.3	18.8	28.1

Marker proteins: ovalbumin (OVA), bovine serum albumin (BSA), lactate dehydrogenase (LDH), catalase (CAT), ferritin (FER), and thyroglobulin (THY).
Sample proteins: aspartate aminotransferase (AAT) isozymes 1, 2, 4, 5 and 7 from the embryo (2n) of two different seeds (I or II) of *Pinus strobus* (see also Fig. 3.16). Taken from [40] by permission of the publisher.

Table 3.6. Slopes (m), standard error of slopes (s(m)), intercepts (b), standard error of intercepts (s(b)) and linear correlation coefficients (r) of the constants of the regression lines in a plot of ln (lnD) vs $t^{-0.5}$ (D [mm]; t [h])

Protein	m	b	s(m)	s(b)	r
OVA	− 0.4621	1.5436	0.0450	0.0230	− 0.9906
BSA	− 0.4971	1.4848	0.0391	0.0184	− 0.9909
LDH	− 0.5848	1.3834	0.0499	0.0235	− 0.9892
CAT	− 0.6441	1.3198	0.0498	0.0234	− 0.9911
FER	− 0.7202	1.0951	0.1389	0.0653	− 0.9485
THY	− 1.5170	0.9717	0.2188	0.1028	− 0.9702
AAT 1, I	− 0.7143	1.4901	0.0197	0.0093	− 0.9989
AAT 1, II	− 0.6887	1.4782	0.0135	0.0063	− 0.9994
AAT 2, II	− 0.6885	1.4893	0.0085	0.0040	− 0.9998
AAT 4, II	− 0.8053	1.4879	0.0146	0.0068	− 0.9995
AAT 5, I	− 1.0504	1.5054	0.0083	0.0039	− 0.9999
AAT 7, II	− 1.2875	1.5290	0.0050	0.0023	− 1.0000

Marker proteins: Ovalbumin (OVA), bovine serum albumin (BSA), lactate dehydrogenase (LDH), catalase (CAT), ferritin (FER), and thyroglobulin (THY). Sample proteins: aspartate aminotransferase (AAT) isozymes 1, 2, 4, 5 and 7 from the embryo (2n) of two different seeds (I or II) of *Pinus strobus* (see also Fig. 3.16)

Table 3.7. Slopes (m), standard error of slopes (s(m)), intercepts (b), standard error of intercepts (s(b)) and linear correlation coefficients (r) of the constants of the calibration lines ln D_{max} vs ln M_r and ln D_{max} vs ln R_s to calculate the molecular mass of AAT isozymes 1–7 from *Pinus strobus*

	m	b	s(m)	s(b)	r
M_r	−0.7428	12.7088	0.0578	0.6993	−0.9881
R_s	−2.0811	7.0118	0.1975	0.3175	−0.9825

Data was calculated by aid of the computer programm MOL-MASS [40].

Table 3.8. Molecular mass and Stokes' radius of aspartate aminotransferase (AAT) isozymes from *Pinus strobus* as estimated by time-dependent PAGGE

Isozymes	Molecular mass (M_r), standard error of M_r, (M_r + −), Stokes' radius (R_s), (R_s + −) and frictional coefficient ($R_s/R_m = f/f_o$)				
	M_r	M_r + −	R_s	R_s + −	R_s/R_m
AAT 1, I	68511	3885	3.44	0.068	1.274
AAT 1, II	73525	2792	3.53	0.047	1.276
AAT 2, II	68860	1650	3.45	0.029	1.274
AAT 4, I	65364	4139	3.39	0.075	1.272
AAT 4, II	69432	2882	3.46	0.051	1.274
AAT 5, I	62497	1496	3.33	0.028	1.271
AAT 7, II	54072	794	3.17	0.017	1.267

Data calculation was performed by aid of the computer program MOL-MASS [40].

These marker proteins may be obtained from Pharmacia (Uppsala, Sweden) and run in one lane. Separation times may be: 1, 2, 4, 8, and 16 h (or longer) [29]. Figures 3.16 and 3.17 exemplify the determination of the molecular mass of aspartate aminotransferase isozymes in 4–27% T gels (see also Tables 3.5 to 3.8).

3.5.3.3.3 Estimation of Frictional Coefficients

The frictional coefficient (f/f_0) relates the hydrodynamic volume of a protein molecule to its molecular mass. It can be calculated from the equation:

$$R_s \,[m] = f/f_0 \times 66.1 \times 10^{-12} \times M_r^{1/3} \tag{11}$$

and the experimentally obtained R_s and M_r values [29]. Extremely high frictional ratios are to be expected for molecules with (a) rod-like or fibrous structures only, which are characterized by a high axial ratio such as fibrinogen or myosin or by (b) bulky and voluminous globular molecules with normal axial ratios, e.g., the spider-like immunoglobulin M, the shell-like apoferritin or the branched α-macroglobulin [29]. "Normal" enzymes do not belong to one of these groups of proteins [29].

In eq. (11) the frictional coefficient of native proteins is assumed to be constant. However, when analyzing the molecular mass (M_r) and Stokes' radius (R_s) of 64 native proteins it turned out that the frictional coefficient increased with increasing protein size [26]. A more precise equation relating R_s and M_r is the following [29]:

$$R_s[m] = M_r^{0.0225} \times 55.1 \times 10^{-12} \times M_r^{0.0142} \times M_r^{1/3}. \tag{12}$$

According to this expression the frictional coefficient of globular proteins $= M_r^{0.0225}$ and increases with molecular masses of 10^3 to 9×10^6 from $f/f_o = 1.17$ to $f/f_o = 1.43$ while the factor 66×10^{-12} of the expression of Siegel and Monty (R_s [nm] $= f/f_o$ 66.1×10^{-12} $M_r^{1/3}$) increases from 61×10^{-12} to 67×10^{-12}.

3.5.3.3.4 Estimation of the Free Electrophoretic Mobility of Proteins

Estimation of the valence (number of unit charges (Z)) in a non-denatured protein requires, besides other physicochemical quantities, knowledge of its free electrophoretic mobility U [m^2 V^{-1} s^{-1}]. This quantity can be calculated from the apparent free electrophoretic mobility (v_{free} [m/s]) which is accessible by extrapolating the apparent migration velocities in a gradient gel (v [m/s]) to a concentration of zero % T [29].

Apparent protein mobilities   The apparent protein velocities can be obtained from the time-dependent migration distances which a protein has migrated in a gradient gel, by applying eq. (13) [29]:

$$v[m/s] = (D_n - D_m)/(t_n - t_m) = dD/dt, \tag{13}$$

where D_n [m] equals the migration distance of a protein at a longer time of electrophoresis (t_n [s]), while D_m [m] equals its migration distance at a shorter time of electrophoresis (t_m [s]).

The apparent migration velocities which a protein acquires at a certain migration distance are now correlated to the corresponding gel concentrations. This is possible if the gel length and the starting and ending gel concentrations are known since, in a linear gel gradient, PAA-concentration (% T) and gel length (D [mm]) are interrelated by eq. (14):

$$T = \alpha D + \beta. \tag{14}$$

Concentration dependent protein mobility   Now the apparent migration velocities "v" are plotted vs the gel concentration which the protein has reached at the corresponding time of electrophoresis (Fig. 3.18). The function by which "v" and "T" are interrelated is best described by an exponential eq. (15) [29]:

$$v \,[m/s] = h \,(T_{max} - T)^\delta, \tag{15}$$

where T = gel concentration reached at a certain time of electrophoresis, T_{max} corresponds to the exclusion limit (corresponding to the maximum migration distance) and h and δ are constants [29]. The maximum gel concentration which a protein can reach is obtainable from the maximum migration distance (cf. Section 3.6.3.3.1) and the course of the gradient, i.e., the constants (α, β) in eq. (16):

$$T_{max} = \alpha D_{max} + \beta. \tag{16}$$

Equation (15) predicts that zero protein mobility (v = 0) is given if the gel concentration (% T) is equal to the exclusion limit (% T_{max}), i.e., if $T = T_{max}$. The apparent free electrophoretic mobility (v_{free} [m/s]) can be computed by extrapolating the apparent protein mobility to zero % T [29]:

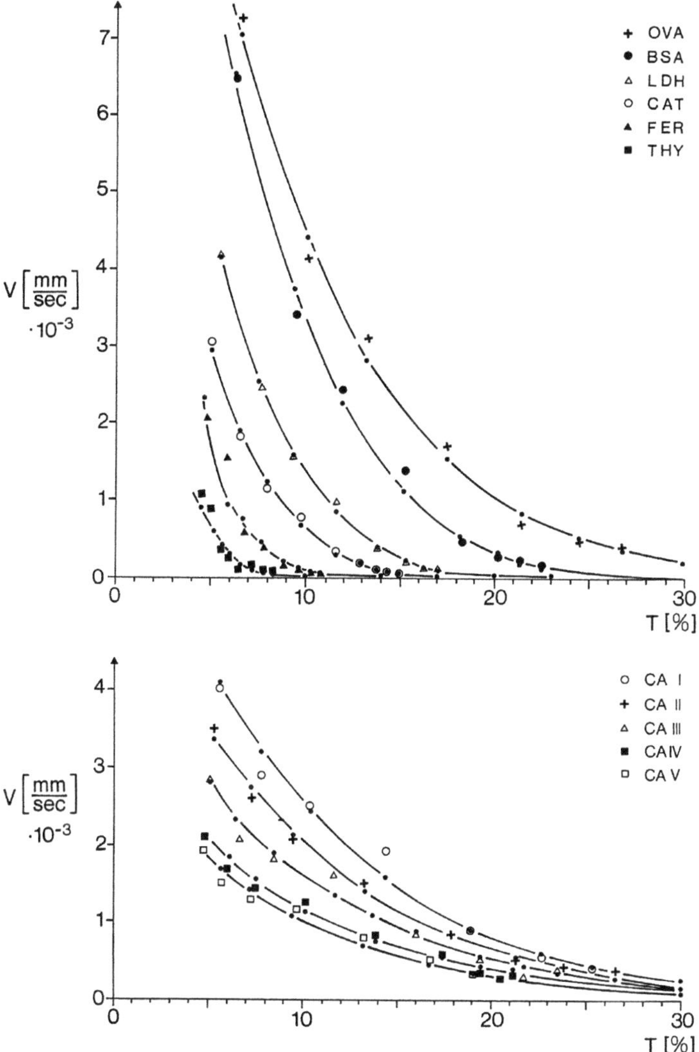

Fig. 3.18. Plot of the migration velocity (v[mm/s]) of non-denatured proteins vs the polyacrylamide concentrations (T[%]) which they had reached after the corresponding times of electrophoresis in a linear polyacrylamide gradient gel ranging from 4 to 27 % T (acrylamide: BIS: 24:1). *Upper panel:* v[mm/s] of the following calibration proteins: *OVA* = ovalbumin; *BSA* = bovine serum albumin; *LDH* = lactate dehydrogenase; *CAT* = catalase; *FER* = ferritin; *THY* = thyroglobulin). *Lower panel:* same plot for various mammalian carbonic anhydrases (CA I – V) [29]. Graphs taken from [29] by permission of the publisher

$$v_{free} \ [m/s] = h \ (T_{max} - 0)^{\delta}$$
(17)

or

$$v_{free} \ [m/s] = h \ T_{max}^{\delta} \ .$$
(18)

The logarithmized form of eq. (15) together with the calculated T_{max}-value can be used to compute the constants h and δ in it [29]:

$$\ln v = \delta \ln (T_{max} - T) + \ln h \ .$$
(19)

Once this data has been obtained a non-linear approximation procedure can be used to calculate the constants more precisely [29].

Free electrophoretic mobility

The free electrophoretic mobility (U $[m^2 \ s^{-1} \ v^{-1}]$) of the protein under consideration results from its apparent free mobility unhindered by the gel matrix ($v_{free} \ [m/s]$) and the electric field strength E $[v \ m^{-1}]$ acting on it [29]:

$$U = v_{free} \ / \ E \ [m/s \ (V/m)^{-1} = m^2 \ s^{-1} \ V^{-1}] \ .$$
(20)

The free electrophoretic mobilities (U) of various marker proteins and five different mammalian carbonic anhydrases were calculated with the described procedures and the values obtained are given in Table 3.9.

Table 3.9. Free electrophoretic mobility (U) and net negative charge (valence, Z; charge, Q) of several marker proteins and carbonic anhydrase (iso)enzymes from mammalia at pH 8.4

Protein	U $[m^2/Vs] \cdot 10^{-9}$		Negative charge	
	$I = 0.529 \cdot 10^3$ $[mol/m^3]$	$I = 0.1 \cdot 10^3$ $[mol/m^3]$	Z	Q [C/molecule] $\cdot 10^{-19}$
OVA	3.45	5.99	13.06	20.92
BSA	4.40	7.85	22.42	35.92
LDH	3.27	6.00	22.63	35.25
CAT	2.60	4.94	21.43	34.33
FER	3.28	6.38	43.81	70.18
THY	2.78	5.62	68.46	109.67
CA I	1.58	2.69	4.93	7.90
CA II	1.17	1.99	3.58	5.74
CA III	1.05	1.79	3.32	5.32
CA IV	0.851	1.46	2.75	4.41
CA V	0.734	1.25	2.31	3.70

Ovalbumin (OVA), bovine serum albumin (BSA), lactate dehydrogenase (LDH), catalase (CAT), ferritin (FER) and thyroglobulin (THY); carbonic anhydrase (iso)enzymes from mammalian erythrocytes: (bovine, I, II), bovine, rabbit (III), rabbit (IV) and canine, human (V).
Conditions of electrophoresis: linear polyacrylamide gradient from 4 to 27% T (acrylamide : BIS = 24:1); gel length: 73 mm; buffer system: 90 mmol l^{-1} Tris, 80 mmol l^{-1} boric acid, 1.25 mmol l^{-1} EDTA-Na$_2$, pH 8.0 (I = 529 $[mol/m^3]$); field strength: 41 V/cm; 4 °C (cf [29]).

3.5.3.3.5 Estimation of the Number of Unit Charges of Proteins

Estimation of the valence (number of unit charges (Z)) in a non-denatured protein requires, besides its free electrophoretic mobility (U), knowledge of its Stokes' radius (R_s), the ionic strength (I) and the viscosity (η) of the electrophoretic buffer system used. Finally Z results from the correlation:

$$Z = \frac{U \, 6 \, \pi \eta \, R_s}{\varepsilon} \cdot \frac{(1 + \kappa \, R_s)}{X_1 \, (\kappa \, R_s)} \tag{21}$$

with: U [m² v⁻¹ s⁻¹], free electrophoretic mobility; η [Pa s], temperature dependent dynamic viscosity of the buffer; R_s [m], Stokes' radius of the protein; Z [C], unit charge (protein charge) = $1.602 \cdot 10^{-19}$; $\pi = 3.14...$; κ [m⁻¹], reciprocal Debye-Hückel thickness and X_1 function of $\kappa \, R_s$ [29].

Time-dependent electrophoresis in a linear polyacrylamide gradient gel of known concentration range can be used to calculate the apparent free electrophoretic mobility of a protein (cf. Sect. 3.6.3.3.4). The free electrophoretic mobility (U) results from the quotient of the apparent free electrophoretic mobility and the field strength (E) applied to the gel [29]. The temperature dependent dynamic viscosity (η) of the buffer system used can be taken from Table 3.10. The Stokes' radius (R_s) of the protein, if not known, can also be calculated from time dependent gradient gel electrophoresis (cf. Sect. 3.6.3.3.3). Kappa (κ), the reciprocal Debye-Hückel thickness of the "ionic cloud" surrounding the protein results from the following equation [29]:

$$\kappa = 1.5906081310^{10} \, [\text{K m/mol}]^{1/2} \, (\text{I/D T})^{1/2} \, [\text{mol/m}^3 \, \text{K}]^{1/2} \, . \tag{22}$$

The ionic strenth I [mol/m³] in this formula results from:

$$I = 1/2 \sum c_i \, Z_i^2 \, [\text{mol/m}^3] \tag{23}$$ **Ionic strength**

where $\sum c_i$ [mol/m³] represents the summed concentrations of the summed ionic species of the buffer multiplied by their squared charges (Z_i).

Hence, taking for example a 90 mmol l⁻¹ Tris, 80 mmol l⁻¹ boric acid, 1.25 mmol l⁻¹ EDTA-Na₂ buffer of pH 8.0, the ionic strength of this buffer adds up to:

Tris (Tris⁺) = 0.09×1^2, plus boric acid ($3H^+ + BO_4^{3-}$) = $3 \times 0.08 \times 1^2 + 0.08 \times (-3)^2$, plus EDTA-Na₂ ($2H^+ + EDTA\text{-}Na_2^{2-}$) = $2 \times 0.00125 \times 1^2 + 0.00125 \times (-2)^2$,

which gives:

$$I = 0.52875 \, [\text{mol/dm}^3] \tag{24}$$

or:

$$I = 0.52875 \times 10^3 \, [\text{mol/m}^3] \, . \tag{25}$$

The temperature dependent dielectric constant of water may be selected from (Table 3.11).

At a temperature of 5 °C (278 K) the dielectric constant of water is 85.90 (Table 3.10). Inserting the values of 278 [K] and 85.90 [N s/m²] into the equation given for κ Eq. (22), yields Eq. (26):

$$\kappa = 1.02930525 \times 10^8 \, [\text{m/mol}]^{1/2} \times \sqrt{I} \, [\text{mol/m}^3]^{1/2} \, . \tag{26}$$ **Kappa**

Table 3.10. Dynamic viscosity (η[N s m^{-2}]) of water as a function of temperature (T[°C]) [29]

T [°C]	[N s m^{-2}] $\times 10^{-3}$	T [°C]	[N s m^{-2}] $\times 10^{-3}$
0	1.787	16	1.109
1	1.728	17	1.081
2	1.671	18	1.053
3	1.618	19	1.027
4	1.567	20	1.002
5	1.519	21	0.9779
6	1.472	22	0.9548
7	1.428	23	0.9325
8	1.386	24	0.9111
9	1.346	25	0.8904
10	1.307	26	0.8705
11	1.271	27	8.8513
12	1.235	28	0.8327
13	1.202	29	0.8148
14	1.169	30	0.7975
15	1.139		

Table 3.11. Dielectric constand (D) of water as a function of temperature t [°C] [29]

T [°C]	D	T [°C]	D
0	87.90	18	80.93
5	85.90	20	80.18
10	83.95	25	78.36
15	82.04	30	76.58

At an ionic strength of I = 0.52875 $\times 10^3$ [mol/m^3], κ becomes:

$$\kappa = 1.02930525 \times 10^8 \ [\text{m/mol}]^{1/2} \times (528.75)^{1/2} \ [\text{mol/m}^{-3}]^{1/2} \tag{27}$$

which rearranges to:

$$\kappa = 2.366842604 \times 10^9 \ [1/\text{m}] . \tag{28}$$

To compute the number of unit charges (Z) in a native protein the value of κR_s and the value of the function X1 of κR_s are needed (cf. Eq. (21)).

The Stoke's radius may also be calculated from the same time dependent gradient gel electrophoretic experiment (Table 3.12).

Taking ferritin as an example which has a Stokes' radius of R_s = 6.20 $\times 10^{-9}$ [m], the value of κR_s = 14.67. The $\log_{10}$ value of κR_s = 1.167 and using this value one obtains from the equation 5 < κR_s < 24 = X1 (κR_s) = 0.2849925 $\times \log_{10}$ (κR_s) + 0.9603828 a value of 1.293 for the function X1 (κR_s). (For the case $\kappa R_s \geq$ 24 and $\kappa R_s \leq$ 5, Fig. 22 and Table 19 in ref. [29] provide solutions.)

Table 3.12. Calculated Stokes' radius (R_s) and molecular mass (M_r) of several marker proteins and carbonic anhydrase (iso)enzymes from mammalia

Protein	R_s [nm]			M_r [g/mol]		
	Ref.	cal.	%	Ref.	cal.	%
OVA	3.05	3.08	+0.9	43000	50181	+16.7
BSA	3.55	3.65	+2.8	67000	80668	+20.4
LDH	4.20	4.32	+2.9	140000	129249	−7.7
CAT	5.25	4.76	−9.3	232000	168653	−27.3
FER	6.10	6.20	+1.6	440000	352110	−20.0
THY	8.50	8.64	+1.7	669000	886379	−32.0
CA I		2.76			37171[a]	
CA II		2.73			35968[a]	
CA III		2.78		38000	37695[a]	
CA IV		2.82			39241	
CA V		2.78		29700	38032	

Marker proteins: OVA = ovalbumin, BSA = bovine serum albumin, LDH = lactate dehydrogenase, CAT = catalase, FER = ferritin, and THY = thyroglobulin.
CA = Carbonic anhydrase: I, II = bovine, III = bovine/rabbit, IV = rabbit, V = canine/human.
Ref.: reference; cal.: calculated value; % : % deviation from the value given in the literature.
[a] The molecular mass of bovine carbonic anhydrase as estimated by sequence analysis was reported to be 28980 [29].
Conditions of electrophoresis: the polyacrylamide gradient increased from 4 to 27% T (acrylamide: BIS = 24:1), the gel length was 73 mm.
Homogeneous buffer system: 90 mmol l⁻¹ Tris, 80 mmol l⁻¹ boric acid, 1.25 mmol l⁻¹ EDTA-Na$_2$, pH 8.0 (I = 529 [mol/m³]. Field strength: 41 V/cm. Maximum running time: 20 h (cf.[29]); 4 °C. Data taken from [29].

From gradient gel electrophoretic results the free electrophoretic mobility of ferritin was calculated to U = 3.28 x 10⁻⁹ [m²/V s] (Table 3.9). Substituting this value, the value of κR_s, the value of X1 (κR_s), R_s and the value for the temperature dependent dynamic viscosity (η [N s/m²]) of water in Eq. (21), the number of unit charges that ferritin acquires under the electrophoretic conditions indicated above can be computed as:

$$Z = \frac{3.28 \times 10^{-9} \times 6 \times \pi \times 1.519 \times 10^{-3} \times 6.20 \times 10^{-9}}{1.602 \times 10^{-19}} \times \frac{(1+14.67)}{1.293} \qquad (29)$$

which works out to:

Surface charge

$$Z = 44.05. \qquad (30)$$

The actual charge of the molecule is given by Z $\times$ ε [C/molecule] = 44.05 $\times$ 1.602 $\times$ 10⁻¹⁹ = 7.057 $\times$ 10⁻¹⁸.

Mammalian carbonic anhydrase is a comparatively small enzyme ($M_r \approx$ 30 000 g/ mol). Nevertheless it migrates extremely slow at pH 8.4 (Table 3.9) which results from its low negative charge. Separation of the various size isomers of carbonic anhydrase is possible because they differ in surface charge (Table 3.9) [29].

3.6 2-D-Electrophoresis under Non-Denaturing Conditions

Two-dimensional (2-D) electrophoresis under non-denaturing conditions can be used to (a) elucidate the total number of non-denatured proteins contained in a given enzyme extract or (b) to determine the total number of isozymic forms contained in a certain enzyme source. The 2-D-system suggested here is simple to perform and efficient. It consists basicly of a polyacrylamide gradient gel which is used in the first dimension and the second dimension after inpregnation with ampholines.

3.6.1 A Pilot Test

Before starting 2-D-electrophoresis the time-length must be found out that is needed to achieve optimum separation of an isozyme system in the first dimension which is a pore-gradient gel electrophoresis under non-denaturing conditions (Sect. 3.5.3). A PAA gradient of 4 to 20 % T (or $\geq 4 \geq 20\,\%$ T), the Tris-EDTA-borate buffer system of pH 8.4 and a voltage gradient of ≥ 40 Vcm^{-1} may be used for that purpose. Separation times can be between 2 and 3 h.

3.6.1 The First Dimension

In the first dimension a vertical electrophoretic apparatus (e.g., Pharmacia, Uppsala, Sweden; Bio Rad, Hercules, CA USA) is needed. Gradient gels are encased in glass cassettes (Sect. 3.5.3.1) while backing them to a polyester film.

At least two glass plates (size $16.5 \times 8 \times 0.15$ cm, Pharmacia, or $10 \times 10 \times 0.1$ cm, Bio Rad) are washed, rinsed with distilled water, dried and wiped off with isopropanol. One of the plates (or more) are dipped into a Repel-Silane (dimethyldichlorosilane solution 2 wt/vol.% in 1,1,1-trichlorethane, LKB, Bromma, Sweden) solution by aid of tweezers and air dried. The remaining glass plate(s) are used as support for a polyester foil (Gel Bond: Marine Colloids, Rockland, MN, USA; Gel Fix: Serva, Heidelberg, Germany) of the same size. The foil is rinsed under running distilled water and placed with its hydrophilic side towards the viewer onto the glass plate. It is firmly attached to the glass plate by aid of a rubber roller (Sect. 3.5.3.1). Afterwards the glass cassette is assembled from this plate, a right and a left distance bar ($100 \times 5 \times 0.5$ mm) placed on the foil, a silanized glass plate and a slot former (Fig. 3.10). The cassette is fixed with self-adhesive tape or clamped together and stored over night at room temperature in a dry and dark place because the Gel Bond foil is light sensitive.

The next day when the foil has completely dried the cassettes are either filled separately with gradient solution (Bio Rad cassetes) or filled all at once in a gel casting device (Pharmacia cassettes) (Sect. 3.5.3.1). The Tris (45 mmol l^{-1}), EDTA-Na$_2$ (1.25 mmol l^{-1}), boric acid (40 mmol l^{-1}) system of pH 8.4 may be used as gel (and electrode) buffer and the polyacrylamide gradient may range from 3.7 % T to 20 % T (acrylamide: BIS = 24:1) (Fig. 3.19). The gradients may be made by an ordinary two chamber gradient mixer or, more reproducible, by an automatic 4 burette system.

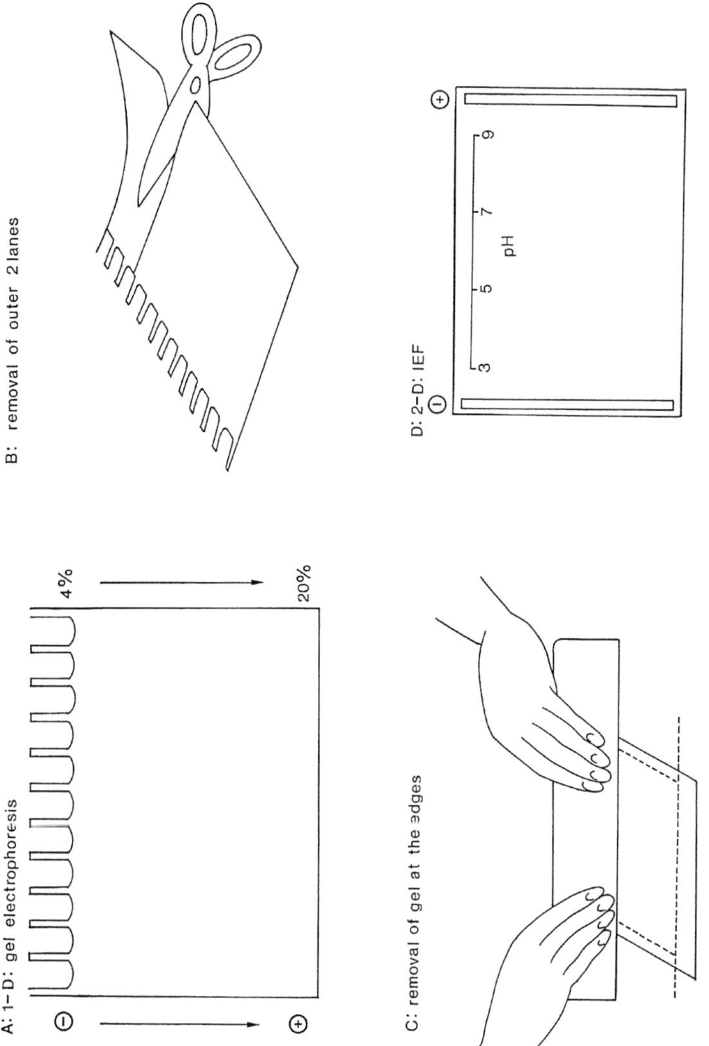

Fig. 3.19. Experimental procedure of 2-D-electrophoresis under non-denaturing conditions. (A) In the first dimension proteins are separated in a 4 to 20 % T PAA gradient gel firmly attached to a polyester foil. Aliquots of sample solution are applied to the center well and one of the outer wells. (B) Before performing the second dimension the gradient gel is taken off the glass cassette and the outer lane containing separated proteins is cut off and stained for the isozymes of interest. (C) A 5 mm strip of gel is removed from all four edges. (D) In the second dimension IEF is performed rectangular to the first dimension and finally the entire gel is stained for the isozymes of interest

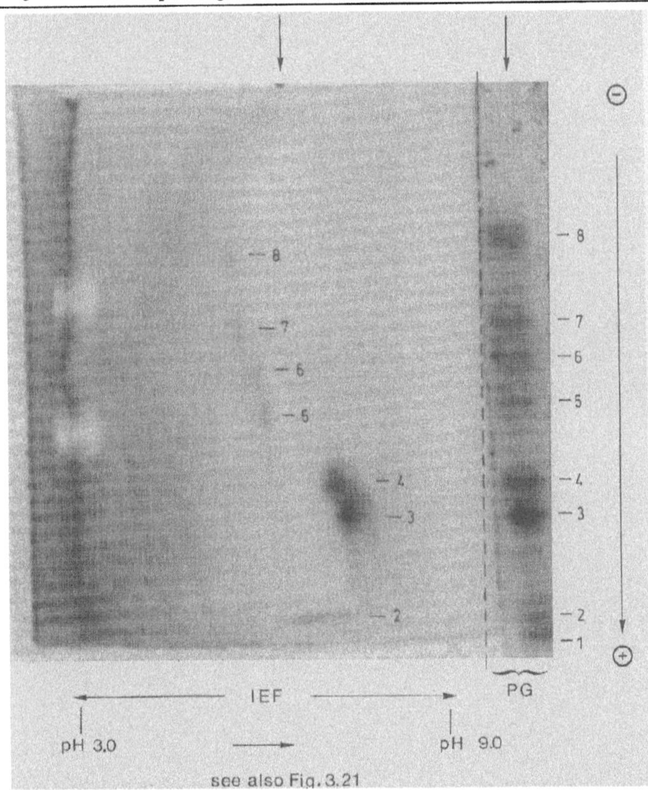

Fig. 3.20. 2-D-electrophoresis of aspartate aminotransferase isozymes from *Pinus strobus* seeds. The left part of the picture shows the separation of aspartate aminotransferase isozymes after IEF (and separation in a 4 to 20 % T PAA gradient before hand). The outer right strip shows the separation pattern after gradient gel electrophoresis. First dimension: electrophoresis in a 4 to 20 % gradient of polyacrylamide (acrylamide: BIS = 24:1); gel and electrode buffer: 45 mmol l⁻¹ Tris, 40 mmol l⁻¹ boric acid, 1.25 mmol l⁻¹ EDTA-Na$_2$, pH 8.5; running time: 3.5 h at 41 V cm⁻¹ and 4 °C. Second dimension: IEF in the same gel used for the first dimension but rectangular to the migration in that gel; range of the pH-gradient: 3.5 to 10 (Ampholine, Pharmacia, Uppsala); duration of IEF: 2 h at 0.5 Watt (constant). Visualization of aspartate aminotransferase isozymes: 48 mg L-aspartate, 28 mg 2-oxoglutarate and 22 mg Fast Blue BB salt in 20 ml of 200 mmol l⁻¹ Na-phosphate buffer of pH 7.5. Extraction of plant material: 4 seeds of *Pinus strobus* were soaked in water for 10 min, separated into endosperm and embryo and crushed in 1 ml of extraction buffer (3.044 g Na$_2$HPO$_4$, 0.207 g NaH$_2$PO$_4$ and 0.5 g Triton X-100 in 100 ml distilled water, pH 7.5). The homogenate was centrifuged for 10 min at 5 000 g and the clear solution between the fatty top layer and the pellet was taken as crude enzyme extract. It was stored on ice until used for electrophoretic studies

Finally, the cassettes are inserted into a vertical electrophoretic apparatus (Pharmacia or Bio Rad). If 10 × 10 cm sized gels are used and if the gel provides 10 sample slots 10 µl of sample are applied to positions 5 and 9 (Fig. 3.19). If 16.5 × 8 cm sized gels are used providing 25 positions of sample application aliquots of 10 µl of sample may be put into the slots number 5, 11, 16 and 24 and the gel may be cut into two halfs following electrophoresis. After the end of the run, the gel is cut between positions 12 and 13 into two equal halfs. Otherwise sample solution is put into the slots number 12 (or any position between number 3 and 21). During the run the gels are cooled to 4 to 6 °C applying a voltage gradient of not more than 40 V cm^{-1}. After termination of electrophoresis the cassettes are taken off and the covering glass plate is carefully removed from the plate to which the polyester foil adheres. The scissors are taken to cut the foil with the gel on top parallele to the course of the gradient, between the sample slots number 23 and 24 (and number 10 and 11 if the gel has been cut into two halfs before). The small strips obtained that way are used to eludicate the position of the separated isozymes by applying one of the histochemical stains listed in Sect. 6 (Fig. 3.20).

3.6.3 The Second Dimension

The remaining gel part with the separated proteins in its center lane is used for an isoelectric focusing. First, from each of the four edges 5 mm of gel are scratched off by aid of a knife. Then the gel is air dried for 5 to 10 min. Afterwards a few drops of kerosene are dropped onto the cooling plate of a horizontal electrophoretic apparatus (Desaga, Heidelberg; Pharmacia-LKB, Bromma) and the foil is then placed on it, carefully avoiding entrapment of air bubbles. Finally, 150 µl of Ampholines, pH 3.5 to 10 (Pharmacia, Uppsala) are dropped onto the middle of the gel (8 × 6 cm) and the spread over the whole gel surface with a round glass stick; a cathodic electrode wick (60 × 5 × 5 mm, containing 0.5 mol l^{-1} NaOH) and an anodic electrode wick

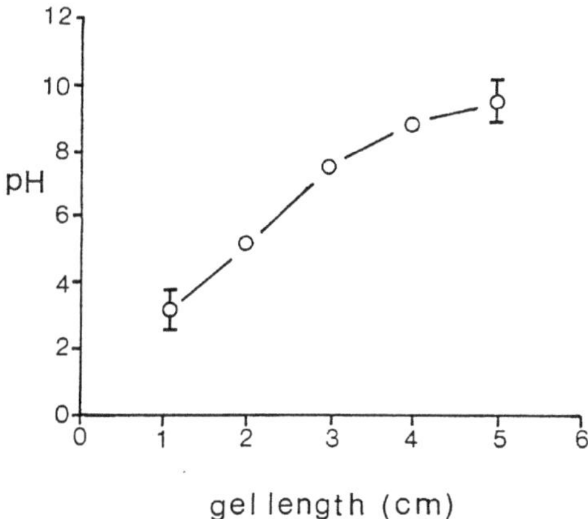

Fig. 3.21. Course of the pH-gradient after 2-D-electrophoresis under non-denaturing conditions. The total gel was cut into squares of 1 × 1 cm and each plate put separately into a few ml of 5 mmol l^{-1} KCl to measure the pH of the resulting solution with a conventional pH-electrode. The mean pH-value of the six parallel rows each of 1 cm distance is indicated (o)

($60 \times 5 \times 5$ mm, containing 0.5 mol l^{-1} H$_3$PO$_4$) are positioned at the opposite edges of the gel representing the former top and bottom of the gradient gel. The electrode wicks are usually made of filter paper, they must be wet but they should not leak and there must be no air bubbles between them and the gel surface. The cathodic (–) and anodic (+) electrodes cover the cathodic and anodic electrode wicks at full length and are firmly pressed onto the wicks with a load. At the end a coolant liquid is circulated through the cooling plate and 85 V (6 mA) are applied to the gel (6×8 cm). After 1 h of IEF the voltage will have reached about 1000 V. The separation is terminated after 2 h. After the end of the run at the upper and lower end of the gel a 5 mm broad strip is cut off and cut into 5 mm long bits. Each bit is put into a numbered glass vessel containing 5 mmol l^{-1} KCl and after a little stirring the pH of each solution is measured (at the temperature used to cool the cooling plate). Otherwise a surface electrode is used to measure the course of the pH-gradient (Fig. 3.21). The remaining gel part is used to stain the separated enzymes by applying one of the protocols given in Chap. 6.

3.7 References

1. Raymond S (1962) Clin Chem 8: 455 – 470
2. Maurer HR (1968) Disk-Elektrophorese, Walter de Gruyter & Co., Berlin
3. McLellan T (1982) Anal Biochem 126: 94 – 99
4. Rodbard D, Chrambach A (1971) Anal Biochem 40: 95 – 134
5. Thomas JM, Hodes ME (1981) Anal Biochem 118: 194 – 196
6. Davis BJ (1964) Annals N Y Acad Sci 121: 404 – 427
7. Ornstein L (1964) Annals N Y Acad Sci 121: 321 – 349
8. Clarke JT (1964) Ann N Y Acad Sci 121: 428 – 436
9. Williams DE, Reisfeld RA (1964) Annals N Y Acad Sci 121: 373 – 381
10. Aronsson T, Grönwall A (1957) Scand J Clin Lab Invest 9: 338 – 341
11. Kleiner H, Schram E (1966) Clin Chim Acta 14: 377 – 389
12. Fitt PS, Fitt EA, Wille H (1968) Biochem 110: 475 – 479
13. Neuhoff V (1968) Arzneim Forsch 18: 35 – 39
14. Neuhoff V, Lezius A (1968) Z Naturforsch (B) 23: 812 – 819
15. Ogilvie JW, Sightler JH, Clark RB (1969) Biochemistry 8: 3557 – 3567
16. McGregor RR, Schnaitman CA, Normansell DE (1974) J Biol Chem 249: 5321 – 5327
17. Yonezawa S, Hori SH (1975) J Histochem Cytochem 23: 745 – 751
18. Willhardt I, Wiederanders B (1975) Anal Biochem 63: 263 – 266
19. Pierce M, Cummings RC, Roth S (1980) Anal Biochem 102: 441 – 449
20. Davies RC, Neuberger A (1973) Biochem J 133: 471 – 492
21. Karpetsky TP, Davies GE, Shriver KK, Levy CC (1980) Biochem J 189: 277 – 284
22. Millard SA, Kubose A, Gal EM (1969) J Biol Chem 244: 2511 – 2515
23. Nimmo HG, Nimmo GA (1982) Anal Biochem 121: 17 – 22
24. Ornstein LB, Davis J (1962) Disc electrophoresis, Distillation Products Ind, Rochester and CANALCO, brochures, Bethesda, unpublished
25. Reisfeld RA, Lewis UJ, Williams DE (1962) Nature 201: 281 – 283
26. Felgenhauer K (1974) Hoppe-Seylers's Z Physiol Chem 355: 1281 – 1290
27. Anderson BL, Berry RW, Telser A (1983) Anal Biochem 132: 365 – 375
28. Neuhoff V (1973) Micromethods in molecular biology. Springer, Berlin, pp 4 – 83
29. Rothe GM (1991) Determination of the size, isomeric nature and net charge of enzymes by pore gradient gel electrophoresis. In: Chrambach A, Dunn MJ, Radola BJ (eds) Advances in electrophoresis vol 4. VCH Publishers, New York, pp 251 – 358

30. Hedrick JL, Smith AJ (1968) Arch Biochem Biophys 126: 155 – 164
31. Rothe GM, Maurer WD (1986) One-dimensional PAA-gel electrophoretic techniques to separate functional and denatured proteins. In: Dunn MJ (ed) Gel electrophoresis of proteins. Wright, Bristol, pp 37 – 140
32. Esposito JJ, Obijeski JF (1976) Prep Biochem 6: 431 – 442
33. Mahadik SP (1975) Anal Biochem 76: 615 – 633
34. Margolis J, Kenrick KG (1968) Anal Biochem 25: 347 – 362
35. Wright GL, Farrell KB, Roberts D (1973) Biochim Biophys Acta 295: 396 – 411
36. Martin RG, Ames BN (1986) J Biol Chem 236: 1372 – 1374
37. Righetti PG, Gelfi C, Gianazza E (1986) Conventional isoelectric focusing and immobilised pH gradients. In: Dunn MJ (ed) Gel electrophoresis of proteins. Wright, Bristol, pp 141 – 202
38. Saravis CA, Cook E (1979) Marine Colloid's Instruction Leaflet, Marine Colloid, Rockland
39. Altland K, Rossmann U (1985) Electrophoresis 6: 314 – 325
40. Rothe GM, Weidmann H (1991) Electrophoresis 12: 703 – 709
41. Görg A, Postel W, Westermeier R (1978) Anal Biochem 89: 60 – 70

4 Sodium Dodecylsulphate Electrophoresis

Sodium dodecylsulphate (SDS) consists of an aliphatic chain of 12 C atoms to which at one end a sulphate residue is bound. It forms complexes with both the polar and non-polar amino acid residues of proteins irrespective of their sizes and shapes, leaving the primary structure uninfluenced. In electrophoresis SDS is used:

a) to separate (enzyme) proteins into their monomeric constituents,
b) to estimate the molecular mass of unfolded (and reduced) polypeptides, and
c) to keep membrane proteins in a solubilized state.

Most non-denatured proteins bind 0.9 to 1.0 g SDS per gram of protein, including the protein core of some glycoproteins [1]. Proteins with reduced disulphide bridges and proteins without disulphide bonds bind 1.4 g SDS per gram of protein in a rather non-specific way. A concentration of SDS above 0.5 mmol l^{-1} is sufficient for binding 1.4 g SDS per gram protein in a primarily hydrophobic way. Binding of the detergent in a micellar form has been excluded [2]. Under the condition that 1.4 g SDS are bound per gram protein, nearly all proteins assume the shape of rod-like particles, the length of which varies unequally with the molecular mass of the protein [3]. Exceptions are papain, pepsin and glucose oxidase [4] and a number of glycoproteins [5] which bind a relatively low amount of detergent. Histones also behave in an "anomalous" way in that SDS cannot compensate for their intrinsic charges. These examples, however, are not sufficient to cast severe doubt on the validity of molecular mass estimations by SDS electrophoresis [6, 7, 8]. It must be pointed out, on the other hand, that protein patterns in SDS electrophoresis must be tested for reproducibility either by performing separations on homogeneous gels of different pore size or by using gradient gels of different slopes [9].

4.1 Homogenous Buffer Systems

By using homogeneous PAA gels and buffer systems containing 0.1 % SDS it is possible to estimate the subunit molecular mass of proteins. Shapiro, Vinuela and Maizel [6] were the first to show that it is possible to estimate the size of 2-mercaptoethanol-reduced SDS-denatured proteins by using 5 % acrylamide gels (acrylamide: BIS = 38:1) and a gel and electrode buffer of 100 mmol l^{-1} phosphate, pH 7.1 with 0.1 % SDS. The proteins which they had used had widely varying intrinsic charges with isoelec-

Composition
of SDS gels

Table 4.1. Continuous buffer systems used in SDS electrophoresis to estimate the subunit size of proteins

Acrylamide (g/100 ml); BIS (g/100 ml) [Ref.]	Gel buffer; pH	Electrode buffer; pH	Molecular mass range ($\times 10^3$)
5; 0.13 [6]	100 mmol l^{-1} P$_i$ 0.1% SDS; 7.1	100 mmol l^{-1} P$_i$ 0.1% SDS; 7.1	15 – 200 [a]
10; 0.27 [10]	100 mmol l^{-1} P$_i$ 0.2% SDS; 7.0	50 mmol l^{-1} P$_i$ 0.1% SDS; 7.0	15 – 200 [b]
15; 0.52	100 mmol l^{-1} P$_i$ 0.1% SDS; 7.2	100 mmol l^{-1} P$_i$ 0.1% SDS; 7.2	10 – 60 [c]
10; 0.34			10 – 100
5; 0.17 [11]			20 – 350

a) Calibration line: log M_r vs distance of migration relative to the α and β chains of haemoglobin. Lysozyme and ribonuclease exhibit an anomalous behaviour. Proteins are denatured for 3 h at 37 °C in 100 mmol l^{-1} phosphate, pH 7.1, containing 1% SDS and 1% 2-mercaptoethanol. Afterwards they are dialyzed for 16 h against 10 mmol l^{-1} phosphate, pH 7.1 with 0.1% SDS and 0.1% 2-mercapto-ethanol. P$_i$: phosphate buffer. Gel dimensions: ø 6 mm, length 75 mm.
b) Calibration line: log M_r vs "m" with "m" representing protein mobility:

$$m = \frac{D \cdot (\text{gel length before staining})}{B \cdot (\text{gel length after destaining})}$$

(D: distance of protein migration, B: distance of the migration of Bromophenol blue).
Proteins are denatured at 37 °C for 2 h in 10 mmol l^{-1} sodium phosphate pH 7.0, containing 1% SDS and 1% 2-mercaptoethanol (0.2 – 0.6 mg protein/ml). Dialysis is performed as described in [6]. A non-linear relationship between log M_r and "m" results if the amount of crosslinker is doubled. Gel dimensions: ø 6 mm, length 100 mm.
c) Calibration line: log M_r vs distance of migration relative to that of chymotrypsinogen. The plot undergoes an inflection if $M_r < 15000$. Proteins are dissolved at about 2 mg per ml and denatured at 45 °C for 30 to 60 min in 1% SDS, 1% 2-mercaptoethanol and 4 M urea. Before electrophoresis they are dialyzed against electrode buffer.

tric points ranging from 4 to 11, but they all migrated anodically in the presence of SDS. By plotting the logarithm of the molecular mass of marker proteins vs the relative distance they migrated at a constant time, the authors obtained a calibration line which they used to calculate the molecular mass of sample proteins. Their results were confirmed by Weber and Osborn [10] and numerous authors thereafter. Dunker and Rueckert [11] extended the accessible molecular range by using gels of 5, 10 and 15% T (Table 4.1).

4.2 Discontinuous Buffer Systems

A discontinuous buffer and gel system was introduced by Laemmli [12]. In this system the sample proteins are sandwiched between the leading chloride ions and the trail-

Table 4.2. Discontinuous gel and buffer systems used to separate denatured proteins by SDS gel electrophoresis

Acrylamide (g/100 ml); BIS (g/100 ml) [Ref.]	Gel buffer; (pH)	Electrode buffer; (pH)
2.5; 0.625	Large pore gel: 200 mmol l^{-1} Tris-SO_4, 0.025 % SDS; (7.8)	Upper buffer: 74 mmol l^{-1} Tris, 0.1 % SDS adjusted to pH 7.8 with concentrated HCl; (7.8)
7.6; 0.4 [15] [a]	Separating gel: 200 mmol l^{-1} Tris-SO_4, 8 mmol l^{-1} urea, 0.025 % SDS; (7.8)	Lower buffer: 200 mmol l^{-1} Tris, 0.04 % SDS, adjusted to pH 7.8 with H_2SO_4
3.0; 0.08	Large pore gel: 125 mmol l^{-1} Tris, adjusted to pH 6.8 with HCl containing 0.1 % SDS	Upper and lower buffer: 192 mmol l^{-1} glycine, 25 mmol l^{-1} Tris, 0.1 % SDS; (8.3)
8; 0.21 10; 0.27 15; 0.40 [12, 16] [b]	Separating gel: 375 mmol l^{-1} Tris, 0.1 % SDS, adjusted to pH 8.8 with HCl	

[a] The calibration curves obtained with this system are linear over a molecular mass range of 2500 – 90 000 when correlating log M_r vs R_f values. The correlation is uninfluenced by acrylamide concentrations in the range 5 – 18 %. Polymerization of the large pore gel is initiated by a final concentration of 0.05 % ammonium persulphate and 0.012 ml Temed per 23.5 ml gel solution; polymerisation of the small pore gel (8 % T) is initiated by a final concentration of 0.0048 % ammonium persulphate and 0.005 ml Temed per 5.0 ml gel solution. As sample buffer serves a 139 mmol l^{-1} Tris solution adjusted to pH 7.8 with glacial acetic acid, containing 0.5 % SDS and 20 % sucrose. Samples are prepared at concentrations of 1 to 5 mg ml^{-1} in 139 mmol l^{-1} Tris, 0.5 % SDS, 20 % sucrose, adjusted to pH 7.8 with glacial acetic acid. Separation conditions: 10 W for 0.75 mm flat gels, around 15 °C. It takes 4 – 5 h for the tracking dye to migrate a distance of 15 cm. R_f = mobility of a protein band relative to Pyronin Y. Omitting urea from the separation gels gives regression lines with inflection points between 10 000 and 20 000 molecular mass. The mobility of calf thymus histones is found to be completely unrelated to their molecular masses.

[b] The separating gels are polymerized by the addition of 0.025 ml Temed and 0.3 ml of 10 % ammonium persulphate solution per 100 ml of gel solution. The required 10 cm gels may be prepared in glass tubes of a total length of 15 cm (ø 6 mm). The length of the stacking gel is 1 cm. It is polymerized in the same way as the separation gel. The samples (0.1 to 0.2 ml) contain in final concentrations: 62.5 mmol l^{-1} Tris-HCl (pH 6.8), 2 % SDS, 10 % glycerol, 5 % 2-mercaptoethanol and 0.001 % Bromophenol blue. Proteins are dissolved by immersing the sample solution for 1.5 min in boiling water. Electrophoresis is carried out at 3 mA per gel until the dye front has reached the bottom of the gel. As marker proteins may be used: bovine serum albumin (68 000), gamma-globulin, heavy chain (50 000), ovalbumin (43 000), gamma-globulin, light chain (23 500) and TMV (17 000).

ing glycine ions. Unstacking occurs in the separation gel when the pH changes and the protein-SDS complexes are sieved according to their apparent molecular masses. Comparing continuous and discontinuous systems the latter are usually superior in their resolving power. It is assumed that a linear correlation exists between the logarithm of the molecular mass of denatured proteins and their relative mobility. But it must be pointed out that: (a) this relationship cannot be applied to glycoproteins when using homogeneous gels [10, 13, 14] and (b) it is limited to a given gel concentration; for 10% gels, the range is 1.1×10^4 to 7×10^4, for 5% gels it is 2.5×10^4 to 20×10^4, and for 3.3% gels it is 20×10^4 to 1×10^6 [10]. Table 4.2 summarizes a number of discontinuous gel and buffer systems used to separate denatured proteins by SDS gel electrophoresis.

4.3 Gradient Gel Systems

In PAA gradient gels bands are much sharper and proteins can be separated over a larger molecular mass range compared with homogeneous gel systems [8, 17]. Some of the gel and buffer systems used most frequently are listed in Table 4.3. Lambin and coworkers [17] demonstrated that a linear relationship between the logarithm of the molecular mass (M_r) of SDS-denatured proteins and the logarithm of the PAA concentration (T) which they reach after a prolongued time of electrophoresis can be set up ($\log M_r = a \log T + b$, where T is the polyacrylamide concentration reached by a protein, "a" is the slope and "b" is the intercept of the regression line [8]). Lambin [8] showed that, in contrast to SDS electrophoresis in homogeneous gels, gradient gel electrophoresis in the presence of SDS offers the possiblility to determine the molecular mass of unreduced, 2-mercaptoethanol reduced proteins and glycoproteins to an accuracy of ± 5.5% (Table 4.4). According to his report, the lowest deviations from literature values were obtained with a crosslinker concentration of 8.4% relative to the total concentration of acrylamide plus BIS per 100 ml of solution. With SDS gradient gel electrophoresis it was also possible to estimate the molecular size of papain and pepsin which bind only traces of SDS [1, 4] with an accuracy of 4 and 11% respectively [8]. In homogeneous SDS-containing PAA gels, ribonuclease and lyzozyme, although binding normal amounts of SDS, migrate anomalously compared to marker proteins [6]. This was not the case when using SDS gradient gels [8]. It has to be mentioned, however, that at shorter times of electrophoresis (t < 1 h at 150 V constant) and with proteins having mol masses below 10 000 or above 300 000, $\log M_r$ is no longer linearly correlated to $\log T$. Rather a sigmoidal correlation between both quantities is found [21].

When using linear polyacrylamide gradient gels in SDS electrophoresis it is possible to estimate the size of proteins before they reach their respective pore limit. A linear correlation is given between the logarithm of the molecular mass (M_r) of proteins and the square root of their respective migration distance (D mm): $\log M_r = -a \sqrt{D} + b$ ("a" = slope, "b" = intercept of the regression line). When using linear gradient gels with a constant ratio of acrylamide to BIS and proteins in the range of 15 000 to 1.000 000 [g/mol] the following parameters do not affect the linearity of this relationship: (a) the buffer system, (b) the concentration of the crosslinker within 1 to 8%

Log M$_r$ = a log T + b

Homogeneous versus gradient gels

Papain, pepsin, ribonuclease, lysozyme

Log M$_r$ = − a √D + b

Table 4.3. Gel and buffer systems used to separate SDS denatured proteins by gradient gel electrophoresis

Gradient range (acrylamide: BIS); [Ref.]	Gel buffer	Electrode buffer
3 – 30 % (25.3:1) [8, 17][a]	10.75 g Tris, 5.04 g boric acid, 0.93 g EDTA-Na$_2$, pH 7.2	10 mmol l^{-1} sodium phosphate, 1 % SDS, 1 % 2-mercoptoethanol, pH 7.2
15 – 40 % (12.6:1) [18][b]	100 mmol l^{-1} sodium phosphate, pH 7.2, 0.1 % SDS, or: 350 mmol l^{-1} Tris sulphate, pH 8.5, 0.1 % SDS, or: 50 mmol l^{-1} Tris-glycine, pH 8.4, 0.1 % SDS, or: 65 mmol l^{-1} Tris borate, pH 9.3, 0.1 % SDS	29 g glycine plus Tris to pH 8.4, 1 g SDS, H$_2$O to 1000 ml
3 – 30 % (28:1) [19][c]	40 mmol l^{-1} Tris, 20 mmol l^{-1} acetate-Na, 20 mmol l^{-1} EDTA, pH 7.4, 0.2 % SDS	same as gel buffer

[a] Gel dimensions: slab gels of a length of 80 mm; conditions of electrophoresis: 40 V, 16 h; plot to estimate M_r: log M_r vs log T (with T: g acrylamide + g BIS to 100 ml), the calibration line is linear over a molecular mass range of 13 000 to 950 000; preparation of proteins: 0.5 mg protein (ml)$^{-1}$ are incubated in 10 mmol l^{-1} phosphate buffer, pH 7.2 with 1 % SDS for a period of 3 min at 100 °C in a water bath; 1 % 2-mercaptoethanol are added to cleave disulphide bridges.

[b] Gel dimensions: micro columns with a length of 15 mm and an internal diameter of 0.43 mm; conditions of electrophoresis: 60 V, 2 h; plot to estimate M_r: log M_r vs R_f, the calibration line is linear over a molecular mass range of 13 000 to 300 000; preparation of proteins: 1 mg protein (ml)$^{-1}$ is incubated in 35 mmol l^{-1} Tris-sulphate, pH 8.6, or 350 mmol l^{-1} Tris-sulphate, pH 8.6, or 100 mmol l^{-1} phosphate pH 7.2 containing 1 % SDS and 1 % 2-mercaptoethanol.

[c] Gel dimensions: slab gels of a length of 80 mm; conditions of electrophoresis: 150 V, 0.5 – 8 h; plot to estimate M_r: log M_r vs $\sqrt{D}$, with D being the migration distance (mm); the calibration line is linear over a molecular mass range of 13 000 to 950 000; re-evaluation of the data pulished by Lambin [8], Lasky [20] and Poduslo and Rodbard [21] confirmed the validity of the linear relationship between log molecular mass and $\sqrt{D}$.

C and (c) the concentration range of the gradient within 3 to 30 % at the commonly used gel length of 8 – 15 cm (Fig. 4.1). The constants "a" and "b", on the other hand, are changed when the parameters (a) to (c) are altered [19].

It has to be mentioned that, with discontinuous buffer systems, partial deloading of SDS-protein complexes may occur, leading to a confusing multitude of bands [7, 18].

With respect to enzyme visualization techniques it is of interest that SDS complexed proteins can be completely freed from the detergent by using SDS-free electrode buffers. The activity of the enzyme β-D-galactosidase (3.2.1.23) denatured with SDS and separated on an SDS-free gradient gel could be restored to around 10 % [18].

Mol mass range of gradient gels

Table 4.4. Molecular mass (M_r) of SDS-denatured proteins as reported in the literature (Lit. value) and per cent deviation of calculated values ($\pm$ % d)[a]

No	Protein Lit. value M_r	D(mm)	$\pm$ % d	T(%)[b]	$\pm$ % d
01	Prealbumin 13745	51.5	− 0.4	20.7	− 4.0
02	Lysozyme 14314	53.5	− 16.5	21.4	− 20.4
03	Ribonuclease B 14700	52	− 10.0	20.9	− 14.0
04	Haemoglobin 15500	51	− 8.6	20.5	− 11.2
05	Avidin 16000	49.5	− 1.7	20	− 4.2
06	Soybean trypsin inhibitor 20095	47	− 6.6	19.2	− 9.2
07	Papain 23426	44.5	− 4.0	18.3	− 4.8
08	α-chain of IgG 23500	–	–	–	–
09	Chymotypsinogen A 25666	43.5	− 5.6	18	− 7.0
10	Carbonic anhydrase B 28739	41	+ 1.8	17.1	+ 2.2
11	Carboxypeptidase A 34409	40	− 8.1	16.7	− 6.3
12	Pepsin 34700	37.5	+ 11.0	15.9	+ 12.5
13	Glycerol-3-phosphate dehydrogenase 35700	37.5	+ 7.9	15.9	+ 9.4
14	Lactate dehydrogenase 36180	37.5	+ 6.4	15.9	+ 7.9
15	Aldolase 38994	36	+ 11.5	15.4	+ 13.1
16	Alcohol dehydrogenase 39805	35.5	+ 13.8	15.2	+ 16.4
17	α_1-Acid glycoprotein 40000	35	+ 18.0	15	+ 21.7
18	Ovalbumin 43000	35.5	+ 5.4	15.2	+ 7.8
19	Fibrinogen chain 47000	–	–	–	–
20	Glutamate oxalacetate transaminase 50000	32.5	+16.7	14.2	+ 19.2
21	Heavy chain IgG 50000	–	–	–	–
22	Fibrinogen β chain 56000	–	–	–	–
23	Catalase 57500	32	+ 5.9	14	+ 9.1

Table 4.4 (continued)

No	Protein Lit. value M_r	D(mm)	± % d	T(%)[b]	± % d
24	Fibrinogen α chain 63500	–	–	–	–
25	Albumin, monomer 66290	31.5	+ 10.7	13.8	+ 14.9
26	Heavy chain IgM 72000	–	–	–	–
27	Transferrin 76000	30.5	– 8.6	13.5	– 6.0
28	Plasminogen 81000	29	– 1.8	13	+ 0.7
29	Phosphorylase b 96800	25.5	+ 14.2	11.8	+ 17.1
30	Cerulosplasmin 124000	23.5	+ 8.8	11.1	+ 11.6
31	Albumin, dimer 132580	24	– 3.3	11.2	+ 1.4
32	Immunoglobulin G 150000	21.5	+ 10.6	10.4	+ 13.4
33	Immunoglobulin A 160000	22	+ 1.6	10.6	+ 0.1
34	Reduced α_2-macroglobulin 190000	–	–	–	–
35	Albumin, trimer 198870	19.75	+ 0.8	9.8	+ 2.6
36	Immunoglobulin A 320000	16	– 3.1	8.5	– 3.5
37	Thyroglobulin 330000	16.5	– 11.6	8.7	– 12.4
38	Fibrinogen 340000	15	+ 3.3	8.2	+ 0.4
39	α_2-Macroglobulin 380000	15.5	– 13.2	8.3	– 13.1
40	Immunoglobulin A, trimer 480000	13	– 4.8	7.5	– 9.5
41	Immunoglobulin M 950000	8.75	– 9.1	6	– 20.1
	Average % deviation:		± 7.8		± 9.6
	Lambin [8]:		–		± 5.9

[a] Literature M_r, migration distances D (mm) of proteins and %T-values as given by Lambin [8]. Calculation of M_r-values was performed by applying the relationship $\log M_r = a \sqrt{D} + b$, with "a" representing the slope of the linear regression line and "b" its intercept; ± % d: deviation of calculated M_r from those given in the literature. The ± %T-scale was constructed with aid of the equation: %T = 0.3425 · D + 3.

A linear correlation also results if a linear 3 – 30 % T gradient is used but with 3.8 % of BIS (instead of 8.4 % BIS) [19].

[b] Electrophoresis was performed without 2-mercaptoethanol.

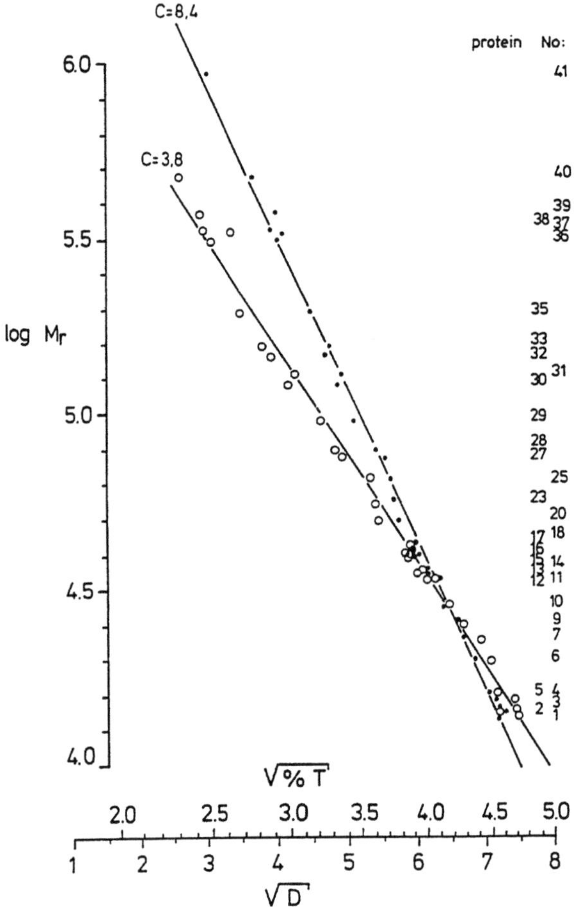

Fig. 4.1. Re-evalution of numerical data as published by Lambin [8]. Migration distances of denatured proteins and protein subunits obtained in SDS-PAGGE were used to set up the log M_r – $\sqrt{D}$ plot. Deviations of the calculated M_r-values from those in the literature are summarized in Table 4.4. Figure taken from [19] by permission of the publisher

4.4 Enzyme Visualization Following SDS Electrophoresis

It is generally considered that only native enzymes can be visualized following electrophoretic separations. Several reports, however, have shown that SDS-denatured proteins can also be visualized provided it is possible to renature their spatial configuration. Nucleases and several other enzymes were included among these [22].

Factors which effect the renaturability of SDS-denatured proteins are essentially the same for water soluble and hydrophobic membrane proteins. Renaturation will be most effective if no disulphide bonds are essential for enzymatic activity and if the native protein is not composed of subunits of different sizes [23]. The renaturation process of many enzymes is favoured by high concentrations of sodium chloride, glycerol, substrates or dithiothreitol [24]. Sulphhydryl agents, however, should be kept at a minimum when using staining solutions which contain phenazine methosulphate (PMS) and a tetrazolium salt, as these compounds are non-enzymatically reduced by free SH-groups [25]. Any denaturing actions of ammonium persulphate or feasible polymerisation products of the gelling procedure are avoided during electrophoresis if the gels are pre-electrophoresed [26]. Cofactors which are essential for activity and which might have been removed during electrophoresis such as Zn^{2+} in alcohol dehydrogenase (1.1.1.1) [23] must be included in the renaturation medium. The degradation of the proteins of interest by proteolytic enzymes occuring in the biological fluid to be investigated, can be avoided by the addition of protease inhibitors like phenylmethanesulphonyl fluoride (PMSF) [27]. Besides, recovery of biological activity may effectively be increased by casting substrates and/or bovine serum albumin into PAA gels prior to electrophoresis [23]. Commercially available SDS preparations differ significantly in their concentration of smaller and longer chain length alkyl sulphate impurities [28 – 30]. The presence of any of these by-products markedly influences the resolution [31], the banding pattern [32] and the renaturability of denatured proteins [22, 33]. It could be demonstrated for RNase (3.1.27.5) that its visualization was prevented by 1-dodecane didodecyl ether (DDE), didodecylsulphate (DDS) and other long-chain hydrocarbons at concentrations of 0.1 wt/vol.% [30].

Upon electrophoresis the SDS present in the sample is replaced by the SDS molecules present in the upper electrophoretic buffer reservoir. Therefore it is the purity of the SDS preparation used in the electrophoretic buffer system which determines the renaturation of a protein. To improve the yield in restored enzyme activities it was also suggested that electrophoresis be carried out in buffer systems containing lower concentrations of contaminating longer chain alkyl sulphates. Similar recoveries were obtained for several nucleolytic enzymes whether, for example, the SDS preparation DX 2490 containing 31 % tetradecylsulphate was used at a 0.05 % concentration or whether the pure SDS preparation from BDH (BDH, Poole, Dorset, England) was used, applying a concentration of 0.2 % [22].

4.4.1 Methods for Renaturing Enzyme Proteins

The key step in a renaturing process is the elimination of SDS in bound and unbound form from the protein of interest. Reactivation, especially of water soluble proteins is best if the denaturant is removed completely. In samples containing hydrophobic proteins, SDS must be replaced by a less denaturing agent such as a non-ionic or zwitter-ionic detergent to keep proteins solubilized. Traces of residual SDS may interfere with the chemical or immunological assay system [22, 34].

The removal of SDS must proceed slowly so that the protein molecules have sufficient time to refold into their native conformation [35]. In order to recover maximum activity several nucleolytic enzymes [22], glucosyltransferases and fructo-

Phosphatases

syltransferase (2.4.1.9) [36] necessitated renaturing times exceeding 12 h. However, activity staining of the membrane enzymes phosphodiesterase (3.1.4.1) and alkaline phosphatase (3.1.3.1) immediately after SDS electrophoresis could be completed within half an hour, probably due to their resistance to SDS [37].

Renaturation of functional proteins following SDS electrophoresis is practicable in two ways:

(a) renaturation is performed after eluting the protein from the gels, or
(b) the protein is renatured directly on the PAA gel.

Three different methods have been described to renature proteins in electrophoretic support media:

(a) washing out of SDS gels with incubation buffer (and positive effectors),
(b) washing of the gels with 25 % isopropanol containing incubation buffer, and
(c) isoelectric focusing of SDS-denatured proteins in the presence of urea and (or) Triton X-100 followed by washing out detergents with incubation buffer (and positive effectors).

4.4.1.1 Renaturation by Washing Out SDS with Buffer

DNases, RNases

DNases (Deoxyribonuclease I, 3.1.21.1) and RNases (Endoribonucleases, 3.1.26. (27.)*) in particular may be renatured simply by replacing the SDS buffer in a PAA gel by a buffer in which the enzyme is active. Visualization may then be performed as usual. DNase I from bovine pancreas and nucleases from *Micrococcus* could be successfully renatured and visualized after electrophoresis in PAA gels containing a SDS-buffer

Protease, amylase, DNA-polymerase

system free of tetradecyl- and hexadecylsulphate [22, 27, 38]. By the same technique, protease, amylase [23], alkaline phosphodiesterase, alkaline phosphatase [37], and DNA-polymerase [39] could also be visualized in situ.

In some cases the sensitivity of enzyme visualization methods could be increased by inclusion of bovine serum albumin (10 g/ml) into a PAA gel prior to polymeriza-

Chymotrypsin, pronase, trypsin, subtilisin, thermolysin

tion. By doing so, lower amounts of nuclease from *Staphylococcus* strains could be visualized while the detection of DNase I remained uninfluenced [28]. Many proteases including trypsin consist of only a single peptide chain. It is not possible to renature them if their proteinogenous disulphide bonds are in a reduced state [23]. Chymotrypsin, pronase, trypsin and subtilisin, for example e.g., could only be visualized after SDS electrophoresis when keeping their disulphide bridges undestroyed [23].

Amylase

On the other hand thermolysin which is free of disulphide bonds [40] could be visualized in situ no matter whether the denaturing medium contained 2-mercaptoethanol or not [23]. α-Amylase from *Bacillus subtilis* contains no cysteine residues [41] while the porcine enzyme contains four disulphide bridges [42]. But both types of enzymes

Lactate dehydrogenase, alcohol dehydrogenase, glucose-6-phosphat dehydrogenase

could be renatured to approximately 20 % after a treatment with 2-mercaptoethanol followed by SDS electrophoresis [23]. Bovine lactate dehydrogenase, porcine liver alcohol dehydrogenase and glucose-6-phosphate dehydrogenase from *Leuconostoc mesenteroides* are composed of different subunits and none of these enzymes contain disulphide bonds [43, 44]. These enzymes could be renatured to a rate of 1 % using individual protein concentrations of up to 5 µg [23].

4.4.1.2 Renaturation by Washing Out SDS with Buffers Containing Isopropanol

If SDS-denatured proteins are solely renatured in situ with buffer, high amounts of enzyme protein are needed. Significantly lower enzyme concentrations can be visualized if SDS is washed out from the gels with an incubation buffer containing 25% isopropanol, followed by a wash with isopropanol-free buffer. The inclusion of isopropanol into the renaturation medium leads to the removal of impurities present in commercial SDS preparations [45, 28, 29]. By using a buffered 25% isopropanol solution for the renaturation process, concentrations of 1 to 5 pg RNase could be detected [28]. The fact that a high molecular weight RNA was polymerized into the PAA gels prior to electrophoresis had no influence on the separation process. In contrast to the native enzyme, SDS-complexed RNase does not bind to its high molecular weight substrate RNA [27]. The sensitivity of the detection of DNA polymerases and nucleases can be increased if an autoradiographical detection method is used instead of staining unhydrolyzed DNA with ethidium bromide [38, 39].

RNase

4.4.1.3 Renaturation by Isoelectric Focusing in the Presence of Urea

Many allosteric enzymes are composed of different subunits [46 – 52]. These enzymes dissociate under the influence of urea, guanidine-HCl or SDS losing their catalytic activity. With the exception of SDS, however, activity is often regained after removal of the detergent. Many urea-denatured enzymes like alkaline phosphatase from the stroma of erythrocytes regain activity when freed from urea [53]. Likewise NADH oxidase and ATPase from membranes of *E. coli* B are to be reactivated after urea treatment [54].

Alkaline phosphatase

NADH oxidase, ATPase

When submitted to SDS electrophoresis the enzyme UDPglucose pyrophosphorylase lost its catalytic activity. Reactivation was possible, however, when the enzyme, after its separation in an SDS-containing gel, was submitted in the second dimension to isoelectric focusing in the presence of urea and the urea finally dialyzed out of the gel [24]. The following enzymes could also be reactivated when submitted to SDS electrophoresis in the first dimension and isoelectric focusing in the presence of urea and Triton X-100 in the second dimension, followed by a wash out of the detergent with an effector containing buffer medium: UDPglucose pyrophosphorylase (2.7.7.9, an octameric enzyme with subunits of a $M_r = 55\,000$), alcohol dehydrogenase (1.1.1.1, a dimeric enzyme with subunits of a $M_r = 22\,000$), and lactate dehydrogenase (1.1.1.27, a tetrameric enzyme having subunits of a $M_r = 35\,000$) [24]. To reactivate UDPglucose pyrophosphorylase the IEF gels were washed in 10 ml of 25 mmol l^{-1} tricine buffer, pH 7.5 containing 10% glycerol, 200 mmol l^{-1} NaCl, and 1 mmol l^{-1} uridine. The gels were washed for a period of four hours changing the renaturation solution every hour. To reactivate alcohol dehydrogenase the IEF gels were washed with 10 ml of a 50 mmol l^{-1} phosphate buffer of pH 7.5 containing 10% glycerol, 200 mmol l^{-1} NaCl, 1 mmol l^{-1} β-NADP$^+$ and 1 mmol l^{-1} dithiothreitol. The washing solution was renewed twice within a period of 2 h. Afterwards the gels were washed twice for 1 h using the same solution but omitting dithiothreitol. Then they were stained using the formazan method. Lactate dehydrogenase was renatured like alcohol dehydrogenase but replacing the phosphate buffer by a 75 mmol l^{-1} Tris-HCl buffer of pH 8 [24].

UDP glucose pyrophosphorylase

Alcohol dehydrogenase, lactate dehydrogenase

Reactivation buffer

To sum up, urea denatured dehydrogenases can be renatured by substituting urea successively with high salt concentrations, substrate and a sulphhydryl reagent. Reassociation to the native structure of heteromeric proteins is clearly impossible as the polypeptide chains are at different locations on a PAA gel after SDS electrophoresis or IEF. Besides the renaturation in situ, several methods have been described to restore the catalytic activity of enzymes after elution of the proteins from the media in which they were separated [55].

4.5 References

1. Pitt-Rivers R, Impiombato FSA (1968) Biochem J 109: 825 – 830
2. Reynolds JA, Tanford C (1970) Proc. Natl Acad Sci USA 66: 1002 – 1007
3. Reynolds JA (1970) J Biol Chem 245: 5161 – 5165
4. Nelson CA (1971) J Biol Chem 246: 3895 – 3901
5. Segrest JP, Jackson RL, Andrews EP, Marchesi VT (1971) Biochem Biophys Res Comm 44: 390 – 395
6. Shapiro AL, Vinuela E, Maizel JV (1967) Biochem Biophys Res Comm 28: 815 – 820
7. Wachneldt TV (1971) Anal Biochem 43: 306 – 312
8. Lambin P (1978) Anal Biochem 85: 114 – 125
9. Maizel JV (1971) Methods in Virology 5: 179 – 246
10. Weber K, Osborn M (1969) J Biol Chem 244: 4406 – 4412
11. Dunker AK, Rueckert RR (1969) J Biol Chem 244: 5074 – 5080
12. Laemmli UK (1970) Nature 227: 680 – 685
13. Grefrath SP, Reynolds JA (1974) Proc Natl Acad Sci USA 71: 3913 – 3916
14. Barker GA, Cotman CW (1972) J Biol Chem 247: 5856 – 5861
15. Anderson BL, Berry RW, Telser A (1983) Anal Biochem 32: 365 – 375
16. King J, Laemmli U (1971) J Mol Biol 62: 465 – 477
17. Lambin P, Rochu D, Fine JM (1976) Anal Biochem 74: 567 – 575
18. Rüchel R, Mesecke S, Wolfrum DI, Neuhoff, V (1974) Hoppe-Seyler's Z Physiol. Chem 355: 997 – 1020
19. Rothe GM (1982) Electrophoresis 3: 255 – 262
20. Lasky M (1978) Protein molecular weight determination using polyacrylamide gradient gels in the presence and absence of sodium dodecyl sulfate. In: Catsimpoolas N (ed), Electrophoresis '78, North Holland, Amsterdam, pp 195 – 210
21. Poduslo JF, Rodbard D (1980) Anal Biochem 101: 394 – 406
22. Lacks SA, Springhorn SS, Rosenthal AL (1979) Anal Biochem 100: 357 – 363
23. Lacks SA, Springhorn SS (1980) J Biol Chem 255: 746 – 773
24. Manrow RE, Dottin RP (1980) Proc Natl Acad Sci 77: 730 – 734
25. Dottin RP, Manrow RE, Fishel BR, Ankermann SL, Culleton, I.L (1979) Localization of enzymes in denaturing polyacrylamide gels. In: Wu R (ed) Methods in enzymology vol 68. Academic Press, New York London Toronto Sydney San Francisco, pp 513 – 529
26. Weber K, Kuter DD (1971) J Biol Chem 246: 4504 – 4509
27. Rosenthal AL, Lacks SA (1977) Anal Biochem 80: 76 – 90
28. Blank A, Sugiyama RH, Dekker CA (1982) Anal Biochem 120: 267 – 275
29. Blank A, Silber JR, Thelen MP, Dekker CA (1983) Anal Biochem 135: 423 – 430
30. Thelen MP, Blank A, McKeon TA, Dekker CA (1982) Fed Proc 41: 1203
31. Matheka HD, Enzmann PJ, Bachrach HL, Migel B (1977) Anal Biochem 81: 9 – 17
32. Margulies MM, Tiffany HL (1984) Anal Biochem 136: 309 – 313
33. Dohnal JC, Garvin IE (1979) Biochim Biophys Acta 576: 393 – 403
34. Zaman Z, Verwilghen RL (1979) Anal Biochem 100: 64 – 69
35. Hager DA, Burgess RR (1980) Anal Biochem 109: 76 – 86
36. Russell RRB (1979) Anal Biochem 97: 173 – 175

37. Dulaney JT, Touster O (1970) Biochim Biophys Acta 196: 29 – 34
38. Huet J, Sentenac A, Fromageot P (1978) FEBS Letters 94: 28 – 32
39. Spanos A, Sedgwick SG, Yarranton GT, Hübscher U, Banks GR (1981) Nucleic Acids Res 9: 5919 – 5925
40. Ohta Y, Oguva Y, Wada A (1966) J Biol Chem 241: 5919 – 5925
41. Takagi T, Toda H, Isemura T (1971) Bacterial and mold amylases. In: Boyer PD (ed) The Enzymes 3rd ed. Vol 5, Academic Press, New York, pp 235 – 271
42. Thoma JA, Spradlin JE, Dyget S (1971) Plant and animal amylases. In: Boyer PD (ed) The Enzymes 3rd ed. Vol 5, Academic Press, New York, pp 115 – 189
43. Olive C, Levy HR (1971) J Biol Chem 246: 2043-2046
44. Appella E, Markert CL (1961) Biochem Biophys Res Commun 6: 171 – 176
45. Blank A, Dekker CA (1982) Biochem 20: 2261 – 2267
46. Anfinsen CB (1962) Brookhaven Symp Biol 15: 184 – 198
47. Martin CJ (1964) Biochemistry 3: 1635 – 1643
48. Westhead EW (1964) Biochemistry 3: 1062 – 1068
49. Kaufmann BT (1963) Biochem Biophys Res Commun 10: 449 – 453
50. Kaufmann BT (1968) J Biol Chem 243: 6001 – 6008
51. Perkins JP, Bertino JR (1965) Biochemistry 4: 847 – 853
52. Stadtman ER (1960) Advan Enzymol 28: 41 – 154
53. Schneidermann LJ (1965) Biochem Biophys Res Commun 20: 763 – 767
54. Weinbaum G, Markman R (1966) Biochim Biophys Acta 124: 207 – 209
55. Rothe GM, Maurer WD (1986). One-dimensional PAA-gel electrophoretic techniques to separate functional and denatured proteins. In: Dunn MJ (ed) Electrophoresis of Proteins . Wright, Bristol, pp 37 – 140

5 Chemistry of Enzyme Visualization

The basic principle of enzyme visualization in situ is to present an enzyme with a solution containing an enzyme specific substrate. Demonstration of an enzyme is achieved if the catalytic action of the enzyme on this substrate produces a coloured reaction product. Often, however, the primary reaction products are colourless and require coupling with a visualizing agent to generate a coloured, preferably insoluble, final reaction product.

The demonstration of many oxidative enzymes uses tetrazolium salts. These salts, which are colourless in diluted solutions and water-soluble, accept hydrogen released from a substrate by enzyme action, and on reduction form highly-coloured water-insoluble microcrystalline deposits known as formazans.

Visualization of hydrolytic enzymes is mostly performed by use of diazonium salts. These will react with enzymatically-released naphthol from corresponding esters or glycosides to produce an intensely-coloured insoluble azo dye. The rate of coupling of diazonium salts to naphthol derivatives depends upon their chemical nature and the pH of the incubating medium. These factors determine to a large extent the choice of diazonium salts. There is an optimal concentration in the incubating solution; if the concentration is less, diffusion of the primary reaction product may occur. If the concentration is greater than optimal, then inhibition of enzyme activity occurs, and the risk of non-specific staining increases.

Diazonium salts are either used in simultaneous or post-coupling methods. Simultaneous coupling means the simultaneous presence of substrate and diazonium salt in a suitable buffer. The enzyme enclosed in the electrophoretic separation medium hydrolyzes the substrate to form an invisible primary reaction product which immediately couples with the diazonium salt to produce a coloured final reaction product. The method is performed at the pH at which the enzyme exhibits maximum activity. In the post-coupling method the colourless primary reaction product is coupled to the diazonium salt in a separate solution and at a different pH-value. This method avoids any deleterious effects that diazonium salts may have on the enzyme-substrate reaction, and takes advantage of the different enzyme activity and coupling pH-optima.

5.1 Methods for Visualizing Oxidative Enzymes

Oxidoreductases

Oxidoreductases catalyze redox reactions according to the general equation: $AH_2 + B = A + BH_2$, with AH_2 being the substrate and B the acceptor molecule for hydrogen. B can be a co-substrate, O_2 or an artificial indicator molecule. Oxidoreductases which transfer hydrogen (electrons) to oxygen are commonly called "oxidases", while those which reduce a pyridine nucleotide coenzyme are named "dehydrogenases"[1–3].

Oxidases and dehydrogenases

5.1.1 The Assay of Dehydrogenases

When dehydrogenases are to be visualized, two different methods are feasible: (a) observation of the change in fluorescence of the pyridine nucleotides NAD(P) or $NAD(P)H_2$ participating in the enzymic reaction or (b) visualization of the oxidation reaction by including an electron acceptor into the incubating medium which turns coloured when reduced.

5.1.1.1 Direct Visualization of Pyridine Nucleotides

$NAD(P)H_2$ fluorescence

The decrease in fluorescence of $NAD(P)H_2$ indicating the action of dehydrogenases can be observed under long-wave ultra-violet light. Enzymes which have been visualized by this method are: alcohol dehydrogenase (1.1.1.1), glutamate dehydrogenase (1.4.1.3), hydroxyl coenzyme A dehydrogenase (1.1.1.35), lactate dehydrogenase (1.1.1.27), glutathione peroxidase (1.11.1.9) and glyceraldehyde-3-phosphate dehydrogenase (1.2.1.12). The decrease in $NAD(P)H_2$-fluorescence is also applied when transferases, lyases, isomerases or ligases are to be localized in a coupled enzyme reaction with a dehydrogenase. The high solubility of NAD(P) and $NAD(P)H_2$ may be disadvantageous when using large-porous electrophoretic support media. Therefore, if possible, a redox dye of the tetrazolium salt type is used to localize dehydrogenases in the oxidation reaction.

Transferases, lyases, isomerases, ligases

5.1.1.2 The Tetrazolium Salt Method

Two types of tetrazolium salts have been developed [4]: monotetrazolium salts and ditetrazolium salts (Fig. 5.1). Monotetrazolium salts accepting two electrons are preferred as compared to ditetrazolium salts demanding four electrons for complete reduction.

In histochemical assays it is sufficient to incubate a tissue together with an enzyme specific substrate and a tetrazolium salt to obtain a colour formation. However, if the same enzyme reaction is to be visualized in vitro, a hydrogen transferring substance must also be included into the incubating medium to substitute for the cell inherent cytochrome system [5]. As electron transfer substance the molecule phenazine methosulfate (N-methyldibenzopyrazine methyl sulphate) (PMS) is mostly used [6]. PMS is capable of accepting the hydrogen from $NAD(P)H_2$ and passing it over to the final redox molecule, the tetrazolium salt (Fig. 5.2). However, PMS has two disadvantages; (a) it is subject to autoxidation and (b) it is extremely sensitive to light

PMS

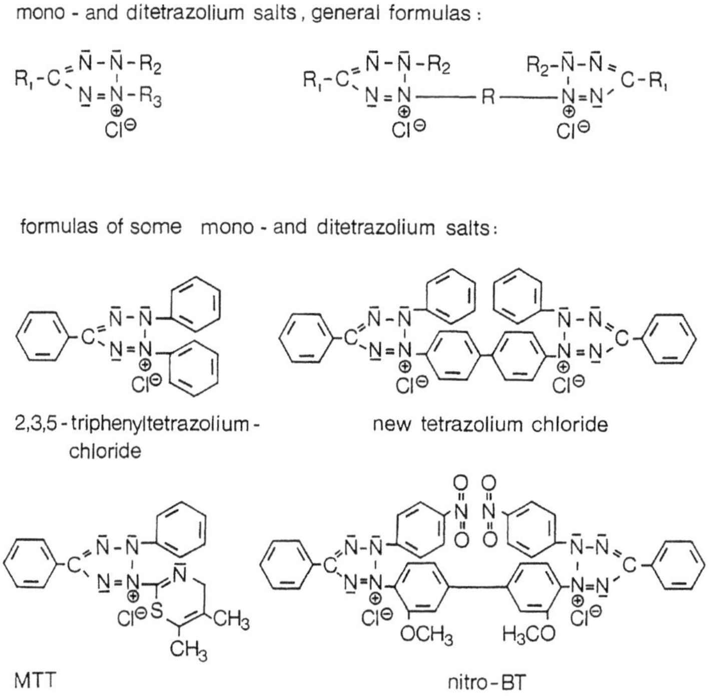

mono - and ditetrazolium salts, general formulas :

formulas of some   mono - and ditetrazolium salts:

2,3,5 - triphenyltetrazolium - chloride

new tetrazolium chloride

MTT

nitro - BT

Fig. 5.1. Formulas of some mono- and ditetrazolium salts

(for which reason it is used for the assay of superoxide dismutase (1.15.1.1)). Meldola's blue (8-dimethylamino-2,3-benzophenoxazine), which is light insensitive, can also be used as electron carrier [7, 8]. Meldola's blue has been proved advantageous when visualizing glucose-6-phosphate isomerase (5.3.1.9) in starch gels [9]. But PMS is usually preferred when staining dehydrogenases in electrophoretic support media (Table 5.1), and the interference of light can easily be avoided by incubating the support media in a moist chamber in the dark. Mg^{2+}-ion concentrations exceeding 10 mmol l^{-1} inhibit the electron transferring capability of PMS [10]. Therefore, where Mg^{2+} is needed to activate a dehydrogenase or an auxilliary enzyme, e.g., in the test system to detect galactosyltransferase from milk, PMS must be replaced by the enzyme diaphorase. Tetrazolium salts can also be reduced non-enzymatically. The most reactive tetrazolium salts are readily reduced by cysteine, reduced glutathione, 2-mercaptoethanol, and other compounds with free SH-groups. A spontaneous reduction takes place in alkaline solutions. Sulphhydryl reagents are often added to enzyme extraction media in order to protect free SH-groups of the enzymes from oxidation by oxygen. Dithiothreitol and 2-mercaptoethanol are uncharged at pH-values below 8 which means that they do not migrate into the electrophoretic support medium at this pH-value. But in starch gel electrophoresis they travel to the cathode by

Meldola's blue

Glucose-6-phosphate isomerase

Galactosyl-transferase

non-enzymatic reduction of tatrazolium salts

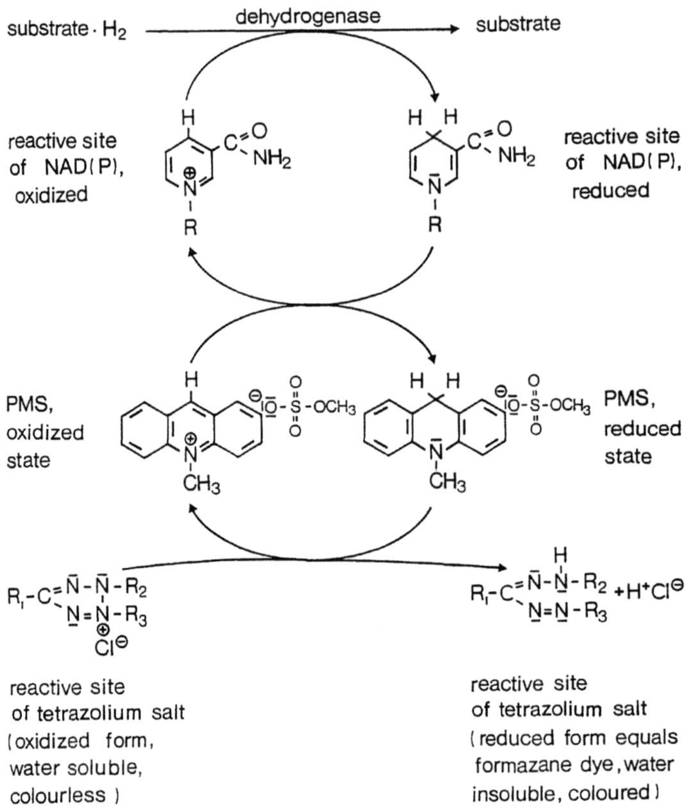

Fig. 5.2. Determination of dehydrogenases using PMS and a tetrazolium salt

electroendosmosis. Unspecific colour reactions may therefore occur at the cathodic side of the gel when sulphhydryl reagents meet a tetrazolium salt.

Dehydrogenases are enzymes which transfer two electrons. Therefore the amount of formazane produced is directly correlated to the amount of dehydrogenated substrate when mono-tetrazolium salts are used. But two substrate molecules must be dehydrogenated to reduce one molecule of a di-tetrazolium salt. This may, at the beginning of the reaction, lead to incompletely reduced di-tetrazolium molecules. The colour of these partially reduced molecules is different from that of the completely reduced di-tetrazolium molecules. Nitro-BT, for example, is red when having accepted two electrons, while the totally reduced dye is bluish black. Therefore, when assaying small amounts of dehydrogenases, the more sensitive mono-tetrazolium salts like MTT 3-(4,5-dimethyl-thiazolyl-2)-2,5- diphenyltetrazolium bromide should be used (Fig. 5.1). Reduced tetrazolium salts can only be re-oxidized by strong oxidation reagents such as lead tetraacetate, HNO_2 or the enzyme superoxide dismutase (1.15.1.1) [11].

Table 5.1. Dehydrogenases which have been visualized in electrophoretic support media by inclusion of phenazine methosulphate and a tetrazolium salt in the incubating medium. For details see Chap. 6: A Compilation of Protocols to Visualize Enzymes Following Electrophoretic Separation

Enzyme

Name	EC number	Source
Alanine dehydrogenase	1.4.1.1	Bacteria
Alcohol dehydrogenase	1.1.1.1	Mammalia, plants
Aldehyde dehydrogenase	1.2.1.3	*Pseudomonas aeroginosa*
Dihydrouracil dehydrogenase (NADP+)	1.3.1.2	Rat
D-Fructose-5-phosphate dehydrogenase	1.1.99.11	*Gluconobacter industrius*
L-Fucose dehydrogenase	1.1.1.122	Swine (liver)
Glucose dehydrogenase	1.1.1.47	Man
Glucose-6-phosphate dehydrogenase	1.1.1.49	Mammalia, plants, insects
L-Glutamate dehydrogenase (NADP+)	1.4.1.4	*Bacillus subtilus*, insects
Glyceraldehyde-3-phosphate dehydrogenase	1.2.1.12	Man, mouse, insects
Glycerol-3-phosphate dehydrogenase (NAD+)	1.1.1.8	Man (muscle)
Histidinol dehydrogenase	1.1.1.23	*Neurospora crassa*
Homoserine dehydrogenase	1.1.1.3	*E. coli*
3(-) Hydroxybutyrate dehydrogenase	1.1.1.30	Plants
3α-Hydroxysteroide dehydrogenase	1.1.1.50	*Pseudomonas testosteroni*
β-Hydroxysteroide dehydrogenase	1.1.1.51	Man (placenta)
Isocitrate dehydrogenase (NADP+)	1.1.1.42	Man, plants, insects
Lactate dehydrogenase	1.1.1.27	Man, mouse, plants
L-Leucine dehydrogenase	1.4.1.9	Bacteria
Lysine 2-monooxygenase	1.13.12.2	*Pseudomonas fluorescence*
Malate dehydrogenase	1.1.1.37	Man, plants, insects
Malate dehydrogenase (oxaloacetate-decarboxylating)	1.1.1.40	*Fusarium oxysporum*, mammalia
Mannitol dehydrogenase	1.1.1.67	*Absidia glauca*
Oestradiol 17-ß-dehydrogenase	1.1.1.62	Man
Phosphogluconate dehydrogenase (decarboxylating)	1.1.1.44	Man, animals, insects, plants, bacteria
Retinole dehydrogenase	1.1.1.105	Mammalia
Shikimate dehydrogenase	1.1.1.25	*E. coli*, plants
Sorbitol dehydrogenase (L-Iditol dehydrogenase)	1.1.1.14	Man
Succinate dehydrogenase	1.3.99.1	Mammalia
Tartrate dehydrogenase	1.1.1.93	*Pseudomonas putida*
Testosterone 17-β-dehydrogenase (NADP+)	1.1.1.64	Pig
Xanthine dehydrogenase	1.2.1.37	*Drosophila*

5.1.1.3 The Nothing Dehydrogenase Reaction

When visualizing dehydrogenases by using the tetrazolium salt method it sometimes happens that a positive reaction is observed even if no substrate is included in the incubating solution. This phenomenon was called the "nothing dehydrogenase reaction". It was found, e.g., for animal lactate dehydrogenase (Fig. 5.3) [12]. The nothing dehydrogenase reaction was not observed with purified lactate dehydrogenase (LDH), or when tissue extracts were treated with N-ethylmaleimide or jodacetamide [13-15]. The false lactate dehydrogenase reaction can also be avoided when pyruvate is included in the indicator system. The nothing dehydrogenase reaction was also observed when staining for the enzymes alcohol dehydrogenase (ADH) [16], glutamate dehydrogenase and malate dehydrogenase [17]. The inhibitor "agrazole" can be used to stop the false malate dehydrogenase reaction. In some cases the "nothing dehydrogenase" effect is due to alcohol dehydrogenase. Like ADH, "nothing dehydrogenase" is inhibited by p-hydroxymercuribenzoate. It has been established that several of the materials used, even when highly purified, contain traces of alcohol. This applies not only to reagents such as PMS but also to galactose in the galactose dehydrogenase assay and even to the starch used in starch gel electrophoresis. It has been shown that drying starch and other materials at 80 °C may eliminate the "nothing dehydrogenase" reaction [18].

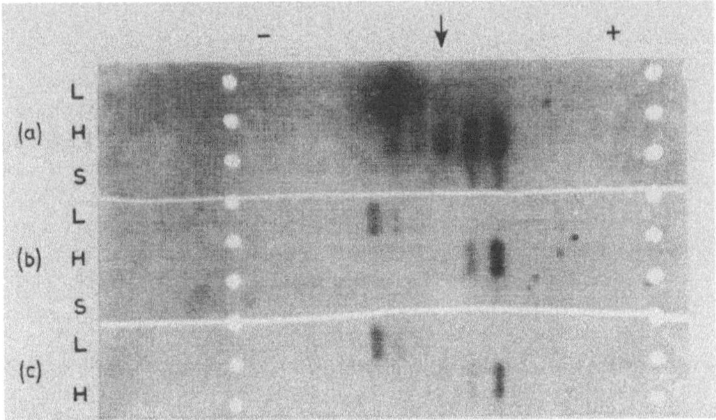

Fig. 5.3. Cellulose acetate electrophoresis of human liver (L) and heart (H) extracts and a human serum (S) stained to show the "nothing dehydrogenase" effect. The Beckmann "Microzone" cell was used and the unstained foil was cut into three parts: (a) stained with the complete tetrazolium system with lactate as substrate; (b) as (a), except hat substrate was omitted; (c) as (a) except that ethanol replaced lactate as substrate. The arrow indicates the point of application of the samples. Note the congruity of LD_1 and LD_2 of the heart extracts with the corresponding "nothing dehydrogenase" and alcohol dehydrogenase bands, and also that the liver extract exhibits an extra "nothing dehydrogenase" band migrating further towards the cathode than LD_5. This band is coincident with the most cationic alcohol dehydrogenase component. Taken from [18] by permission of the author and the publisher

5.1.1.4 The Use of DCPIP in Combination with Tetrazolium Salts

In some cases PMS has to be replaced by 2,6-dichlorophenol-indophenol (DCPIP) when visualizing dehydrogenases via the tetrazolium salt method. DCPIP cannot be reduced by NAD(P)H$_2$ and therefore it is used as electron carrier where NAD(P)H$_2$ is needed as cosubstrate and MTT is intended to be used as redox dye. DCPIP is less sensitive than PMS but it may be reduced by sulphhydryl reagents such as reduced glutathione. Dehydrogenases which have been visualized by use of DCPIP or by DCPIP and a tetrazolium salt are: amine dehydrogenase (1.4.99.3; bacteria, plants, mammalia), dihydrolipoamide reductase (NAD$^+$) (1.6.4.3; bacteria, plants, mammalia), glutathione reductase (1.6.4.2; man), diaphorase (1.8.1.4; plants) and NADH dehydrogenase (1.6.99.3; man) (Chap. 6). The electron flow when using DCPIP and MTT to visualize a NADPH$_2$-dependent dehydrogenase is demonstrated for the enzyme glutathione reductase (1.6.4.2) in Fig. 5.4 [19].

Glutathione reductase

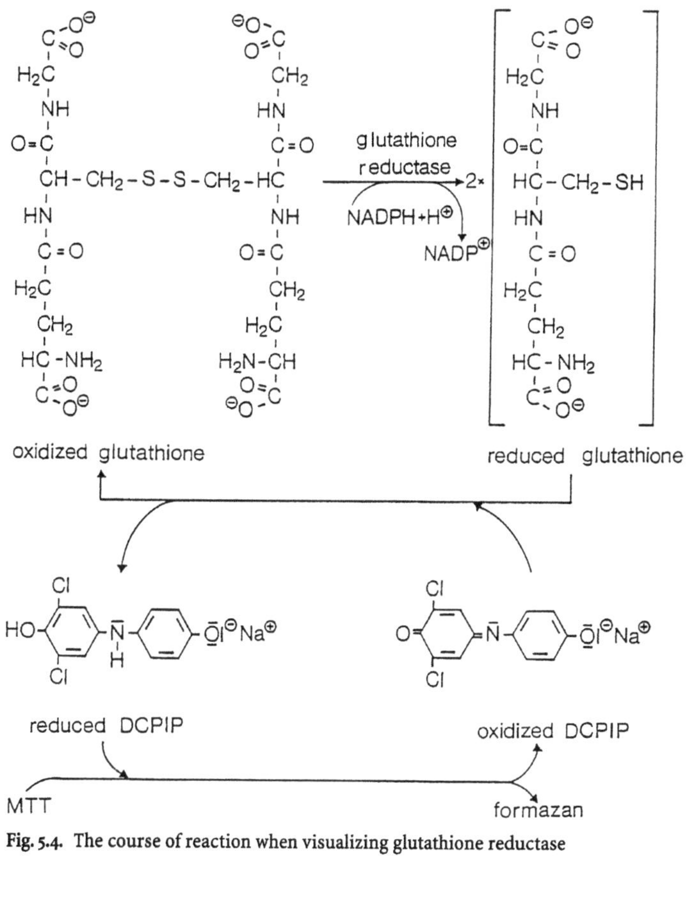

Fig. 5.4. The course of reaction when visualizing glutathione reductase

5.1.2 The Assay of Oxidases

5.1.2.1 Peroxidases

Peroxidases are wide-spread, especially in plants, and appear in a large number of isozymes [21–24].

In the presence of H_2O_2 they catalyze the oxidation of phenols (AH_2) and aromatic amines (AH_2) according to the equation:

$$\text{enzyme-}H_2O_2 + AH_2 = \text{enzyme} + A + 2\,H_2O.$$

The determination of peroxidases and oxidases are often performed with benzidine or o-dianisidine, but as these compounds have proved to be carcinogenic they should no longer be used. The polyphenol pyrogallol together with H_2O_2 can be used as substrate for peroxidases. By enzymatic oxidation and a following non-enzymic reaction step, the red-coloured purpurogallin is formed. The compound 3-amino-9-ethyl-carbazol can also be used as electron donor, but it too should be handled with care [19]. After its oxidation a brown dye is obtained (Fig. 5.5) which, in contrast to purpurogallin is only slightly water-soluble [22]. 3,3',5,5'-Tetramethylbenzidine has been recommended as substrate for peroxidases since it has been shown to be non-carcinogenic in animal tests [25]. Visualization of plant peroxidase isozymes in gels of PAA is strongly influenced by the concentration of H_2O_2 used in the staining with benzidine dihydrochloride as an electron donor. The more slowly migrating isozymes of *Phaseolus aureus*, for example, stain more intensely at higher H_2O_2 concentrations (0.0075%) than the faster migrating bands for which the optimum H_2O_2 concentration is 0.0015%.

3,3',5,5'-tetra-methylbenzidine

Carbazol method

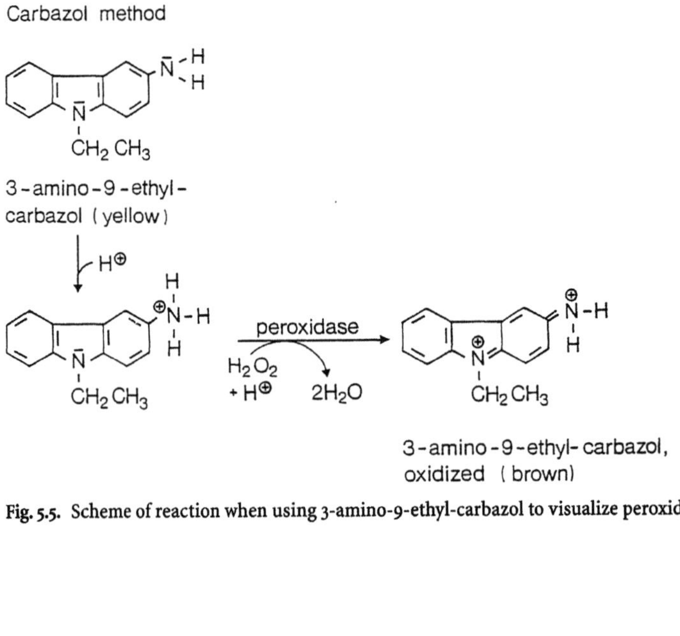

Fig. 5.5. Scheme of reaction when using 3-amino-9-ethyl-carbazol to visualize peroxidases

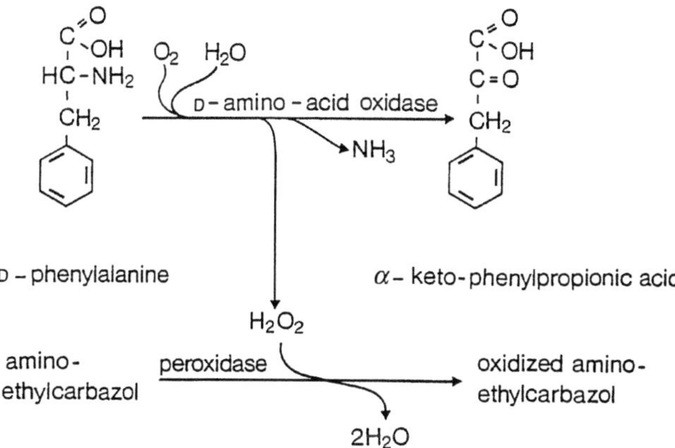

Fig. 5.6. Scheme of reaction when using peroxidase as auxiliary enzyme in enzyme visualization

5.1.2.2 Peroxidase Used as Auxiliary Enzyme

Enzyme reactions in which H_2O_2 is liberated may be visualized by the addition of peroxidase and a chromogenic hydrogen donator like 3-amino-9-ethyl-carbazol. Coupled enzyme tests of this kind have been set up, e.g., for D-(L)-amino acid oxidase (cf. Fig. 5.6) [19, 26], glucose oxidase [26] glycollate oxidase [19] and xanthin oxidase [26], β-D-fructofuranosidase [27, 28], peptidases [29, 30, 31] and acid β-glucosidase [19].

Oxidases, glycosidases, peptidases

5.1.2.3 Phenolases

Phenolases are a group of enzymes comprising, e.g., catechol oxidase (1.10.3.1), laccase (1.10.3.2), ascorbate oxidase (1.10.3.3), o-aminophenol oxidase (1.10.3.4), and monophenol monooxygenase (tyrosinase, 1.14.18.1). The plant enzyme catechol oxidase (synonym o-diphenoloxidase, o-diphenolase, tyrosinase) accepts as substrate 3,4-dihydroxyphenyl alanine (DOPA) (Fig. 5.7) [32]. It has been pointed out, however, that DOPA is also consumed by peroxidase when H_2O_2 is present, and that H_2O_2 is produced when DOPA is oxidized by catechol oxidase [33]. Consequently the possibility arises of compulsorily comprehending peroxidase besides polyphenoloxidase. Caffeic acid and m-phenylenediamine at pH 5.5 were considered to be specific substrates for laccase only [33] (Fig. 5.7). Another way to differentiate between catechol oxidase and peroxidase is to use the inhibitor diethyldithiocarbamate. As catechol oxidase is a copper-containing enzyme, it is completely inhibited by 10^{-2} mol l^{-1} diethyldithiocarbamate, whereas the iron-containing peroxidases remain unaffected [11].

Catechol oxidase

Peroxidases versus phenolases

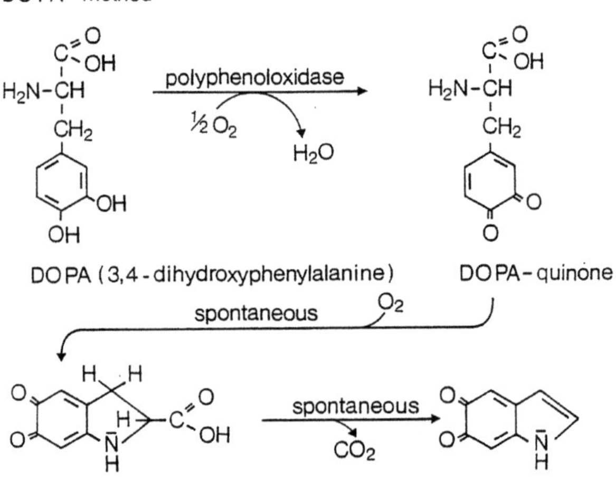

Fig. 5.7. Scheme of DOPA-oxidation by polyphenoloxidase; reaction scheme when using caffeic acid and m-phenylenediamine to visualize laccase (syn. polyphenol oxidase)

5.1.2.4 The Use of Phenazine Methosulphate and a Tetrazolium Salt for the Determination of Oxidases

PMS and the tetrazolium salt MTT can also be used to localize oxidases such as aldehyde oxidase (Fig. 5.8). In a coupled enzyme assay the method has also been used to visualize purine nucleoside phosphorylase (2.4.2.1) [34, 35] and adenosine deaminase (3.5.4.4) [36, 37] (Fig. 5.8).

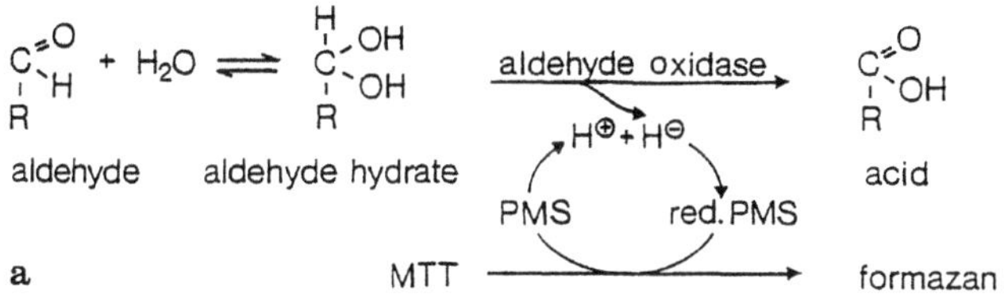

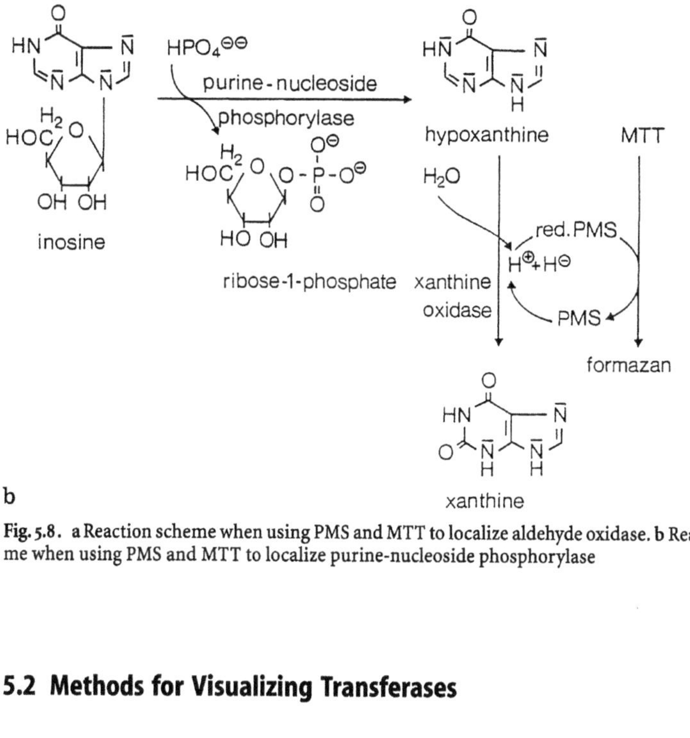

Fig. 5.8. a Reaction scheme when using PMS and MTT to localize aldehyde oxidase. b Reaction scheme when using PMS and MTT to localize purine-nucleoside phosphorylase

5.2 Methods for Visualizing Transferases

5.2.1 General Aspects

There are no specific methods for localizing the catalytic activities of transferases. The methods used rely on assays originally developed for the detection of oxidoreductases or hydrolases. Transferases are mostly visualized by use of a so-called coupled assay, i.e., one of the reaction products of a transferase reaction is used as substrate for a coupled dehydrogenase reaction which is made visible. Transferases which liberate phosphoric acid or a phosphate-containing compound may be localized by the so-called metal salt method. By this method phosphoric acid is precipitated as insoluble calcium or lead phosphate depending on the pH of the incubating medium. The milky precipitates are directly visible in transparent polyacrylamide

Coupled assay

Ca- or Pb-phosphate precipitates

gels while in starch gels they are not. Therefore, in starch gels the insoluble metal salts of phosphoric acid are converted into the black coloured sulphides. These steps are time-consuming because they need a complete wash-out of the unprecipitated Ca- or Pb-salts. Because the metal salts are directly visible in acrylamide gels, these are prefered with this visualization method.

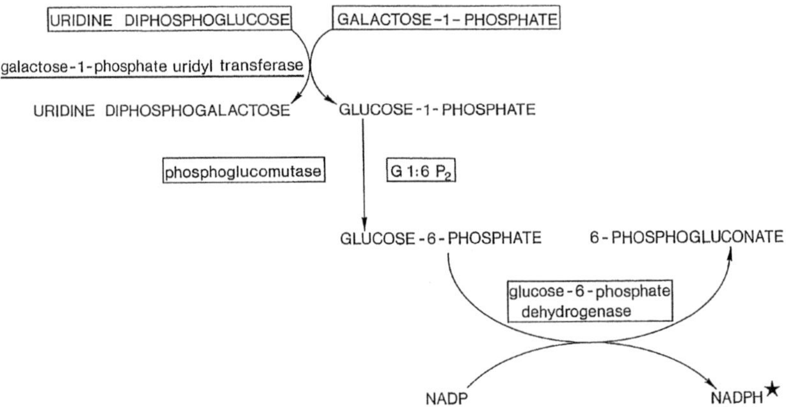

Fig. 5.9. Reaction scheme of the assay of hexose-1-phosphate uridylyl transferase (syn.: galactose-1-phosphate uridyl transferase) (2.7.7.12) using a coupled test with phosphoglucomutase and glucose-6-phosphate dehydrogenase as auxiliary enzymes; ★ visualization product

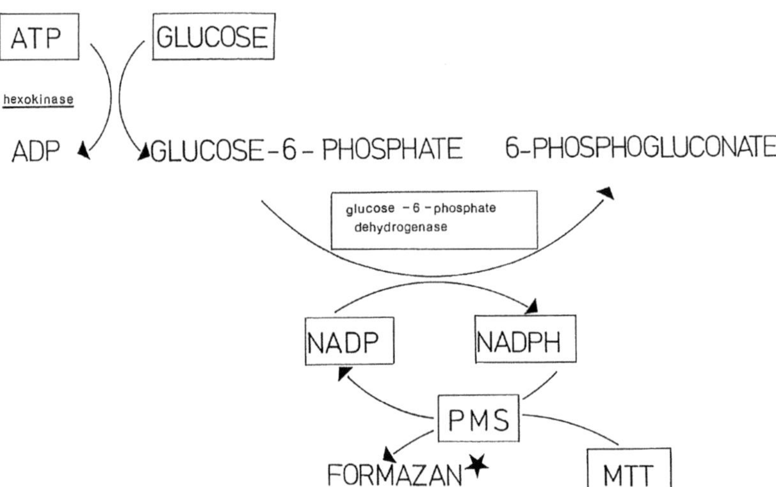

Fig. 5.10. Reaction scheme of the assay of hexokinase (2.7.1.1) using a coupled test with glucose-6-phosphate dehydrogenase as auxiliary enzyme, ★ visualization product

Table 5.2. Transferases which have been detected with an enzyme linked staining method using a dehydrogenase and its coenzyme to finally visualize their location. For details see Chap. 6: A Compilation of Protocols to Visualize Enzymes Following Electrophoretic Separation

Enzyme

Name	EC number	Source
Adenylate kinase	2.7.4.3	Man
Alanine aminotransferase	2.6.1.2	Mammalia
Aspartate aminotransferase	2.6.1.1	Man
Bisphosphoglyceromutase	2.7.5.4	Man
Creatine kinase	2.7.3.2	Mammalia
Fructokinase	2.7.1.4	*Drosophila*
Glucose-1-phosphate uridylyltransferase	2.7.7.9	Man
Glycine aminotransferase	2.6.1.4	Rat
Guanylate kinase	2.7.4.8	Man
Hexokinase	2.7.1.1	Man
Hypoxanthine phosphoribosyltransferase	2.4.2.8	Mammalia
Leucine aminotransferase	2.6.1.6	Mammalia
Nucleosidetriphosphate-adenylate kinase	2.7.4.10	Mammalia
6-Phosphofructokinase	2.7.1.11	Man
Phosphoglucomutase	2.7.5.1	Man
Phosphoglycerate kinase	2.7.2.3	Man
Phosphoglyceromutase	2.7.5.3	Man
Purine-nucleoside phosphorylase	2.4.2.1	Man
Pyridoxyl kinase	2.7.1.35	Man
Pyruvate kinase	2.7.1.40	Man
Ribosephosphate pyrophosphokinase	2.7.6.1	Mammalia
Transaldolase	2.2.1.2	*Bifidobacteria*
Transketolase	2.2.1.1	*Bifidobacteria*
Tyrosine aminotransferase	2.6.1.5	*Bifidobacteria*
UDPglucose-hexose-1-phosphate uridylyltransferase	2.7.7.12	Man
Uridinemonophosphate kinase	2.7.4.*	Man

* not further specified

5.2.2 The Assay of Transferases in a Coupled Test with a Dehydrogenase

The principle of a coupled enzyme system to localize transferases with a dehydrogenase as auxiliary enzyme is shown in Figs. 5.9 and 5.10 where the reaction schemes for hexose-1-phosphate uridylyltransferase (syn. galactose-1-phosphate uridyltransferase) (2.7.7.12), and hexokinase (2.7.1.1) are shown. More test systems for transferases based on a coupled test with a dehydrogenase are listed in Table 5.2.

Hexokinase

5.2.3 The Assay of Transferases Applying the Metal-Salt Method

Ornithine carbamoyltransferase (2.1.3.3) has been visualized by means of the metal-salt method [38]. The scheme of the reaction is presented in Fig. 5.11. More transferases which have been localized in electrophoretic support media using the metal-

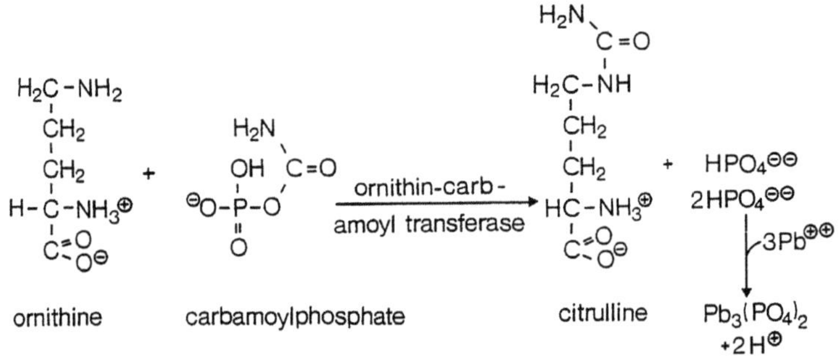

ornithine   carbamoylphosphate   citrulline $Pb_3(PO_4)_2$
                                                                                      $+2H^{\oplus}$

Fig. 5.11. Reaction scheme of the assay of ornithine carbamoyl transferase

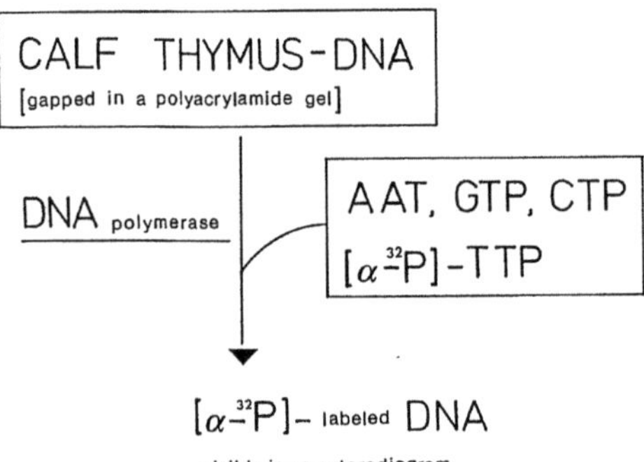

Fig. 5.12. Reaction scheme in the assay of DNA-polymerase. DNA polymerases are assayed after (SDS-) polyacrylamide gel electrophoresis using gels with gapped DNA. If SDS gel electrophoresis has been performed the gels are washed free from SDS after the separation of DNA polymerases renaturing enzyme activity. Afterwards the gels are incubated in an assay mixture which contains one of the four deoxyribonucleoside triphosphates radioactively labeled (usually [α - ³²P]TTP) at low concentration and high specific activity. After washing the gel with trichloroacetic acid and drying, incorporation of the label into the DNA within the gel is detected by autoradiography

salt method are: 3-enolpyruvoylshikimate-5-phosphate synthase (2.5.1.19; *Neurospora crassa*), glucose-1-phosphate uridylyltransferase (2.7.7.9; yeast), ornithine carbamoyl-transferase (2.1.3.3; man) and RNA nucleotidyltransferase (= RNA polymerase (2.7.7.6, *E. coli*) (Chap. 6). A transferase which uses a high molecular weight substrate is DNA-nucleotidly transferase (syn. DNA polymerase) (2.7.7.7) [39]. The reaction scheme to stain for this enzyme is shown in Fig. 5.12.

5.3 Methods for Visualizing Hydrolases

Hydrolases catalyze the hydrolytic cleavage of C–O, C–N, C–C, and some other bonds, including those of phosphoric anhydrides. According to the substrates used, hydrolases are also named esterases, glycosidases, amylases, etc. [1]. Three types of methods are taken to detect hydrolytic enzymes: (a) the use of substrates which result in coloured (e.g., *p*-nitrophenol, indigo) or fluorescent (e.g., methylumbelliferone) compounds when enzymatically hydrolyzed, (b) application of substrates being a naphthol derivative and coupling the hydrolyzed naphthol to a diazo-compound to produce a coloured (diazo-)dye and (c) precipitation of liberated ortho-phosphate by Ca^{2+} or Pb^{2+}-ions depending on the pH-value of the incubating medium. Pb-phosphate can be transformed to PbS, which has a black colour, while $Ca_3(PO_4)_2 \cdot (H_2O)_n$ must first be transformed to CaS that can be made visible as black CoS.

Three different methods to detect hydrolases

5.3.1 The Indigo Method

A simple method for detecting non-specific esterases or glycosidases is to use indoxyl esters or indoxyl glycosides as substrates such as 3-indoxyl phosphate, 3-indoxyl sulphate, 3-indoxyl acetate, indoxyl-1,3-diacetate or indoxyl-β-D-glucoside. Hydrolases which can use one of these substrates liberate the indoxyl moiety which is then nonenzymatically oxidized to indigo in the presence of O_2 and a catalyst such as Cu^{2+}-acetate or potassium ferrocyanide plus potassium ferricyanide (Fig. 5.13). While indoxyl esters and glycosides are colourless, the indigo dye has a faint blue colour [40 – 42].

Esterases, glycosidases

Indoxyl method

indoxyl acetate
syn. 3-acetoxindole

acylesterase, H_2O → H_2O, O_2 (Cu^{++})

Indigo Blue

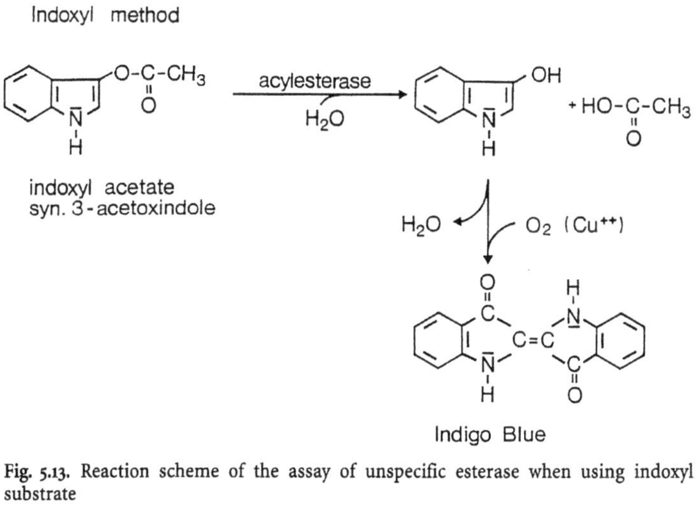

Fig. 5.13. Reaction scheme of the assay of unspecific esterase when using indoxyl acetate as substrate

5.3.2 The Phenolphthalein Method

Phenolphthalein diphosphate may be used to detect phosphatases, phenolphthalein disulphate is substrate of sulphatases while phenolphthalein mono-β-D-glucuronic acid is substrate of β-D-glucuronidase.

Phenolphthalein is colourless at acidic pH-values but turns red in alkaline solutions. The dye is water soluble and therefore tends to diffuse so that, in general, staining methods which lead to water insoluble stains and which are less toxic are

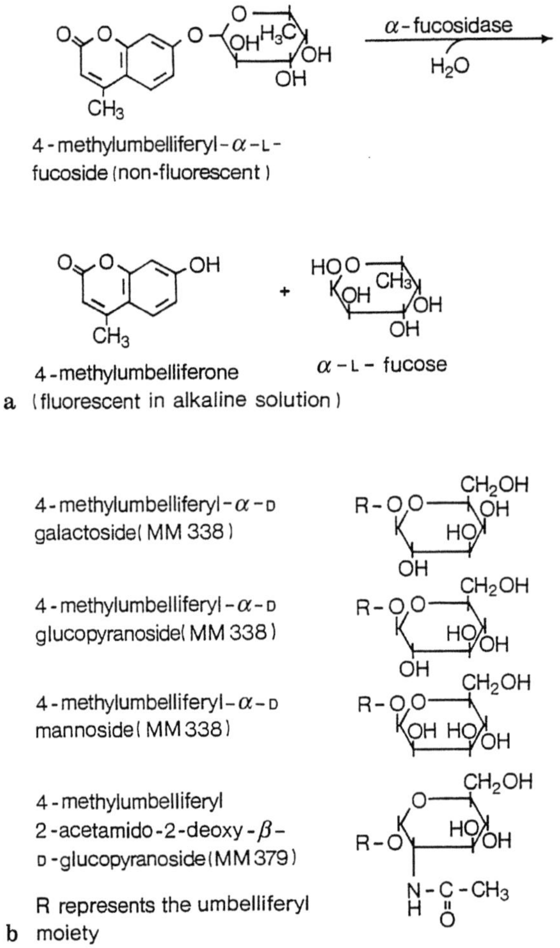

4-methylumbelliferyl-α-L-fucoside (non-fluorescent)

α-fucosidase, H_2O

4-methylumbelliferone (fluorescent in alkaline solution)

α-L-fucose

a

4-methylumbelliferyl-α-D galactoside (MM 338)

4-methylumbelliferyl-α-D glucopyranoside (MM 338)

4-methylumbelliferyl-α-D mannoside (MM 338)

4-methylumbelliferyl 2-acetamido-2-deoxy-β-D-glucopyranoside (MM 379)

R represents the umbelliferyl moiety

b

Fig. 5.14. a Course of the chemical reaction when using 4-methylumbelliferyl-α-L-fucoside to detect the enzyme α-fucosidase. b Structure of 4-methylumbelliferyl derivatives used to stain for glycosidases

preferred [19, 43]. If staining for acid phosphatase is intended then at least two incubations at 4 to 5 °C for 15 min each in 0.1 mol l⁻¹ acetate buffer of pH 5.0 are necessary prior to incubation with substrate in order to adjust the pH of the gel to an acidic milieu. If this step is not carried out, the stain for acid phosphatase will lead to the demonstration of alkaline phosphatase since electrophoresis is generally performed with alkaline buffers.

5.3.3 The Umbelliferone Method

A very sensitive and convenient method for detecting hydrolases is the use of esters or glycosides of 4-methylumbelliferone. The 4-methylumbelliferone moiety set free by enzymatic hydrolysis is visualized under long-wave UV-light. A disadvantage of the method, however, is the fact that 4-methylumbelliferone fluoresces exclusively at alkaline pH-values, while many hydrolases exhibit their pH-optimum at an acidic pH. In such cases the enzyme reaction is carried out at an acidic pH-value while the detection of 4-methylumbelliferone is performed after having alkalified the electrophoretic support medium either by exposing it to ammonia vapours or by incubating it into an alkaline buffer solution [44 – 52]. The course of the chemical reaction when using a 4-methylumbelliferone derivative is exemplified in Fig. 5.14 where the reaction scheme for the enzyme α-fucosidase is shown. Methylumbelliferone is highly soluble so that it diffuses rapidly out of large pore gels. Besides, it is unstable in alkaline solutions which results in a background colouring. Therefore, it is recommended to evaluate stained gels immediately. Enzymes which have been visualized by using a methylumbelliferone derivative as substrate have been compiled in Table 5.3. To detect active enzyme bands in electrophoretic support media they must be viewed under long wave UV-light (around 350 nm). It is also possible to photograph the gels. For this

4-methyl-umbelliferone substrates

α-fucosidase

Photography under UV-light

Table 5.3. Transferases which were detected *in situ* by applying fluorogenic derivatives of umbelliferone as substrate. For details see Chap. 6: A Compilation of Protocols to Visualize Enzymes Following Electrophoretic Separation

Enzyme

Name	EC number	Source
β-N-Acetyl-D-glucosaminidase	3.2.1.30	Man
β-N-Acetyl-D-hexoseaminidase	3.2.1.52	Man
Acid phosphatase	3.1.3.2	Man
Arylsulphatase	3.1.6.1	Man
Carboxylesterase	3.1.1.1	Man
α-D-Galactosidase	3.2.1.22	Man
β-D-Galactosidase	3.2.1.23	Man
α-D-Glucosidase	3.2.1.20	Man
β-D-Glucuronidase	3.2.1.31	Man
α-L-Fucosidase	3.2.1.51	Man
α-D-Mannosidase	3.2.1.24	Man
Phosphodiesterase I	3.1.4.1	Mammalia
Triacylglycerol lipase	3.1.1.3	Man

purpose the camera has to be equiped with a yellow filter, e. g., with the following filters: B + W 49 ES and B + W (threefold) (Filterfabriken, Wiesbaden, Germany) [53].

5.3.4 The Metal-Salt Method

Detection of phosphate, pyrophosphate or CO_2-liberating enzymes

Enzymes which hydrolyze phosphate ester bonds (e. g., acid or alkaline phosphatase) may be visualized by the so-called metal-salt method. Lead or calcium ions are used to precipitate enzymatically liberated ortho-phosphate. At acidic pH-values lead precipitates phosphate while Ca^{2+} ions precipitate P_i at alkaline pH-values (Fig. 5.15) [54, 55]. The method can also be used to detect pyrophosphate or CO_2-liberating enzymes in electrophoretic support media [56]. In contrast to polyacrylamid gels, the precipitates are not directly visible in starch gels. In starch gels the Pb precipitates must be transformed to the black lead sulphides by washing the gels with ammonium sulphide while the Ca-precipitates are transformed to the corresponding cobalt salts which are then transformed to the black coloured CoS. Enzymes which have also been localized by use of the metal-salt method are ornithine carbamoyltransferase (2.1.3.3) [19] and 5'-nucleotidase (3.1.3.5) (Figs. 5.11 and 5.15) [42]. The calcium precipitation method,

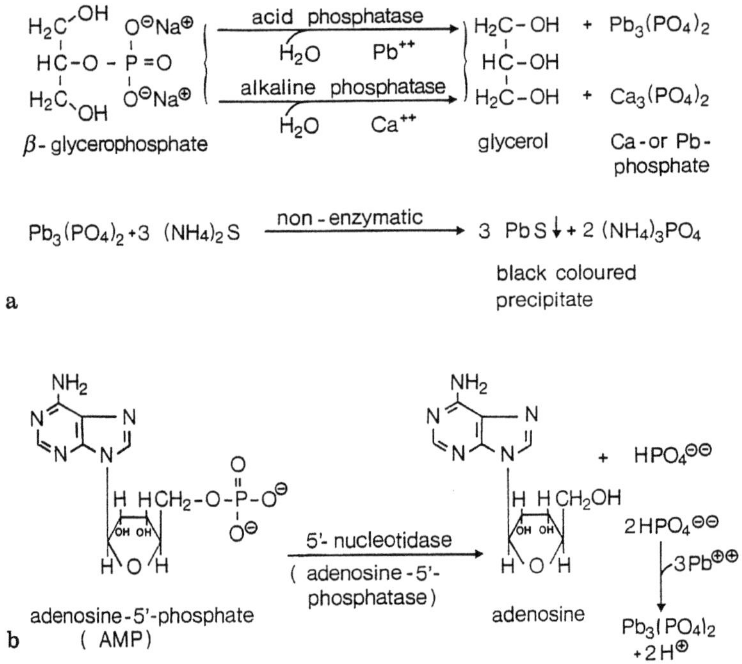

$$Pb_3(PO_4)_2 + 3\ (NH_4)_2S \xrightarrow{\text{non-enzymatic}} 3\ PbS\downarrow + 2\ (NH_4)_3PO_4$$

black coloured precipitate

a

b

Fig. 5.15. a Reaction scheme when visualizing acid or alkaline phosphatases by applying the metal-salt method. b Reaction scheme to stain for 5'-nucleotidase applying the metal-salt method

when using polyacrylamide gels, is very sensitive but limited to the following conditions [56]: (a) the pH-value of the enzyme reaction must be between pH 5 and 10 and (b) the Ca^{2+}-concentration in the test solution must exceed 5 mmol l^{-1}. Hydrolases which have been visualized by the metal-salt method are: acid phosphatase (3.1.3.2; wheat), adenosinetriphosphatase (3.6.1.3; *Micrococcus luteus*), alkaline phosphatase (3.1.3.1; various sources except higher plants), 3′:5′-cyclic-nucleotide phosphodiesterase (3.1.4.17; rat), fructose-bisphosphatase (3.1.3.11; ox) and 5′-nucleotidase (3.1.3.5; plants) (Chap. 6).

Another method uses ammonium molybdate to localize ortho-phosphate in electrophoretic support media. Inorganic pyrophosphatase (3.6.1.1) has been localized by this method [19, 57]. The dye molybdene blue is formed when pyrophosphate reacts with ammonium molybdate in the presence of sulphuric and ascorbic acid. This assay

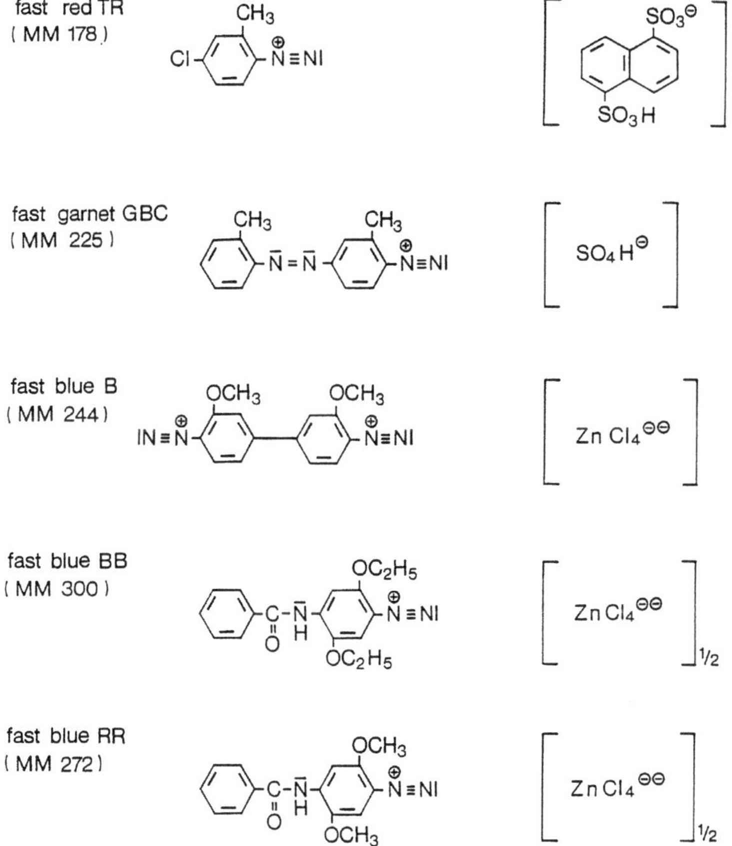

Fig. 5.16. Diazonium salts as used for the determination of naphthols

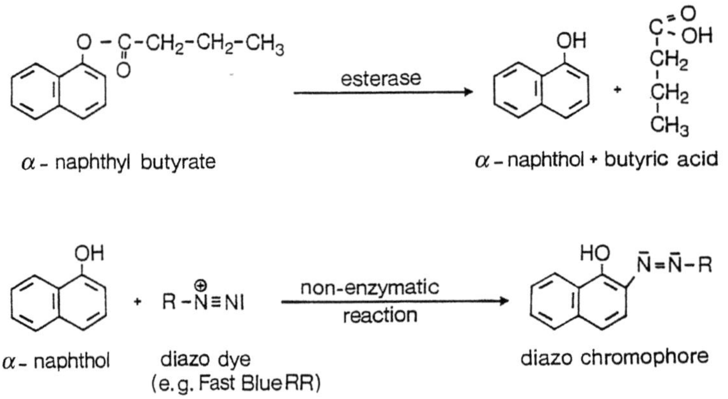

Fig. 5.17. **a** Reaction scheme when staining for unspecific esterase and applying the method of azo-coupling. **b** Reaction scheme when detecting leucine amino peptidase with L-leucine-β-naphthylamide and coupling the liberated β-naphthol to a diazonium salt

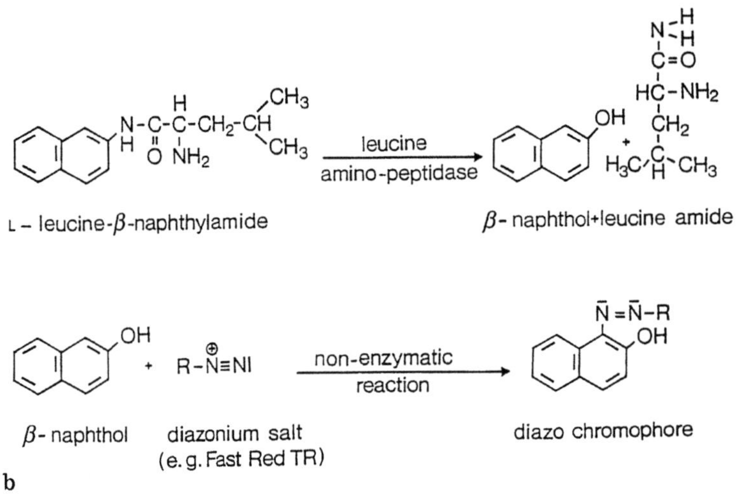

has been utilized to localize the enzyme inosine triphosphatase (3.6.1.19) in starch gels [44].

5.3.5 The Method of Azo-Coupling

A further sensitive method to visualize hydrolyzing enzymes is to use derivatives of α- or β-naphthol as substrates. During the enzymatic hydrolysis of such substrates

Table 5.4. Hydrolytic enzymes which were detected *in situ* by application of the azo-coupling method. For details see Chap. 6: A Compilation of Protocols to Visualize Enzymes Following Electrophoretic Separation

Enzyme

Name	EC number	Source
Acetylcholinesterase	3.1.1.7	Animals, insects
β-N-Acetyl-D-Glucosaminidase	3.2.1.30	Man
Acid phosphatase	3.1.3.2	Man
Alkaline phosphatase	3.1.3.1	Man
Aminopeptidase (cytosol)	3.4.11.1	Man, rat
Carboxylesterase	3.1.1.1	Man
Cathepsin B	3.4.22.1	Man
Cholinesterase II (unspecific)	3.1.1.8	Mammalia, insects
Chymotrypsin	3.4.21.1	Mammalia, insects
Cystyl aminopeptidase	3.4.11.3	Man
Dipeptidase	3.1.3.11	Rat
Fructose-bisphosphatase	3.1.3.11	Swine
α-D-Glucosidase	3.2.1.20	*Saccharomyces cerevisiae*
Phosphodiesterase I	3.1.4.1	Plants
Thrombin (Fibrinogenase)	3.4.21.5	Man
Trypsin	3.4.21.4	Bovine, insects

α- or β-naphthol and the corresponding acid or amide are liberated. By the process of azo-coupling the non-coloured naphthol compound is bound non-enzymatically to a diazonium salt. The basic structure of diazonium salts is R – N$^+$= N which on formation to azo groups (–N≡N–), confer colour. Azo-dyes are water-insoluble and their colour depend both on the diazonium salt used and the compound to be coupled to it (Fig. 5.16). When aminopeptidase (leucine aminopeptidase (3.4.11.1)) liberates β-naphthol from L-leucyl-4-methoxy-β-naphthylamide and the naphthol compound is coupled to the diazonium salt Fast blue B, a red coloured dye is formed (Fig. 5.17). A bluish colour, however, develops if α-naphthol couples to the same diazonium salt. α-Naphthol is liberated, e.g., when o-naphthylnanoate or α-naphthylpelargonate are hydrolyzed by the enzyme triacylglycerol lipase (3.1.1.3) [58, 59]. Table 5.4 summarizes hydrolases which have been visualized by applying the azo-coupling method. Another enzyme which can be localized by the method of azo-coupling is aspartate aminotransferase (synonym glutamic-oxalacetic transaminase) (2.6.1.1). Here, enzymatically liberated oxalacetate is coupled to the diazonium salt Fast Violet B to produce a coloured insoluble dark blue dye [23, 60].

α- and β-naphthol substrates

Diazonium salts

Triacylglycerol lipase

Aspartate aminotransferase

5.3.5.1 Simultaneous and Post-Coupling Methods

The coupling of enzymatically liberated naphthol derivatives to diazonium salts can be performed by two different methods: (a) the diazonium salt is directly included in the test system (simultaneous coupling), or (b) it is added after the formation of sufficient α- or β-naphthol (post-coupling). Post-coupling must be applied when the coupling salt contains a heavy metal which inhibits the enzyme reaction or, if op-

timum enzyme activity occurs above pH 6, since diazonium salts are rapidly hydrolyzed in aquous solutions of this pH-range. Fast Blue B is provided as Zn-salt and
since Zn-ions inhibit the activity of proteases these enzymes are visualisized by the
post-coupling method [61]. Fast Garnet GBC (diazotized 2-aminoazotoluole-ZnCl$_2$)
also contains Zn. Although it couples at optimum rates at pH 6.5, a pH-value at which
many proteases react, its use is limited by its low water solubility at this pH-value.
Post-coupling offers an advantage over simultaneous coupling in that enzyme bands
can be studied fluorometrically if 4-methoxy-β-naphthylamine derivatives are used
as substrate. The liberated methoxy naphthylamine can be examined under UV-light
before gels are reacted with a suitable diazonium salt.

Fast Blue B,
Fast Garnet GBC

5.3.5.2 The Detection of Peptidases

The azo-coupling method is often used to visualize peptidases. In the course of corresponding investigations it has been found that α-naphthylamine couples much
faster to diazonium salts than β-naphthylamine. Substituting the β-naphthol-ring by
a methoxy-group improves the reactivity with diazonium salts e.g. L-leucyl-4-
methoxy-β-naphthylamide couples, after hydrolyses by aminopeptidase at pH 6.5, 40
times faster to Fast Blue B or Fast Garnet GBC than β-naphthylamine [58]. By this fact
it is possible to detect even low aminopeptidase concentrations [62, 63].

Aminopeptidases

Peptidases are not as unspecific as is often assumed. They hydrolyze only a limited number of peptide bonds in a given peptide because the amino acid residues which
are close to the peptide bond to be hydrolyzed influence their hydrolytic ability. Substituting only a single amino acid in the neighbourhood of the peptide bond which is
to be hydrolyzed can inhibit their activity [61]. It is obvious therefore that the modification of naphthylamides has increased the sensitivity of peptidase assays considerably. Cathepsin B (3.4.22.1) is better visualized by use of α-N-benzoyloxycarbonyl-
L-arginyl-L-arginin-4-methoxy-β-naphthylamide than by taking the formerly used
benzoyl-arginine-β-naphthylamide [64]. Further amino acid derivatives of 4-me-
thoxy-β-naphthylamine which have been used to localize mammalian peptidases can
be taken from Table 5.5.

Cathepsin B

Diazo compounds react with free SH-groups. Therefore, the post-coupling
method must also be applied when reagents with free SH-groups such as cysteine,
2-mercaptoethanol, reduced glutathione or dithiothreitol have to be added into the
incubating solution for example to protect peptidases from denaturation [65]. Therefore, IEF gels were incubated with α-N-benzyloxycarbonyl-L-arginyl-L-arginin-4-
methoxy-β-naphthylamide and cysteine at pH 6 to detect human cathepin B (3.4.22.1)
which liberates the fluorescent 1-methoxy-3-naphthylamine [64]. After the formation
of a sufficient amount of the fluorescent 1-methoxy-3-naphthylamine the gels were put
into a solution acidified to pH 4 that contained, besides Fast Garnet GBC, mersalyl
acid to prevent the reactive group of cysteine from reacting with the diazonium salt.
At acid pH-values mersalyl acid is superior to compounds of similar reactivity such
as p-chloromercuribenzoic acid [62]. At those sites where enzymatically liberated 1-
methoxy-3-naphthylamine reacts with Fast Garnet GBC a dark blue colour is formed.
The most effective synthetic substrates for cathepsin B contain double basic residues
at the point of cleavage such as the substrate N-benzoyloxycarbonyl-Ala-Arg-
4-methoxy-β-naphthylamide [66]. Upon enzymatic hydrolysis the liberated 4-

Thiol reagents
reacting
with diazo dyes

Mersalyl acid

Table 5.5. Amino acid derivatives of 4-methoxy-β-naphthylamine (4 MβNA) that have been used to histochemically localize peptidase enzymes in tissues of mammalia [61]. BZ: benzyl, CBZ: benzyloxycarbonyl

Substrate	Enzyme
Lys-Ala-4MβNA	Diaminopeptidase II
BZ-Val-Lys-Lys-Arg-4MβNA	Cathepsin B 1a
CBZ-Arg-Arg-4MβNA-2HCl	Cathepsin B 1
N-CBZ-Arg-Arg-Arg-4MβNA-diacetat	Trypsin
L-arginyl-L-Arg-4MβNA-3HCl	Diaminopeptidase III
Pro-Arg-4MβNA	Diaminopeptidase I, Cathepsin C
N-α-BZ-Phe-Val-Arg-4MβNA	Thrombin
γ-L-glutamyl-4MβNA	γ-Glutamyl transpeptidase
Leu-4MβNA	Aminopeptidase
Phe-Pro-Ala-Met-4MβNA	Cathepsin B 1b
Gly-Pro-4MβNA	Diaminopeptidase IV
CBZ-Pro-Ala-Gly-Pro-4MβNA	Collagenase
His-Ser-4MβNA	Diaminopeptidase 1, Cathepsin C
N-CBZ-L-Pro-L-Phe-L-His-L-Leu-L-Leu-L-Val-L-Tyr-L-Ser-4MβNA	Renin
Ser-Tyr-4MβNA	Diaminopeptidase I

methoxy-β-naphthylamide can be visualized under ultra-violet light. 4-Methylcoumarin derivatives may also be used to detect cathepsin B or cathepsin H. To stain for cathepsin B Z-Phe-Arg-4-methyl-7-coumarylamid can be used as chromogenic substrate, while Arg-4-methyl-7-coumarylamid is a chromogenic substrate of cathepsin H. Upon enzymic hydrolysis the highly fluorescent 7-amino-4-methylcoumarin is liberated. Its fluorescence at 365 nm is 6.7 times higher than that of the 4-methoxy-2-naphthylamide [67, 68]. Besides, the coumaryl derivative has the advantage of not being carcinogenic as are the 2-naphthol derivatives. Prolin endopeptidase from rat brain could be detected with the chromogenic substrate 7-(N-succinyl-Gly-Pro)-4-methyl-coumarinamide [69]. Leucine aminopeptidase is another example of an enzyme that can be visualized by a fluorogenic methyl-coumaryl derivative, namely 7-L-leucyl-4-methylcoumarinyl-amide [70]. Another fluorogenic substrate is o-aminobenzoyl-glycyl-p-nitro-L-phenylalanyl-L-proline which can be used to detect the depeptidyl carboxypeptidase angiotensin-I-converting enzyme (3.4.15.1). Enzymatically liberated o-aminobenzoyl-glycin fluoresces at 360 nm [71]. Five different peptide derivatives were used to stain for elastase (3.4.21.11), a chymotrypsin-like enzyme [67]. Each substrate comprised the same peptide sequence, namely Me-O-Suc-Ala-Ala-Pro-Val-R but differed in the residue R which was either a -4-nitroanilide or a -thiobenzyl- ester, or a -o-ethyl-4-methyl-7-coumarylamide, or a -1-methoxy-3-naphthylamide. Elastase was most active towards the thiobenzylester derivative and 2.4 pM (0.072 ng/ml) of human elastase could be detected with this substrate. Enzyme activity can be visualized by the formation of 3-carboxy-4-nitrothiophenoxide which rises

Cathepsin H

Prolin endopeptidase, leucine aminopeptidase

Angiotensin-I-converting enzyme, elastase

when the enzymatically liberated benzyl-mercaptane reacts with 4,4′-dithiodipyridine (or Ellman's reagent). The formed dye has its maximum absorbance at 412 nm ($\varepsilon = 13\,600$ [l/mol × cm]).

To detect proteases, a cellulose acetate membrane (0.2 μm pore size) previously impregnated with a fluorogenic substrate was applied to the surface of a processed ultrathin IEF gel [72]. The technique was used to detect the following enzymes when taking the substrates indicated in parentheses: trypsin (Z-Gly-Gly-Arg-AFC) at pH 8.2, glandular kallikrein (D-Val-Leu-Arg-AFC) at pH 8.2, dipeptidyl aminopeptidase IV (Gly-Pro-AFC) at pH 7.6, cathepsin B (BZ-Val-Lys-Lys-Arg-AFC) at pH 6.4 and dipeptidyl aminopeptidase II (Lys-Ala-AFC) at pH 5.6. The peptide substrates are acrylamide derivatives of 7-amino-4-trifluoromethylcoumarin (AFC). They are unique in that the acylated derivatives have a blue fluorescence, while the free liberated product is wavelength-shifted into the green region of the spectrum. Enzymic activity was monitored by an ultraviolet (UV) lamp (long-wavelength, UV SL-25; Ultra-Violet Products, Inc., San Gabriel, CA, USA). Cellulose acetate membranes came from Gelman and Sartorius. The binding of AFC derivatives to the membrane is through the hydrophobic region of the substrate. Hydrolysis of AFC substrates does not lead to bands of yellow-green flourescence, which distinguishes this fluorophore in solution reactions, but rather, bright blue-green fluorescent bands are formed. Cellulose nitrate membranes could not be used to stain proteases because synthetic peptide substrates bind through the peptide moiety to the membrane, making it unrecognizable by proteases. The consequence is that, in general, no hydrolysis occurs whether the substrate is limited to one amino acid or to an extended oligopeptide. Therefore, peptide substrate impregnation is effectively limited to the use of cellulose diacetates and regenerated hydrate membranes.

5.3.6 The Dithiothreitol Formazan Method

Hydrolases which catalyze reactions leading to an increase in the pH-value of the reaction medium can be visualized by the dithiothreitol formazan method. It uses an

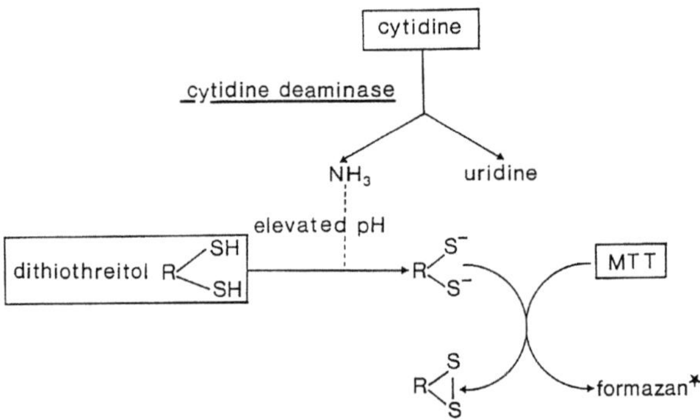

Fig. 5.18. Determination of cytidine deaminase using dithiothreitol and Nitro BT as staining agents, * visualization product

increase in the OH⁻-ion concentration to transform the dithiol compound dithiothreitol from the non-dissociated to the dissociated form. The dissociated molecule is
capable of reducing a tetrazolium salt whereby it is oxidized forming an intramole-
cular disulphide bond (Fig. 5.18). With this detection method the less sensitive tetrazolium salt nitro BT is preferably used instead of the more sensitive MTT. The dithiothreitol formazan method has been applied to visualize the enzymes arginase (3.5.3.1)
and cytidine deaminase (3.5.4.5) [19].

Arginase,
cytidine
deaminase

5.3.7 The Coupled Test with a Dehydrogenase or Peroxidase

Several hydrolases were also visualized by coupling one of their reaction products to
a dehydrogenase and its coenzyme. These are: adenosylhomocysteinase (3.3.1.1; man),
alkaline phosphatase (3.1.3.1; animals, bacteria), arginase (3.5.3.1; man), cytidine
deaminase (3.5.4.5; man), dipeptidase (3.4.13.11; man) and tripeptide aminopeptidase
(3.4.11.4; man) (Chap. 6).

Hydrolases which were detected by including a peroxidase and a chromogenic
hydrogen donor in the staining solution are: aminoacylase (3.5.1.14; mammalia), β-D-
fructofuranosidase (3.2.1.26; plants), α-D-glucosidase (3.2.1.20; man) and peptidases
(3.4.11(13)*; mammalia) (Chap. 6).

5.3.8 Enzymes which Hydrolyze High-Molecular Weight Substrates

High-molecular weight substrates diffuse extremely slowly into electrophoretic support media. Two different principles have been suggested to surmount this problem:
(a) the substrate is mixed with fluid agar and poured onto the support medium so that
after the gelatination an agar overlay is formed; during the visualization process a
certain number of enzyme molecules diffuse into the agar overlay where they hydrolyze the high-molecular weight substrate, and (b) high-molecular weight substrates
may be polymerized directly into polyacrylamide gels before starting electrophoresis;
after electrophoresis the conditions are changed so that the separated hydrolases are
able to attack the hight-molecular weight substrate.

5.3.8.1 Nucleases

DNA containing polyacrylamide gels may be used to visualize deoxyribonuclease I
(DNase) (3.1.21.1) [73]. After electrophoresis the DNA-containing gels are incubated in
substrate buffer and finally stained for non-hydrolyzed DNA with methylene green;
deoxyribonuclease I active sites appear as white bands on a green background. This
sort of staining is called negative staining. The method was also used to visualize
DNases from mammalia and *Drosophila melanogaster*. The detection limit for pancreatic DNase I was found to be at 250 pg enzyme protein. The sensitivity of the
method can be increased significantly by staining the DNA with ethidium bromide
and inspecting the gels under long-wave UV-light [74]. The use of labeled DNA or RNA
substrates further increases the sensitivity so that the following enzymes could be
detected: DNases from human blood serum [53, 75] DNases and RNases from tobacco

DNases

RNases

leaves [76] and RNases from virus infected chinese cabbage [77]. Positively charged nucleases, however, cannot be treated in this way because they already bind at the start to the negatively charged nucleic acids [78]. One of the possible expedients to this problem is to include SDS into the electrophoretic separation system and finally to renature the enzyme protein in situ (cf. Chap. 4). By doing so, enzyme concentrations in the pg-range could be visualized.

The different nucleases can be classified according to their substrate specificity into five different groups [78]:

(a) DNases which hydrolyze only native DNA,
(b) DNases which hydrolyze specifically denatured DNA,
(c) DNases which hydrolyze both native and denatured DNA,
(d) Nucleases which hydrolyze only single stranded DNA or RNA, and
(e) RNases which hydrolyze specifically RNA.

5.3.8.2 Proteases

By using one of the available β-naphthylamide- or 4-methylcoumaryl peptides a large number of peptidases can be detected (Table 5.5). Dipeptidases, tripeptidases [79], carboxy peptidases [80] and amino peptidases could also be visualized in a coupled test with L-amino-acid oxidase and peroxidase as indicator reaction [79].

Methods for detecting proteinases, however, are limited. Whereas trypsin- and chymotrypsin-like proteinases can be detected with chromogenic substrates of the β-naphthyl-, amide-, and β-naphthoic ester-type, it is not possible to localize proteinases of the thermolysine type [81] and other metal-salt dependent proteinases [82] with one of these substrates. In order to detect thermolysin, trypsin and chymotrypsin a denatured protein is used as substrate. Provided the protein substrate does not migrate under the used electrophoretic conditions it can be included in a polyacrylamide gel before gelatinization. Denatured haemoglobin is a suitable substrate for proteinases no matter whether the electrophoretic separation is performed at acidic or alkaline conditions. Casein, however, can only be used under acidic conditions since it migrates anodically under alkaline pH-values. Following electrophoresis, haemoglobin-containing PAA gels may be fixed with trichloroacetic acid leading to the formation of colourless proteinase bands on a brown background. Casein-containing gels are stained with one of the usual protein stains. Proteinase active zones are then visible as colourless bands on a blue background. When protein-containing gels are used to separate trypsin or chymotrypsin these enzymes can be differentiated from enzymes of similar activities by the use of specific inhibitors. Chymotrypsin-like enzymes are inhibited by L-1-tosylamido-2-phenylethylchloromethylketone [83], while trypsin-like proteases are inhibited by 1-chloro-3-tosylamido-7-amino-2-heptanone [84]. The proteinase plasmin (3.4.21.7) can be assayed with the protein substrates fibrin and casein, or with synthetic substrates with only one hydrolyzable bond such as esters and amides of basic amino acids. One such amide, containing lysine as the carboxyl terminus, benzyl-oxycarbonyl-Ala-Ala-Lys-4-methoxy-2-naphthylamine (CBZ-Ala-Ala-Lys-MβNA) was found to be a better substrate for plasmin than for other mammalian serine proteases [85]. Sites of enzymic activity could be detected by observing under ultra-violet light electrophoretic separation media containing plasmin and its substrate CBZ-Ala-Ala-Lys-MβNA, since 4-

methoxy-2-naphthylamine set free upon enzymic hydrolysis is highly fluorescent. The naphthylamine compound may also be coupled to Fast Blue B resulting in a coloured diazo dye [85]. Plasmin occurs in blood as a zymogen (plasminogen) which can be activated by proteolytic enzymes of high specificity. These are known as plasminogen activators and they occur in body fluids and many different tissues [86].

5.3.8.3 Amylases

There exist three different types of amylases: (a) α-amylase (3.2.1.1) acting on endogeneous 1,4-α-D- glucosidic bonds in polysaccharides containing three or more 1,4-α-linked-D-glucose units, (b) β-amylase (3.2.1.2) which hydrolyzes 1,4-α-D-glucosidic linkages in polysaccharides so as to remove successive maltose units from the non-reducing ends of the chains and (c) exo-1,4-α-D-glucosidase (3.2.1.3) which hydrolyzes terminal 1,4-linked α-D-glucose residues successively from the non-reducing ends of the chains with release of β-D-glucose [1]. α-Amylases are generally visualized by observing their action on soluble starch. Since amylases bind to starch, it is not possible to separate them in starch gels. Polyacrylamide or other separation media are used for that purpose. Afterwards these gels are incubated in soluble starch and finally non-hydrolyzed starch is visualized with the iodine staining technique. α-Amylases are then indicated as faint bands on a blue background (Chap. 6).

5.4 Methods for Visualizing Lyases

There are no specific visualization techniques for lyases, and methods which were originally used for the detection of oxidoreductases or hydrolases were modified to detect them. Lyases were visualized (a) in a coupled test system using a dehydrogenase as indicator reaction (Table 5.6), (b) by following the colour change of 2,6-

Table 5.6. Lyases which were detected with an enzyme-linked staining method using a dehydrogenase to finally visualize their location. For details see Chap. 6: A Compilation of Protocols to Visualize Enzymes Following Electrophoretic Separation

Enzyme

Name	EC number	Source
Aconitate hydratase	4.2.1.3	Man
Anthranilate-5-phosphoribosylpyro-phosphate phosphoribosyl-transferase multi enzyme complex		*Neurospora crassa*
Anthranilate synthase	4.1.3.27	*Neurospora crassa*
Argininosuccinate lyase	4.3.2.1	Man
Enolase	4.2.1.11	Rabbit
Fructose-bisphosphate aldolase	4.1.2.13	Man, bacteria
Fumarate hydratase	4.2.1.2	Mammalia

dichlorophenolindophenol (citrate-(si)-synthase (4.1.3.7), glyoxalase I (4.4.1.5)), (c) staining of liberated phosphate (chorismate synthase (4.6.1.4; *Neurospora crassa*), cystathionine *β*-synthase (4.2.1.22; yeast), pyruvate decarboxylase (4.1.1.1; yeast), phospho-2-keto-3-deoxyheptonate aldolase (= DAHP-synthase, 4.1.2.15; *Neurospora crassa*)), or (d) by other methods (carbonate dehydratase (4.2.1.1; mammalia), d-TDP-glucose 4,6-dehydratase (4.2.1.46; *E. coli*), threonine dehydratase (4.2.1.16; bacteria)).

5.5 Methods for Localizing Isomerases and Ligases

The test systems published to visualize isomerases are limited. The following enzymes have been visualized in electrophoretic support media: glucosephosphate isomerase (5.3.1.9), mannosephosphate isomerase (5.3.1.8) and triosephosphate isomerase (5.3.1.1) [19]. The location of these enzymes was made visible by using several auxiliary enzymes and a dehydrogenase as reaction indicator (Chap. 6).

Isomerase

Very few staining systems have been described to localize ligases in electrophoretic support media. These are limited to the enzymes tryptophanyl synthetase (6.1.2) [19] and glutamine synthetase (6.3.1.2) [87].

5.6 Visualization of Enzymes by Different Methods

Many enzymes can be visualized by more than one method. Dehydrogenases for example can be detected by either observing the change in fluorescence of the reduced pyridine nucleotides or by observing the formation of a coloured formazane. With some exceptions both methods are also applicable to coupled test systems. For the detection of certain hydrolytic enzymes a large number of test systems can be used. Some enzymes can be visualized by up to five different methods such as the umbelliferone-, the naphthol ester-, the *p*-nitrophenyl-, or the metal-salt method. An overview on enzymes which have been visualized with different methods is provided Table 5.7.

Table 5.7. Enzymes which have been visualized by a number of different staining methods. For details see Chap. 6: A Compilation of Protocols to Visualize Enzymes Following Electrophoretic Separation

Enzyme [Ref.]	Brief reaction scheme	Recognition product
β-N-Acetyl-D-glucos-aminidase [19]	1. enzymatic hydrolysis of 4-methylumbelliferyl-2-acetamido-2-deoxy-β-D-glucopyranoside	4-methylumbelliferone (fluorescent)
[19]	2. enzymatic hydrolysis of naphthol AS BI 2-acetamido-2-deoxy-β-D-glucopyranoside and non-enzymatic coupling of liberated naphthol to Fast Garnet GBC	diazo dye

Table 5.7 (continued)

Enzyme [Ref.]	Brief reaction scheme	Recognition product
Adenylate kinase [57, 88]	1. coupled enzyme test: (1) ADP + ADP = ATP + AMP, (2) ATP + glucose = ADP + glucose-6-phosphate, (3) glucose-6-phosphate + NADP = 6-phosphogluconate + $NADPH_2$, $NADPH_2$ + PMS + MTT = formazan; (1) adenylate kinase, (2) hexokinase, (3) glucose-6-phosphate dehydrogenase	formazan
[19]	2. coupled enzyme test: (1) ATP + AMP = ADP + ADP; (2) ADP + PEP = ATP + pyruvate; (3) pyruvate + $NADH_2$ = lactate + NAD; (1) adenylate kinase, (2) pyruvate kinase, (3) lactate dehydrogenase	NAD (non-fluorescent)
Alcohol dehydrogenase [19]	1. ethanol + NAD = acetaldehyde + $NADH_2$; $NADH_2$ + PMS + MTT = formazan	formazan
[19]	2. acetaldehyde + $NADH_2$ = ethanol + NAD	NAD (non-fluorescent)
Alkaline phosphatase [19]	1. enzymatic hydrolysis of β-naphthyl-phosphate and non-enzymatic coupling of liberated 2-naphthol to Fast blue B (or Fast blue RR)	diazo dye
[89, 90]	2. enzymatic hydrolysis of AMP and formation of a yellow dye in the presence of ammonium molybdate, H_2SO_4 and ascorbic acid	ammonium molybdato-phosphate
[90]	3. enzymatic hydrolysis of p-nitrophenyl phosphate; visualization of liberated p-nitro phenol at alkaline pH	p-nitrophenol
[90]	4. coupled enzyme test (1) AMP = adenosine + Pi; (2) Pi + inosine = ribose-1-phosphate + hypoxanthine; (3) hypoxanthine + H_2O_2 + PMS + nitro BT = PMS + formazan; (1) alkaline phosphatase, (2) inosine phosphorylase, (3) xanthine oxidase	formazan
Aminopeptidase [91]	1. enzymatic hydrolysis of L-leucyl-4-methoxy-2-naphthylamide-HCl and non-enzymatic coupling of liberated 2-naphthol to Fast blue B;	diazo dye
[92]	2. enzymatic hydrolysis of L-leucine-p-nitroanilide and inspection of fluorescent p-nitroanilin	p-nitroanilin (fluorescent)

Table 5.7 (continued)

Enzyme [Ref.]	Brief reaction scheme	Recognition product
Arginase [93]	1. increase in pH: (1) arginine + H_2O = ornithine + urea, (2) urea = $2 NH_3 + CO_2$ (increase in OH^-); non-enzymatic reactions: OH^- + dithiothreitol = dithiotreitol-SH; dithiotreitol-SH + nitro BT = dithiotreitol + formazan; (1)arginase, (2) urease	formazan
[94]	2. coupled enzyme test: (1) arginine + H_2O = ornithine + urea; (2) urea + $2 NH_3 + CO_2$; (3) $NH_3 + NADH_2$ + 2-oxoglutarate = L-glutamate + H_2O + NAD; (1) arginase, (2) urease, (3) glutamate dehydrogenase	NAD (non-fluorescent)
Aspartate aminotransferase [19]	1. coupled enzyme test: (1) aspartate + 2-oxoglutarate = oxaloacetate + L-glutamate; (2) oxaloacetate + $NADH_2$ = malate + NAD; (1) aspartate aminotransferase, (2) malate dehydrogenase	NAD (non-fluorescent)
[19]	2. 2-oxoglutarate + L-aspartate = L-glutamate + oxaloacetate, oxaloacetate + Fast blue B = diazo dye	diazo dye
Carbonate dehydratase [95]	1. enzymatic hydrolysis of 4-methyl-umbelliferyl-acetate; inspection of liberated 4-methylumbelliferone under UV-light	4-methylumbelliferone (fluorescent)
[19]	2. enzymatic hydrolysis of fluorescein-diacetate; inspection of liberated fluorescein under UV-light	fluorescein (fluorescent)
Carboxyl esterase [19]	1. enzymatic hydrolysis of 4-methyl-umbelliferyl-acetate and inspection of 4-methylumbelliferone under UV-light	4-methylumbelliferone (fluorescent)
[23]	2. enzymatic hydrolysis of α-naphthyl-acetate and coupling of liberated α-naphthol to Fast blue B	
[42, 96]	3. enzymatic hydrolysis of 5-bromo-indoxylacetate and non-enzymatic oxidation of liberated 5-bromoindoxyl to blue indigo in the presence of Cu_2^+-ions	indigo (blue dye)

Table 5.7 (continued)

Enzyme [Ref.]	Brief reaction scheme	Recognition product
Catalase [19]	1. $H_2O_2 + H_2O_2 = 2H_2O + O_2$; 2. $H^+ + H_2O_2 + 2Fe^{3+} = 2Fe^{2+} + 2H_2O$; Fe^{2+} + K-hexacyanoferrate III = Turnbull's blue	"white patches" on a blue background when using starch gels
[97]	2. $H_2O_2 + H_2O_2 = 2H_2O + O_2$; H_2O_2 + diaminobenzidine = $2 H_2O$ + brown benzidine	"white patches" on a brown background
[44, 98]	3. $H_2O_2 + H_2O_2 = 2H_2O + O_2$; $2H^+ + H_2O_2 + 2J^- = J_2 + 2H_2O$; J_2 stains starch blue	"white patches" on a blue background with starch gel
Creatine kinase [57, 99]	1. coupled enzyme test: (1) phosphocreatine + ADP = ATP + creatine; (2) ATP + glucose = glucose-6-phosphate + ADP; (3) glucose-6-phosphate + NADP = 6-phosphogluconate + $NADPH_2$; $NADPH_2$ + PMS + MTT = formazan; (1) creatine kinase, (2) hexokinase, (3) glucose-6-phosphate dehydrogenase	formazan
Creatine kinase	2. coupled enzyme test: (1) ATP + creatine = ADP + phosphocreatine; (2) ADP + PEP = ATP + pyruvate; (3) pyruvate + $NADH_2$ = lactate + NAD; (1) creatine kinase, (2) pyruvate kinase, (3) lactate dehydrogenase	NAD (non-fluorescent bands on a fluorescent background)
3':5'-Cyclic-nucleotide phosphodiesterase [100]	1. coupled enzyme test: (1) 3':5'-cyclic AMP + H_2O = 5'-AMP; (2) 5'-AMP + ATP = 2ADP; (3) ADP + PEP = ATP + pyruvate; (4) pyruvate + $NADH_2$ = lactate + NAD; (1) 3':5'-cyclic-nucleotide phospho-diesterase, (2) adenylate kinase, (3) pyruvate kinase, (4) lactate dehydrogenase	NAD (non-fluorescent)
[101]	2. enzymatic hydrolysis of 3',5'-cyclic AMP to 5'-AMP and precipitation of 5'-AMP with Ca^{2+}-ions	Ca-precipitate of 5'-AMP
[101, 102]	3. enzymatic hydrolysis of 1'-N_6-etheno-2-aza-adenosine 3',5'-phosphate	2-aza-ε-AMP (fluorescent)

Table 5.7 (continued)

Enzyme [Ref.]	Brief reaction scheme	Recognition product
Cytidine deaminase [103]	1. increase in pH: cytidine + H_2O = uridine + NH_3; NH_3 causes an increase in the OH^--ion concentration and triggers the following non-enzymatic reactions: OH^- + dithiothreitol = dithiotreitol-SH; dithiotreitol-SH + nitro BT = dithiotreitol + formazan	formazan
[94]	2. coupled enzyme test: (1) cytidine + H_2O = uridine + NH_3; (2) NH_3 + 2-oxoglutarate + $NADH_2$ = L-glutamate + H_2O + NAD; (1) cytidine deaminase, (2) glutamate dehydrogenase	NAD (non-fluorescent)
Deoxyribonuclease I [104]	1. DNA is polymerized into a PAA gel. At the sites of enzymatic activity DNA is digested. Finally the entire gel is stained for DNA with ethidium-bromide	non-fluorescent bands on a fluorescent gel
[105]	2. same procedure as in 1. but the gels are stained with methyl green	"white patches" on green gels
[106]	3. same procedure as in 1. but after electrophoresis the gels are incubated in a buffered solution containing dGTP, dCTP, dATP and $[\alpha\text{-}^{32}P]$dTTP. Unreacted substrates are washed out after a while and incorporation of labled dTTP is localized by autoradiography.	autoradiography of labeled DNA
Enolase [19]	1. coupled enzyme test: (1) 3-phosphoglycerate = 2-phosphoglycerate; (2) 2-phosphoglycerate = PEP + H_2O; (3) PEP + ADP = pyruvate + ATP; (4) pyruvate + $NADH_2$ = lactate + NAD; (1) phosphoglycerate phosphomutase, (2) enolase, (3) pyruvate kinase, (4) lactate dehydrogenase	NAD (non-fluorescent bands on a fluorescent background)
[107]	2. coupled enzyme test: (1) 2-phosphoglycerate = PEP + H_2O; (2) PEP + ADP = pyruvate + ATP; (3) ATP + glucose = glucose-6-phosphate + ADP; (4) glucose-6-phosphate + NADP = $NADPH_2$ + PMS + MTT = NADP + PMS + formazan; (1) enolase, (2) pyruvate kinase, (3) hexokinase, (4) glucose-6-phosphate dehydrogenase	formazan

Table 5.7 (continued)

Enzyme [Ref.]	Brief reaction scheme	Recognition product
β-D-Fructofurano-sidase [108, 109]	1. Coupled enzyme test: (1) sucrose = glucose + fructose; (2) glucose + O_2 = gluconate + H_2O_2; H_2O_2 + 3,3′-diaminobenzidine = oxidized benzidine + H_2O; (1) fructofuranosidase, (2) glucose oxidase	oxidized benzidine
[110]	2. sucrose = glucose + fructose; glucose + $Fe^{3+}Cl_3$ = Fe^{2+} + gluconic acid; Fe^{2+} + $K_3[Fe^{3+}(CN)_6]$ = blue dye	Turnbull's blue
Galactokinase [111]	1. Coupled enzyme test: (1) ATP + galactose = ADP + galactose-1-phosphate; (2) ADP + PEP = pyruvate + ATP; (3) pyruvate + $NADH_2$ = lactate + NAD; (1) galactokinase, (2) pyruvate kinase, (3) lactate dehydrogenase	NAD (non-fluorescent bands on a fluorescent background)
[113]	2. Enzymatic hydrolysis of 4-methyl-umbelliferyl-β-D-galactoside and detection of 4-methylumbelliferone	4-methylumbelliferone (fluorescent)
[113]	3. Coupled enzyme test: (1) ATP + galactose = ADP + galactose-1-phosphate; (2) UDPglucose + galactose-1-phosphate = glucose-1-phosphate + UDPgalactose; (3) glucose-1-phosphate = glucose-6-phosphate; (4) glucose-6-phosphate + NADP = 6-phosphogluconate + $NADPH_2$; (1) galactokinase, (2) UDPglucose hexose-1-phosphate uridylyltransferase, (3) phosphoglucomutase, (4) glucose-6-phosphate dehydrogenase	$NADPH_2$ (fluorescent)
[114]	4. ^{14}C-D-galactose + ATP = ADP + ^{14}C-galactose-1-phosphate; the labled product is precipitated with lanthane and localized by autoradiography	^{14}C-labeled product
α-D-Glucosidase [19]	1. Enzymatic hydrolysis of 4-methyl-umbelliferyl-α-D-glucopyranoside and inspection of fluorescent 4-methylumbelliferone	4-methylumbelliferone (fluorescent)
[19]	2. Coupled enzyme test: (1) maltose + H_2O = glucose; (2) glucose + O_2 = gluconate + H_2O_2; H_2O_2 + 3-amino-9-ethylcarbazol =	oxidized carbazol

Table 5.7 (continued)

Enzyme [Ref.]	Brief reaction scheme	Recognition product
α-D-Glucosidase [19]	$2H_2O$ + oxidized carbazol (brown); (1) glucosidase, (2) gluconate oxidase	oxidized carbazol
[115]	3. Enzymatic hydrolysis of 6-bromo-2-naphthyl-α-D-glucoside and coupling of Fast blue B to liberated 2-naphthol	diazo dye
Glutamate dehydrogenase [115]	1. glutamate + H_2O + NADP = NH_3 + 2-oxoglutarate + $NADPH_2$; $NADPH_2$ + PMS + MTT = PMS + NADP + formazan	formazan
[115]	2. 2-oxoglutarate + NH_3 + $NADPH_2$ = glutamate + H_2O + NADP	NADP (non-fluorescent)
Glutathione reductase [19]	1. $NADPH_2$ + glutathione$_{ox.}$ = NADP + 2 glutathione$_{red.}$; glutathione$_{red.}$ + DCPIP = glutathione$_{ox.}$ + DCPIP$_{red.}$; DCPIP$_{red.}$ + MTT = DCPIP + formazan	formazan
[57]	2. $NADPH_2$ + glutathione$_{ox.}$ = NADP + 2 glutathione$_{red.}$; glutathione$_{red.}$ + 2-nitrobenzoic acid = nitro dye	nitro dye
Lactate dehydrogenase [57]	1. lactate + NAD = pyruvate + $NADH_2$; $NADH_2$ + PMS + MTT = NAD + PMS + formazan	formazan
[19]	pyruvate + $NADH_2$ = lactate + NAD	NAD (non-fluorescent)
Lactoyl-glutathione lyase [20]	1. glutathione$_{red.}$ + methylglyoxal = S-D-lactoyl-glutathione. Following incubation, DCPIP and MTT are added to the incubating medium. Glutathione$_{red.}$ reduces DCPIP which reduces MTT.	"white patches" on a blue background
[116]	2. Starch gels are incubated with a solution containing glutathione$_{red.}$ and methylglyoxyl. At the end of the incubation period the gels are overlayered with a J_2/KJ-containing agar solution. Sites of enzyme activity stain blue since here J_2 is not reduced to J⁻ by glutathione.	starch stained with J_2
α-D-Mannosidase [48]	1. Enzymatic hydrolysis of 4-methylumbelliferyl-α-D-mannoside and inspection of liberated 4-methylumbelliferone.	4-methylumbelliferone (fluorescent)
[117]	2. Enzymatic hydrolysis of p-nitrophenyl-α-D-mannopyranoside and inspection of liberated yellow p-nitrophenole	p-nitrophenol (yellow)

Table 5.7 (continued)

Enzyme [Ref.]	Brief reaction scheme	Recognition product
NADH-Diaphorase (Cytochrome b_5 reductase) [19]	1. $NADH_2 + DCPIP = NAD + DCPIP_{red.}$; $DCPIP_{red.} + MTT = DCPIP + formazan$	formazan
[118]	2. $NADH_2 + DCPIP = NAD + DCPIP_{red.}$	NAD (non-fluorescent)
Nucleoside triphosphatase [19]	1. Precipitation of enzymatically liberated orthophosphate with lead ions	precipitate of $Pb_3(PO_4)_2$
[120, 121]	2. Staining of enzymatically liberated otho-phosphate as blue molybdato-complex	
Phosphodiesterase I [122]	1. Enzymic hydrolysis of 4-methylumbelliferyl-5′-thymidylate and inspection of fluorescent 4-methylumbelliferone	4-methylumbelliferone (fluorescent)
[123]	2. Enzymic hydrolysis of α-naphthyluridine-5′-phosphate and coupling of Fast red TR to liberated α-naphthol	diazo dye
Phosphoglyceromutase [19]	1. Coupled enzyme test: (1) 2,3-bis-phosphoglycerate + 2-phosphoglycerate = 3-phosphoglycerate + 2,3-bis-phosphoglycerate; (2) 3-phosphoglycerate + ATP = ADP + 3-phosphoglyceroylphosphate; (3) 3-phosphoglyceroylphosphate + $NADH_2$ = glyceraldehyde-3-phosphate + P_i + NAD; (1) phosphoglyceromutase, (2) phosphoglycerate kinase, (3) glyceraldehyde phosphate dehydrogenase (phosphorylating)	NAD (non-fluorescent)
	2. Coupled enzyme test: (1) 3-phosphoglycerate + 2,3-bis-phosphoglycerate = 2,3-bis-phosphoglycerate + 2-phosphoglycerate; (2) phosphoglycerate = PEP + H_2O; (3) PEP + ADP = pyruvate + ATP; (4) pyruvate + $NADH_2$ = lactate + NAD; (1) phosphoglyceromutase, (2) enolase, (3) pyruvate kinase	NAD (non-fluorescent)
Superoxide dismutase [19]	1. $MTT + 2O_2^- + 2H^+ = 2O_2 + formazan$; formazan + PMS = MTT + PMS	"white patches" on a blue background
[124]	2. Upon exposure to light the two catalysts Temed and riboflavin form radicals which reduce Nitro BT to formazan. Superoxide dismutase hinders this reaction sequence and white bands appear on a dark blue background.	"white patches" on a blue background

Table 5.7 (continued)

Enzyme [Ref.]	Brief reaction scheme	Recognition product
Triosephosphate isomerase [125]	1. Coupled enzyme test: (1) glyceraldehyde-3-phosphate = dihydroxyacetone phosphate; (2) dihydroxyacetone phosphate + $NADH_2$ = α-glycerol-3-phosphate + NAD; (1) triosephosphate isomerase, (2) glycerol-3-phosphate dehydrogenase	NAD (non-fluorescent)
[126]	2. Coupled enzyme test: (1) α-glycerophosphate + NAD = $NADH_2$ + dihydroxyacetone phosphate; (2) dihydroxyacetone phosphate = glyceraldehyde-3-phosphate; (3) glyceraldehyde-3-phosphate + NAD = $NADH_2$ + 3-phosphoglycerate; $NADH_2$ + PMS + MTT = NAD + PMS + formazan; (1) glycerol-3-phosphate dehydrogenase, (2) triosephosphate isomerase, (3) glyceraldehyde-phosphate dehydrogenase	formazan

5.7 References

1. Bielka H, Horecker BL, Jakoby WB, Karlson P, Keil B, Liebecq C, Lindberg B, Webb EC (1979) Enzyme Nomenclature Academic Press, New York
2. Dawson CR, Tarpley WB (1951) The copper oxidases. In: Sumner IB, Myrbäck K (ed) The Enzymes vol II, Academic Press, New York, pp 454 – 498
3. Theorell H (1951) Flavin-containing enzymes. In: Sumner JB, Myrbäck K (ed) The Enzymes vol II, Academic Press, New York, pp 335 – 356
4. Kuhn R, Jerchel, D (1941) Ber Deutsch Chem Ges 748: 941 – 948
5. Hunter R, Markert C (1951) Science 125: 1294 – 1295
6. Latner AL, Skillen AW (1961) The Lancet 2: 1286 – 1288
7. Burd JF, Usategui-Gomez M (1973) Clin Chim Acta 46: 223 – 227
8. Möllering H, Wahlefeld AW, Michael G (1978) Visualisierung NAD(P)-abhängiger Reaktionen. In: Bergmeyer HU in collaboration with Gawehn K (eds) Principles of Enzymatic Analysis, Verlag Chemie Weinheim Germany, pp 145 – 153
9. Melrose TR, Brown CG, Sharma RD (1981) Res Vet Sci 29: 298 – 304
10. Pierce M, Cummings RC, Roth S (1980) Anal Biochem 102: 441 – 449
11. Graumann W, Neumann K (1960) Handbuch der Histochemie vol VIII Enzyme Teil I Gustav Fischer Verlag Stuttgart Germany
12. Silverstein E, Geller H (1974) J Chromatography 101: 327 – 337
13. Zimmermann H, Pearse AGE (1959) J Histochem Cytochem 7: 271 – 275
14. Anderson H (1965) Acta Histochem 21: 120 – 134
15. Katz AM, Kalow W (1965) Can J Biochem 43: 1653 – 1659
16. Shaw CR, Koen AL (1968) Ann N Y Acad Sci 151: 149 – 156
17. Ressler N, Stitzer K (1967) Biochim Biophys Acta 146: 1 – 9
18. Wilkinson HJ (1970) Isoenzymes, Chapman and Hall Ltd. London
19. Harris H, Hopkinson DA (1976) Handbook of enzyme electrophoresis in human genetics North Holland Publ Comp Amsterdam Amer Elsevier Pub Comp Inc New York

20. Kömpf I, Bissbort S, Gussman S, Ritter H (1975) Humangenetik 27: 141 – 143
21. McCune D (1961) Ann NY Acad Sci 94: 723 – 730
22. Graham RC, Lundholm U, Karnovsky MJ (1965) J Histochem Cytochem 13: 150 – 152
23. Brewbaker JL, Upadhya MD, Mäkinen Y, McDonald T (1968) Phys Plantarum 21: 930 – 940
24. Mäder M, Meyer Y, Bopp M (1975) Planta 122: 259 – 268
25. Holland VR, Saunders BC, Rose FL, Walpole AL (1974) Tetrahedron 30: 3299 – 3302
26. Tsuge H, Nakanishi Y (1980) Activity staining for flavoprotein oxidases. In: Mc Cormick DB, Wright LD (eds) Methods in Enzymology 66 E (Vitamins and coenzymes). Academic Press, New York San Francisco London, pp 473 – 483
27. Faye L (1981) Anal Biochem 112: 90 – 95
28. März L, Barna J, Ebermann R (1976) J Chromatogr 123: 495 – 496
29. Van Someren H, van Henegouwen HB, Los W, Wurzer-Figurelli E, Doppert B, Veroloet M, Meera Khan P (1974) Humangenetik 25: 189 – 201
30. Lewis WHP, Harris H (1967) Nature (London) 215: 351 – 355
31. Baker I.P (1974) Biochem Genet 12: 199 – 201
32. Stafford HA, Galston AW (1970) Plant Physiol 46: 763 – 767
33. Van Loon LC(1971) Phytochemistry 10: 503 – 507
34. Edwards YH, Hopkinson DA, Harris H (1971) Ann Hum Genet London 34: 395 – 408
35. Turner BM, Fisher RA, Harris H (1975) Post-translational alterations of human erythrocyte enzymes. In: Markert CL (ed) Isozymes I Academic Press, New York, pp 781 – 795
36. Spencer N, Hopkinson DA, Harris H (1968) Ann Hum Genet 32: 9 – 14
37. Meuwissen HJ, Pollara B, Pickering RJ (1975) J Pediat 86: 169 – 181
38. Baron DN, Buttery JE (1972) J Clin Pathol 25: 415 – 421
39. Spanos A, Sedgwick SG, Yarranton GT, Hübscher U, Banks GR (1981) Nucleic Acids Res 9: 1825 – 1839
40. Barrett RJ, Seligman (1951) AM Science 114: 579 – 582
41. Holt SJ, Withers RFJ (1952) Nature 170: 1012 – 1014
42. Rothe GM (1972) Beitr Biol Pflanzen 48: 433 – 444
43. Swallow DM, Povey S, Harris H (1973) Ann Hum Genet 37: 31 – 38
44. Shaw CR, Prasad R (1970) Biochem Genet 4: 297 – 320
45. Rattazzi MC, Marks JS, Davidson RG (1973) Am J Hum Genet 25: 310 – 316
46. Dreyfus JC, Alexandre Y (1972) Biochem Biophys Res Commun 48: 914 – 920
47. Ho MW, O'Brien JS (1971) Clin Chim Acta 32: 443 – 450
48. Poenaru C, Dreyfus JE (1973) Clin Chim Acta 43: 439 – 442
49. Hooghwinkel GJM, Veltkamp WA, Overdijk B, Lisman JJW (1972) Z Physiol Chem 353: 839 – 841
50. Robinson D, Stirling JL (1968) Biochem J 107: 321 – 327
51. Chern CJ, Croce CM (1976) Am J Hum Genet 28: 350 – 356
52. Franke W (1940) Die Enzyme der Desmolyse. In: Nord FF, Weidenhagen R (eds) Handbuch der Enzymologie I Akademische Verlagsgesellschaft Becker & Erler, Leipzig, pp 673 – 845
53. Zöllner EJ, Müller WEG, Zahn RK (1973) Z Naturforsch 28c: 376 – 378
54. Takamatsu H (1938) Trans Soc Path Japan 29: 429 – 498
55. Gomori G (1939) Proc Soc Exp Biol NY 42: 23 – 26
56. Nimmo HG, Nimmo GA (1982) Anal Biochem 121: 17 – 22
57. Brewer GJ, Sing CF (1970) An introduction to isozyme techniques Academic Press, New York
58. Nachlas MM, Goldstein TP, Rosenblatt DH, Kirsch M, Seligman (1959) AM J Histochem Cytochem 7: 50 – 65
59. Abe M, Kramer SP, Seligman AM (1964) J Histochem Cytochem 12: 364 – 383
60. Schwartz MK, Nisselbaum JS, Bodansky O (1963) Techn Bull Regist Med Techn 33: 101 – 104
61. Smith RE, Van Frank RM (1975) The use of amino acid derivatives of 4-methoxy-β-naphthylamine for the assay and subcellular localization of tissue proteinases. In: Dingle JT, Dean RT (eds), Lysosomes in Biology and Pathology vol 4, North Holland, Amsterdam, pp 19 – 264
62. Barrett AJ (1973) Biochem J 131: 809 – 822
63. Takahashi K, Isemura M, Ikenaka T (1979) J Biochem 85: 1053 – 1060
64. Mort JS, Leduc M (1982) Anal Biochem 119: 148 – 152
65. Barrett AJ (1972) Anal Biochem 47: 280 – 293

66. Quinn PS, Judah JD (1978) Biochem J 127: 301 – 309
67. Castillo MJ, Nakajima K, Zimmermann M, Powers JC (1979) Anal Biochem 99: 53 – 64
68. Barrett AJ (1980) Biochem J 187: 909 – 912
69. Kato T, Nakano T, Kojima K, Nagatsu T, Sakakibara S (1980) J Neurochem 35: 527 – 535
70. Saifuku K, Sekine T, Namihisa T, Kanaoka Y (1978) Clin Chim Acta 84: 85 – 91
71. Carmel A, Ehrlich-Rogozinsky S, Yaron A (1979) Clin Chim Acta 93: 215 – 220
72. Smith RE (1984) J Histochem Cytochem 32: 1265 – 1274
73. Boyd JB, Mitchell HK (1965) Anal Biochem 13: 28 – 42
74. Grdina DJ, Lohman PHM, Hewitt RR (1973) Anal Biochem 51: 255 – 264
75. Strong JE, Hewitt RR (1975) Investigation of human plasma DNAse(s) by DNA-polyacrylamide gel electrophoresis. In: Markert CL (ed) Isozymes vol 3 (Developmental biology) Academic Press, New York, pp 473 – 483
76. Van Loon LC (1975) FEBS Letters 51: 266 – 269
77. Randles JW (1968) Virology 36: 556 – 563
78. Rosenthal AL, Lacks SA (1977) Anal Biochem 80: 76 – 90
79. Lewis WHP, Harris H (1967) Nature (London) 215: 351 – 355
80. Ward CW (1975) Aust J Biol Sci 28: 1 – 23
81. Blackwood CE, Erlanger BF, Mandel I (1965) Anal Biochem 12: 128 – 136
82. Morihara K, Tsuzuki H, Oka T (1968) Arch Biochem Biophys 123: 572 – 588
83. Schoellmann G, Shaw E (1963) Biochemistry 2: 252 – 255
84. Shaw E, Mares-Guia M, Coken W (1965) Biochemistry 4: 2219 – 2224
85. Clavin SA, Bobbitt JL, Shuman RT, Smithwick EL (1977) Anal Biochem 80: 355 – 365
86. Astrup T (1966) Fed Proc 25: 42 – 51
87. Miller RE, Shelton E, Stadtman ER (1974) Arch Biochem Biophys 163: 155 – 171
88. Fildes RA, Harris H (1966) Nature 209: 261 – 263
89. Fisher RA, Turner BM, Dorkin HL, Harris H (1974) Ann Hum Genet 37: 341 – 353
90. Klebe RJ, Schloss S, Mock L, Link CR (1981) Biochem Genet 19: 921 – 927
91. Nachlas MM, Moris B, Rosenblatt D, Seligman AM (1960) J Biophys Biochem Cytol 7: 261 – 264
92. Strongin AY, Azarenkova NM, Vaganova TI, Levin ED, Stepanov VM (1976) Anal Biochem 74: 597 – 599
93. Farron F (1973) Anal Biochem 53: 264 – 268
94. Nelson RL, Povey MS, Hopkinson DA, Harris H (1977) Biochem Genet 15: 1023 – 1035
95. Hopkinson DA, Peters J, Harris H (1974) Ann Hum Genet 37: 477 – 484
96. Uriel J (1961) Annals l'ínstitute Pasteur 101: 104 – 110
97. Gregory EM, Fridovick I (1974) Anal Biochem 58: 57 – 62
98. Gruft H, Gaafar HA (1974) Am Review Resp Disease 110: 320 – 323
99. Eppenberger ME, Eppenberger HM, Kaplan NO (1967) Nature 214: 239 – 241
100. Monn E, Christiansen RO (1971) Science 173: 540 – 541
101. Tsou KC, Lo KW, Yip KF (1974) FEBS Letters 45: 47 – 49
102. Yip KF, Tsou KC (1973) Tetrahedron Letters 33: 3087 – 3090
103. Teng Y-S, Anderson JE, Giblett ER (1975) Am J Hum Genet 27: 492 – 497
104. Kim HS, Liao T-H (1982) Anal Biochem 119: 96 – 101
105. Porter ACG (1981) Anal Biochem 117: 28 – 31
106. Spanos A, Sedgwick SG, Yarranton GT, Hübscher U, Banks GR (1981) Nucleic Acids Res 9: 5919 – 5925
107. Hullin DA, Thompson RJ (1977) Anal Biochem 82: 240 – 242
108. Faye L (1981) Anal Biochem 112: 90 – 95
109. Chung PLY, Trevithick JR (1970) J Bacteriol 102: 423 – 429
110. März L, Barna J, Ebermann R (1976) J Chromatogr 123: 495 – 49
111. Walker DG, Khan HH (1968) Biochem J 108: 169 – 175
112. Ho MW, O'Brien JS (1969) Science 165: 611 – 613
113. Tedesco TA, Diamond R, Orkwiszewski KG, Boedecker HJ, Croce CM (1974) Proc Natl Acad Sci USA 71: 3483 – 3486
114. Nichols EA, Elsevier SM, Ruddle FH (1974) Cytogenet Cell Genet 13: 275 – 278
115. Van der Helm HJ (1962) Nature 194: 773
116. Parr CW, Bagster IA, Welch SG (1977) Biochem Genet 15: 109 – 113

117. Gabriel O, Wang S-F (1969) Anal Biochem 27: 545 – 554
118. West CA, Gomperts BD, Huehns ER, Kessel I, Ashby JR (1967) Br Med J 4: 212 – 214
119. Abrahams A, Baron C (1967) Biochemistry 6: 225 – 229
120. Selwyn MJ (1967) Biochem J 105: 279 – 288
121. Wachstein M, Meisel E (1957) Am J Clin Pathol 27: 13 – 23
122. Lerch B (1968) Experientia 24: 889 – 890
123. Hawley DH, Tsou KC, Hodes ME (1981) Anal Biochem 117: 18 – 23
124. Beauchamp C, Fridovitch I (1971) Analyt Biochem 44: 276 – 287
125. Kaplan JC, Teeple L, Shore N, Beutler E (1968) Biochem Biophys Res Commun 31: 768 – 773
126. Scopes RK (1964) Nature 201: 924 – 925

6 A Compilation of Protocols to Visualize Enzymes Following Electrophoretic Separation

The following tables supply protocols for the electrophoretic separation as well as for the succeeding visualization of enzymes. Separation systems and enzyme visualization procedures are adjusted to each other.

To some extent several different electrophoretic systems have been published so far for the separation of certain enzymes; those are also listed in the following.

If an enzyme is to be determind, it is best to consult first "The Index to Enzyme Recipes and Separation Systems" (sect. 6.1). In an alphabetical order all enzymes are listed here for which a histochemical assay may be found under "Protocols to Visualize Enzymes Following Electrophoretic Separation" (sect. 6.2).

Besides, the index Table 6.1 informs on the electrophoretic system(s) the enzyme (system) concerned has been separated with.

When performing enzyme separation on Cellogel[R] (C), the respective buffer system is given in Table 6.2.1, polyacrylamide gel electrophoretical systems (D) are listed in Table 6.2.2, separations performed by applying isoelectric focusing (I) are given in Table 6.2.3, and buffer systems used for the separation in starch gels (S) can be taken from Table 6.2.4.

6.1 Index to Enzyme Recipes and Separation Systems

Name of enzyme	EC number	System[1]			
		C	D[a)]	I	S
Acetylcholinesterase	3.1.1.7		1 (1)		1,2
α-N-Acetyl-D-glucosaminidase	3.2.1.50		1		
β-N-Acetyl-D-glucosaminidase	3.2.1.30	1	1		3,4
β-N-Acetyl-D-hexosaminidase	3.2.1.52	2	1,2		3
Acid phosphatase	3.1.3.2		1 (2)	5	
Aconitate hydratase	4.2.1.3				6

1 C= Cellogel[R], D = polyacrylamide gel electrophoresis (PAGE), I = isoelectric focusing, S = starch gel electrophoresis. For more information see Tables 6.2.1 to 6.2.4. a) Remarks.

Name of enzyme	EC number	System[1]			
		C	D[a)]	I	S
Acylphosphatase	3.6.1.7			1	
Adenosine deaminase	3.5.4.4	3			7
Adenosinetriphosphatase	3.6.1.3		1 (3)		
Adenosylhomocysteinase	3.3.1.1		1 (4)		
Adenylate kinase	2.7.4.3	3			5
Alanine aminotransferase	2.6.1.2	4			4
Alanine dehydrogenase	1.4.1.1				8
Alcohol dehydrogenase	1.1.1.1			2	4
Aldehyde dehydrogenase	1.2.1.3		1 (5)		9
Alkaline phosphatase	3.1.3.1	5	1 (6)		10
Amine dehydrogenase	1.4.99.3		1 (7)		
Amine oxidase (copper-containing)	1.4.3.6		1 (8)		11
D-Amino-acid oxidase	1.4.3.3				12
L-Amino-acid oxidase	1.4.3.2		1 (9)	1	
Aminoacylase	3.5.1.14	6			
Aminopeptidase (cytosol)	3.4.11.1		1 (10)		13
Aminotransferases	2.6.1.(1 – 6)	2			
AMP deaminase	3.5.4.6				14
α-Amylase	3.2.1.1	7	1		
Anthranilate phosphoribo-syltransferase	2.4.2.18		1 (11)		
Anthranilate synthase	4.1.3.27		1 (11)		
Arginase	3.5.3.1				15, 16
Argininosuccinate lyase	4.3.2.1				17
Arylsulphatase	3.1.6.1	8	1		
Aspartate aminotransferase	2.6.1.1	4			4
D-Aspartate oxidase	1.4.3.1				12
Carbonate dehydratase (NADP+)	4.2.1.1		1		18
Carboxylesterase	3.1.1.1		1		
Catalase	1.11.1.6		1 (12)		20
Catechol oxidase	1.10.3.1		1 (13)		1
Cathepsin B	3.4.22.1		1		
Cellulase	3.2.1.4		1 (14)		
Cholinesterase	3.1.1.8				1,2
Chymotrypsin	3.4.21.1	9	3 (15)		
Citrate (si)-synthase	4.1.3.7	10			
Creatine kinase	2.7.3.2				21
3',5'-Cyclic-nucleotide phosphodiesterase	3.1.4.17				22
Cystathionine β-synthase	4.2.1.22		1 (16)		

Name of enzyme	EC number	System[1]			
		C	D[a]	I	S
Cystyl aminopeptidase	3.4.11.3		4 (17)		
Cytidine deaminase	3.5.4.5				23
Deoxyribonuclease I	3.1.21.1			3	
Dihydrouracil dehydro-genase (NADP⁺)	1.3.1.2		1 (18)		
Dipeptidase	3.4.13.11		1 (18)		7
endo-β-N-Acetylglucos-aminidase	3.2.1.96	11			
Enolase	4.2.1.11	12			24
β-D-Fructofuranosidase	3.2.1.26		1 (19)		
Fructokinase	2.7.1.4				22
Fructose-bisphosphatase	3.1.3.11		1 (18)		
Fructose-bisphosphate aldolase	4.1.2.13				12
L-Fucose dehydrogenase	1.1.1.122		1 (18)		
α-L-Fucosidase	3.2.1.51				25
Fumarate hydratase	4.2.1.2	13			
Galactokinase	2.7.1.6				26
α-D-Galactosidase	3.2.1.22	14	1 (20)		27
Glucose dehydrogenase	1.1.1.47				28
Glucose oxidase	1.1.3.4		1 (20)	1	
Glucose-1-phosphate uridylyltransferase	2.7.7.9	13	1		
Glucose-6-phosphate dehydrogenase	1.1.1.49	15			29
Glucose-phosphate isomerase	5.3.1.9	15			30
α-D-Glucosidase	3.2.1.20			1	3
β-D-Glucuronidase	3.2.1.31				4
L-Glutamate dehydro-genase (NADP)	1.4.1.4		5 (29)		17
Glutamine synthetase	6.3.1.2		1 (21)		
Glutaminyl-peptide γ-glutamyltransferase	2.3.2.13				
Glutathione peroxidase	1.11.1.9				13
Glutathione reductase	1.6.4.2	16			21
Glyceraldehyde-phosphate dehydrogenase	1.2.1.12		1		31

Name of enzyme	EC number	System[1]			
		C	D [a)]	I	S
Glycerol-3-phosphate dehydrogenase (NAD)	1.1.1.8				4,32
Glycollate oxidase	1.1.3.1		1		
Guanine deaminase	3.5.4.3				33
Guanylate kinase	2.7.4.8	17			22
Hexokinase	2.7.1.1				12
Homoserine dehydrogenase	1.1.1.3		6 (22)		
3-Hydroxybutyrate dehydrogenase	1.1.1.30				34
β-Hydroxysteroid dehydrogenase	1.1.1.51		7 (23)		
3-Hydroxyacyl-CoA dehydrogenase	1.1.1.35	18			
3α-Hydroxysteroid dehydrogenase	1.1.1.50	19			
Hypoxanthine phospho-ribosyltransferase	2.4.2.8	20			29
L-Iditol dehydrogenase	1.1.1.14				35
Inorganic pyrophosphatase	3.6.1.1				36
Isocitrate dehydrogenase (NADP)	1.1.1.42	21			37
Lactate dehydrogenase	1.1.1.27	21			38
Lactose synthase	2.4.1.22		8 (24)		
Lactoyl-glutathione lyase	4.4.1.5				39
Leucine dehydrogenase	1.4.1.9				40
Lysine 2-mono oxygenase	2.13.12.2		1 (25)		
Malate dehydrogenase	1.1.1.37	22			4
Malate dehydrogenase (oxalo-acetate-decarboxylating, (NADP))	1.1.1.40	23			41
Mannitol dehydrogenase	1.1.1.67		1 (26)		
Mannosephosphate isomerase	5.3.1.8	24			7
α-D-Mannosidase	3.2.1.24		1 (27)		4
Melilotate 3-mono-oxygenase	1.14.13.4		1 (28)		
Monophenol mono-oxygenase	1.14.18.1		1 (29)		
NADH dehydrogenase	1.6.99.3				42
NADPH dehydrogenase	1.6.99.1		9 (30)		

Name of enzyme	EC number	System[1]			
		C	D[a)]	I	S
NAD(P) nucleosidase	3.2.2.6	22			
Nitrate reductase (NADH)	1.6.6.1		10 (31)		
Nitrogenase	1.18.2.1		11 (32)		
Nucleoside triphosphatase	3.6.1.15		12		
Nucleosidetriphosphate-adenylate kinase	2.7.4.10				43
Nucleosidetriphosphate pyrophosphatase	3.6.1.19				29
5'-Nucleotidase	3.1.3.5		1 (31)		
Oestradiol 17β-dehydrogenase	1.1.1.62		1 (23)		
Penicillinase	3.5.2.6				44
Pepsin	3.4.23.3				
Peptidases	3.4.11(13)	17			45
Peroxidase	1.11.1.7		2	6	
Phosphodiesterase I	3.1.4.1	24	1 (31)		
6-Phosphofructokinase	2.7.1.11				46
Phosphoglucomutase	2.7.5.1	13			4
Phosphogluconate dehydrogenase (decarboxylating)	1.1.1.44	15			47
Phosphoglycerate kinase	2.7.2.3	25			7
Phosphoglyceromutase	2.7.5.3				4
Phosphorylase	2.4.1.1		13 (33)		
Polyribonucleotide nucleotidyl-transferase	2.7.7.8		1 (34)		
Purine-nucleoside phosphorylase	2.4.2.1	20			29
Pyridoxal kinase	2.7.1.35				48
Pyruvate kinase	2.7.1.40	3			49
Retinol dehydrogenase	1.1.1.105				50
Ribonuclease (pancreatic)	3.1.27.5		14		
Ribosephosphate pyrophosphokinase	2.7.6.1	26			
RNA nucleotidyltransferase	2.7.7.6		15 (34)		
Sucrose phosphorylase	2.4.1.7		1 (35)		
Superoxide dismutase	1.15.1.1	15			48
Testosterone 17β-dehydrogenase (NADP+)	1.1.1.64			4	

Name of enzyme	EC number	System[1]			
		C	D[a]	I	S
Tetrahydrofolate dehydrogenase	1.5.1.3	26			
Threonine dehydratase	4.2.1.16		1 (36)		
Thrombin	3.4.21.5			5	
Transaldolase	2.2.1.2				51
Transketolase	2.2.1.1				51
Triacyl glycerol lipase	3.1.1.3	27			
Triosephosphate isomerase	5.3.1.1	28			52
Tripeptide aminopeptidase	3.4.11.4	25			
Trypsin	3.4.21.4		16 (18)		
UDPglucose-hexose-1-phosphate uridylyl-transferase	2.7.7.12				53
Urease	3.5.1.5		1 (35)		
Xanthine oxidase	1.2.3.2			1	23

6.2 Protocols to Visualize Enzymes Following Electrophoretic Separation

Acetylcholinesterase (3.1.1.7)

(syn: serum cholinesterase, pseudocholinesterase)

Reaction scheme:

An acetylcholine + H_2O = choline + a carboxylic anion (α-naphthyl acetate + H_2O = acetate + α-naphthol, α-naphthol + Fast red TR = coloured *diazo dye*)

Electrophoresis:

Technique	pH	System	Staining	Source
starch gel 5 V/cm, 5 h	8.6	S_2	FM	human

Recipe:

Dissolve 1% α-naphthyl acetate in 50% aqueous acetone, add 0.5 ml of this solution slowly under stirring to 25 ml of a 200 mmol l^{-1} phosphate buffer of pH 7.1, add 10 mg of Fast red TR. Filtrate if necessary. Impregnate starch gel until red zones are visible. In serum or plasma, cholinesterase isozymes are the principal esterases. In extracts of human tissues various esterases are also detectable by this staining procedure. The major isozyme C_4 and the minor isozymes C_1, C_2 and C_3 are best separated at pH 8.0 to 8.6, while C_5 appearing in about 10% in European populations is best separated from C_4 at pH 5.0. Treatment with neuraminidase results in a marked decrease in anodal mobility [1, 2].

α-N-Acetyl-D-glucosaminidase (3.2.1.50)

Reaction scheme:

o-Nitrophenyl-N-acetyl β-D-glucosaminide + H_2O = N-acetyl-β-D-glucosamine + *o-nitrophenol* (yellow)

Electrophoresis:

Technique	pH	System	Staining	Source
PAGE	8.9	D_1	FM	Jack bean meal

Recipe:

Processed PAA gels are rinsed with water and incubated at room temperature for 30 min in a 200 mmol l⁻¹ acetate buffer of pH 4.9. Subsequently the gels are immerced for 30 min in 25 mmol l⁻¹ citrate buffer of pH 5.5 containing 5 mmol l⁻¹ (= 1.7 mg/ml) o-nitrophenyl-N-acetyl-β-D-glucosaminide. Enzyme activity is indicated by a developing yellow band of o-nitrophenol at an alkaline pH [3].

β-N-Acetyl-D-glucosaminidase (3.2.1.30)

(The enzyme hydrolyzes non-reducing 2-acetamido-2-deoxy-β-D-glucose residues in chitobiose and higher analogues in glycoproteins).

Reaction scheme:

AS BI 2-acetamido-2-deoxy-β-D-glucopyranoside + H_2O = N-acetyl-glucosamine + AS BI naphthol, AS BI naphthol + Fast Garnet GBC = *diazo-dye* (coloured)

Electrophoresis:

Technique	pH	System	Staining	Source
starch gel, 13 V/cm, 4 h	5.7	S_3	OL	human

Recipe:

Dissolve 4 mg naphthol AS BI 2-acetamido-2-deoxy-β-D-glucopyranoside in 2 ml hot absolute ethanol and mix with 8 ml 0.1 mol l⁻¹ citrate buffer of pH 4.5. Finally add 10 mg Fast Garnet GBC. Incubate processed starch gel and wash afterwards with methanol/water/acetic acid (5:5:1) [4 – 6].

β-N-Acetyl-D-glucosaminidase (3.2.1.30)

Reaction scheme:

4-Methylumbelliferyl-2-acetamido-2-deoxy-β-D-glucopyranoside + H_2O = N-acetyl-glucosamine + *4-methylumbelliferone* (fluorescent)

Electrophoresis:

Technique	pH	System	Staining	Source
starch gel 13 V/cm, 5 h	8	S4	FM	human

Recipe:

Dissolve 20 mg 4-methylumbelliferyl-2-acetamido-2-deoxy-β-D-glucopyranoside in 10 ml of a 25 mmol l^{-1} citrate-phosphate buffer of pH 4.0. Drop the solution on the cut surface of a processed starch gel and incubate for 45 min at 37 °C. Then wash the gel with H$_2$O and spray the surface with a 7.4 N NH$_4$OH-solution. Sites of enzyme activity appear as fluorescent bands under UV-light [5, 6].

β-N-Acetyl-D-glucosaminidase (3.2.1.30)

Reaction scheme:

Naphtol AS BI 2-acetamido-2-deoxy-β-D-glucopyranoside + H$_2$O = N-acetyl-glucosamine + AS BI Naphtol, AS BI Naphthol + Fast Garnet GBC = *diazo-dye* (coloured)

Electrophoresis:

Technique	pH	System	Staining	Source
starch gel, 13 V/cm, 4 h	5.7	S$_3$	FA	human

Recipe:

Dissolve 20 mg naphthol AS BI 2-acetamido-2-deoxy-β-D-glucopyranoside in 10 ml hot absolute ethanol. Dissolve 50 mg Fast Garnet GBC in 40 ml 0.1 mol l^{-1} citrate buffer of pH 4.5, mix both solutions and pour over the cut surface of a processed starch gel. Wash the gel after incubation with a mixture of methanol/water/acetic acid (5:5:1) [6].

β-N-Acetyl-D-hexosaminidase (3.2.1.52)

Reaction scheme:

4-Methylumbelliferyl-β-N-acetyl-galactosaminide + H$_2$O = β-N-acetyl-galactosamine + *4-methylumbelliferone* (fluorescent)

Electrophoresis:

Technique	pH	System	Staining	Source
starch gel	5.7	S$_3$	FA	human

Recipe:

Dissolve 4 mg 4-methylumbelliferyl-β-N-acetyl-galactosaminide in 10 ml of a 150 mmol l^{-1} citrate-phosphatebuffer of pH 4.4. Hexosaminidase A from human tissues is almost totally transformed to hexosaminidase D at 50 °C along with a 10 % loss in enzyme activity. The conversion is inhibited by N-acetyl-galactosamine, a competitive inhibitor of hexosaminidase B. Removal of N-acetyl-neuraminic acid alone through the reaction of neuraminidase does not cause the transformation [7].

Acid phosphatase (3.1.3.2)

Reaction scheme:

α-Naphthyl phosphate + H_2O = α-naphthol + P_i
α-naphthol + Fast Blue BB = *diazo-dye* (coloured)

Electrophoresis:

Technique	pH	System	Staining	Source
starch gel, 13 V/cm, 5 h, 4 °C	7	S_5	FM	human

Recipe:

Dissolve 10 mg Fast Garnet GBC (or Fast Blue BB)- salt in 10 ml of a 0.05 mol l⁻¹ citrate-buffer of pH 4.5 and add 0.4 ml α-naphthyl phosphate-Na$_2$ (1% in 50% acetone).

Filtrate, and drop on the cut surface of a processed starch gel. The appearance of blue (or red) bands indicates the presence of active enzyme(s) [5, 6, 8].

Acid phosphatase (3.1.3.2)
(alternative method)

Reaction scheme:

4-Methylumbelliferyl phosphate + H_2O = phosphate + *4-methylumbelliferone* (fluorescent)

Electrophoresis:

Technique	pH	System	Staining	Source
starch gel, 13 V/cm, 5 h, 4 °C	7	S_5	FM	human

Recipe:

Dissolve 20 mg 4-methylumbelliferyl-di-hydrogen phosphate in 100 ml 100 mmol l⁻¹ citrate buffer, pH 6.3 and drop on the cut surface of a processed starch gel. Incubate at 37 °C for 5 – 90 min. Monitor under long wave UV-light. The fluorescence of weak bands may be intensified by making the gel surface alkaline with ammonia vapor. A citrate buffer of pH 5.5 – 6.5 is used to detect "red cell" acid phosphatase isozymes while a citrate buffer of pH 4.5 is used to detect "tissue" acid phosphatase [6].

Acid phosphatase (3.1.3.2)
(alternative method)

Reaction scheme:

α-Glycerophosphate + H_2O = glycerol + orthophosphate, orthophosphate + Ca^{2+} = $Ca_3(PO_4)_2 \cdot (H_2O)_n$ (precipitate)

Electrophoresis:

Technique	pH	System	Staining	Source
PAGE, 3 mA/gel, 2 h	8.6	D_1	FM	cattle

Recipe:

Dissolve 2.2 mg α-glycerophosphate in 10 ml of 100 mmol l^{-1} acetic acid/acetatebuffer of pH 5, containing 222 mg CaCl$_2$. Sites of enzyme activity appear as milky white bands [9].

Aconitate hydratase (4.2.1.3)

(syn: Aconitase)

Reaction scheme:

cis-Aconitate + H$_2$O = isocitrate, isocitrate + NADP (+ isocitrate dehydrogenase) = 2-oxoglutarate + NADPH + PMS = NADP + reduced PMS, reduced PMS + MTT = PMS + *formazane* (blue)

Electrophoresis:

Technique	pH	System	Staining	Source
starch gel, 3.5 V/cm, 17 h, 4 °C	7.5	S_6	AOL	human

Recipe:

Dissolve 75 mg cis-aconitic acid in 20 ml of a 400 mmol l^{-1} Tris-HCl buffer of pH 8 and re-adjust pH to 8.0. Add 5 ml MgCl$_2$-6H$_2$O (122 mg/ml), 1 ml NADP (5mg NADP-Na$_2$/ 1 ml H$_2$O), 0.1 ml isocitrate dehydrogenase (20 U/ml), 1 ml MTT (5 mg/1 ml H$_2$O) and 0.25 ml PMS (5 mg/ml H$_2$O).

Mix the substrate solution with 25 ml of a 2 % agar-solution, cooled down to 45 °C. Pour the solution on the cut surface of a processed starch gel [6].

Acylphosphatase (3.6.1.7)

Reaction scheme:

Acylphosphate + H$_2$O = a fatty acid anion + orthophosphate, acetylphosphate + NH$_2$OH = hydroxamic acid, hydroxamic acid + FeCl$_3$ = *coloured dye* (red to violet)

Electrophoresis:

Technique	pH	System	Staining	Source
IEF	3.5 – 10	I_1	MOL	horse

Recipe:

Place electrophoretic support medium in a mixture of 1 ml acetyl phosphate-K-Li-salt (11 mg/ml), 5 ml of a 100 mmol l^{-1} acetate buffer of pH 5.3 and 4 ml distilled water. After a suitable time the reaction is stopped by discarding the substrate solution and adding 3 ml of stain A, 3 ml H_2O and 9 ml stain B. (Stain A: immediately before use equal volumes of 4 mol l^{-1} NH_2OH-HCl and 3.5 mol l^{-1} NaOH are mixed. Stain B: Mix equal volumes of 5% $FeCl_3$ in 0.1 N HCl, 12% trichloroacetic acid and 3 N HCl.) The intensity of the red to violet coloured enzyme bands may be quantitatively estimated at a wave length of 530 nm [10].

Adenosine deaminase (EC 3.5.4.4)

Reaction scheme:

Adenosine + H_2O = inosine + NH_3, inosine orthophosphate (+ nucleoside phosphorylase) = hypoxanthine + α-D-ribose-1-phosphate, hypoxanthine + H_2O + PMS (+ xanthine oxidase) = xanthine + reduced PMS, reduced PMS + MTT = PMS + *formazane* (blue)

Electrophoresis:

Technique	pH	System	Staining	Source
Starch gel, 3 V/cm, 18 h	6.5	S_7	AOL	human

Recipe:

Add to 25 ml 50 mmol l^{-1} phosphate buffer of pH 7.5, 15 mg adenosine, 25 ml nucleoside phosphorylase (25 U/ml), 1 ml MTT (5 mg/ml H_2O) and 1 ml PMS (5 mg/ml H_2O). Mix the substrate solution with 25 ml of a 2% agar solution, cooled down to 45 °C. The electrophoretic properties of the red cell adenosine deaminase change on storage of haemolysates due to the oxidation of reactive sulphhydryls with oxidized glutathione. The changes can be prevented or reversed by the addition of 10 mmol l^{-1} 2-mercaptoethanol to the lysates [6]. Red cell adenosine deaminase is a single polypeptide chain (M_r: 37 000). High molecular mass forms (M_r: 200 000) result from a combination of the monomeric enzyme with dimeric glycoproteins. Insoluble adenosine deaminase is membrane bound [11].

Adenosinetriphosphatase (3.6.1.3)

Reaction scheme:

ATP + H_2O = ADP + ortho-phosphate, ortho-phosphate + Ca^{2+} = $Ca_3(PO_4)_2 - (H_2O)_n$ (precipitate)

Electrophoresis:

Technique	pH	System	Staining	Source
PAGE 5 mA/gel, 1.5 h, gel length: 65 mm	8.6	D_1	FM	*Micrococcus luteus*

Recipe:

Dissolve in 10 ml 100 mmol l^{-1} Tris-HCl, pH 8.0, 6 mg ATP-Na_2-$3H_2O$ and 5.5 mg $CaCl_2$. The formation of milky white bands indicate areas of enzyme activity [12].

Adenosylhomocysteinase (3.3.1.1)

Reaction scheme:

S-Adenosyl-L-homocysteine + H_2O = adenosine + L-homocysteine, L-homocysteine + dimethylthetin (+ thetin homocysteine methyltransferase) = hypoxanthine + α-D-ribose-1-phosphate, hypoxanthine + H_2O + PMS (+ xanthine oxidase) = xanthine + reduced PMS, reduced PMS + MTT = PMS + *formazane* (blue coloured)

Electrophoresis:

Technique	pH	System	Staining	Source
PAGE 1 – 2 m A/gel, 4 – 6 h	8.7	D_1	FM	calf

Recipe:

To 10 ml of a 50 mmol l^{-1} Tris-HCl buffer of pH 8.2 are added: 3.84 mg 5-adenosyl-L-homocysteine, 7.8 mg dimethylthetin chloride, 15.3 mg disodium hydrogen arsenate-7 H_2O, 0.1 ml thetin: homocysteine methyltransferase (2 000 U/ml), 0.1 ml adenosine deaminase (10 000 U/ml), 0.1 ml nucleoside phosphorylase (10 000 U/ml), 0.1 ml xanthine oxidase (5000 U/ml), 1 mg PMS and 1 mg MTT. The measurement of the hydrolytic reaction requires the use of coupled enzyme systems to remove the inhibitory endproducts [13].

Adenylate kinase (2.7.4.3)

Reaction scheme:

ADP + ADP = ATP + AMP, glucose + ATP (+ hexokinase) = glucose-6-phosphate + ADP, glucose-6-phosphate + NADP (+ glucose-6-phosphate dehydrogenase) = 6-phosphogluconate + NADPH, NADPH + PMS = NADP + reduced PMS, reduced PMS + Nitro BT = PMS + *formazane* (blue coloured)

Electrophoresis:

Technique	pH	System	Staining	Source
starch gel 13 V/cm, 5 h;	7	S_5	AOL	human

Recipe:

Dissolve 10 mg ADP, 7.5 mg NADP, 7.5 mg Nitro BT, 0.5 mg PMS, 10 mg $MgCl_2$-6H_2O and 22.5 mg glucose in 10 ml of a 200 mmol l^{-1} Tris-HCl buffer of pH 8 and add 85 Units of hexokinase and 40 Units of glucose-6-phosphate dehydrogenase.

Sites of enzyme activity are indicated by the appearence of blue bands [5, 6, 14].

Alanine aminotransferase (2.6.1.2)

(syn: Glutamate pyruvate transaminase)

Reaction scheme:

L-Alanine + 2-oxoglutarate = pyruvate + L-glutamate, pyruvate + NADH (+ lactate dehydrogenase) = lactate + *NAD* (non fluorescent)

Electrophoresis:

Technique	pH	System	Staining	Source
Cellogel, 12 V/cm, 2 h	7.8	C_2	FM	rat

Recipe:

Dissolve 30 mg NADH, 40 mg L-alanine and 20 mg α-ketoglutaric acid in 10 ml of 40 mmol l^{-1} Tris-HCl, pH 8.0 and add 300 Units of lactate dehydrogenase. Sites of enzyme activity appear as non-fluorescent bands under long wave UV-light (375 nm) in front of a fluorescent background [15, 16].

Alanine dehydrogenase (1.4.1.1)

Reaction scheme:

L-Alanine + H_2O + NAD^+ = pyruvate + NH_3 + NADH, NADH + PMS = NAD + reduced PMS, reduced PMS + MTT = PMS + *formazane* (blue coloured)

Electrophoresis:

Technique	pH	System	Staining	Source
starch gel, 13 V/cm, 5 h	7.0	S_8	FM	bacteria

Recipe:

Dissolve in 100 ml of a 100 mmol l^{-1} phosphate buffer of pH 7 100 mg DL-alanine, 50 mg NAD, 30 mg MTT and 2 mg PMS. Pour the solution on the cut surface of a processed starch gel. Sites of enzyme activity are indicated by the appearance of blue bands [17].

Alcohol dehydrogenase (1.1.1.1)

Reaction scheme:

An aldehyde or ketone + NADH = an alcohol + *NAD⁺* (non fluorescent)

Electrophoresis:

Technique	pH	System	Staining	Source
starch gel, 13 V/cm, 5 h	8	S_4	AOL	human, plants

Recipe:

To 25 ml of a 50 mmol l^{-1} Tris-phosphate buffer of pH 7 are added 0.1 ml acetaldehyde and 10 mg NADH-Na_2. Mix with 25 ml of a 2 % agar-solution cooled to 45 °C and pour on the cut surface of a processed starch gel. Under long wave UV-light sites of alcohol dehydrogenase activity occur as non-fluorescent bands on a fluorescent background [5, 6, 14].

Alcohol dehydrogenase (1.1.1.1)

Reaction scheme:

An alcohol + NAD = an aldehyde or ketone + NADH, NADH + PMS = NAD + reduced PMS, reduced PMS + Nitro BT = PMS + *formazane* (blue coloured)

Electrophoresis:

Technique	pH	System	Staining	Source
starch gel, 13 V/cm, 5 h	8	S_4	AOL or FM	human, plants

Recipe:

Starch gel electrophoresis: 0.4 ml 95% ethanol (butanol or octanol), 5 mg NAD, 3 mg Nitro BT and 0.2 mg PMS are dissolved in 9.6 ml of a 28 mmol l^{-1} Tris-HCl buffer of pH 8 and put on the cut surface of a processed starch gel [5, 6].

Aldehyde dehydrogenase (1.2.1.3)

Reaction scheme:

Glycol aldehyde + NAD = glycolic acid + NADH, NADH + PMS = NAD + reduced PMS, reduced PMS + Nitro BT = PMS + *formazane* (blue coloured)

Electrophoresis:

Technique	pH	System	Staining	Source
starch gel, 2 mA, 2 h	7.6	S_9	FM	*Pseudomonas aeruginosa*

Recipe:

Gels are incubated for 3 min at room temperature in a reaction mixture of 10 ml of a 100 mmol l^{-1} potassium phosphate buffer of pH 7.2, containing 72 mg NAD and 40 µl 2-mercaptoethanol. To initiate the reaction 0.1 ml of glycolaldehyde (1.2 mg/ml), 0.1 ml of Nitro BT (10 mg/ml) and 0.02 ml of PMS (5 mg/ml) are added per 10 ml of reaction mixture. The formation of dark bands indicates regions of enzyme activity. Bisulphite stabilizes the unstable protein. Excess substrate inhibits the enzyme reaction and a reducing agent such as 2-mercaptoethanol is required for enzymatic activity [18].

Alkaline phosphatase (3.1.3.1)

Reaction scheme:

β-Naphthyl phosphate-Na$_2$ + H$_2$O = orthophosphate + β-naphothol, β-naphthol + Fast blue RR = *diazo-dye* (coloured)

Electrophoresis:

Technique	pH	System	Staining	Source
starch gel, 5 V/cm, 5 h	8.6	S_{10}	OL	human

Recipe:

Dissolve in 10 ml of a 60 mmol l^{-1} borate buffer of pH 9.7, 5 mg β-naphtyl phosphate-Na, 12 mg $MgSO_4$-$7H_2O$ and 5 mg Fast blue B- (or Fast blue RR-) salt. After development of coloured bands gels are washed with methanol/water/acetic acid (5:5:1) [5, 6, 14]. Further methods were described by Fisher et al. [19] and Klebe et al. [20].

Alkaline phosphatase (3.1.3.1)

Reaction scheme:

An orthophosphoric monoester + H_2O = an alcohol + orthophosphate, orthophosphate + Ca^{2+} = $Ca_3(PO_4)_2 - (H_2O)_n$ (precipitate)

Electrophoresis:

Technique	pH	System	Staining	Source
PAGE 3 mA/gel, 2 h, gel length: 75 mm	8.6	D_1	FM	calf

Recipe:

Processed gels are equilibrated in a 50 mmol l^{-1} glycine-KOH buffer of pH 10 for 20 – 30 min at 37 °C. Then they are incubated in 10 ml of a 50 mmol l^{-1} glycine-KOH buffer of pH 10, consisting 5.6 mg p-nitrophenyl phosphate-Na_2, 0.13 mg $ZnCl_2$, 2 mg $MgCl_2$-$6H_2O$ and 11 mg $CaCl_2$. The appearence of white bands locates sites of enzyme activity [21].

Amine dehydrogenase (1.4.99.3)

Reaction scheme:

Methylamine + H_2O + PMS = acetaldehyde + NH_3 + reduced PMS, reduced PMS + dichlorophenolindophenol (blue coloured) = PMS + *reduced dichlorophenolindophenol* (colourless)

Electrophoresis:

Technique	pH	System	Staining	Source
PAGE	8.9	D_1	AOL	*Pseudomonas,* pea, ox, pig

Recipe:

Dissolve in 10 ml of a 200 mmol l^{-1} phosphate buffer of pH 7.5 4.5 mg methylaminehydrochloride and 1 mg PMS. Add to this solution 0.1 ml dichlorophenolindophenol-Na-$2H_2O$ (1.63 mg/ml) and 0.1 ml KCN-solution (6.5 mg/ml). Mix the substrate solution with 10 ml of a 2% agar solution cooled to 45 °C [22].

Amine oxidase (copper-containing) (1.4.3.6)

Reaction scheme:

1,4-Diaminobutane + H_2O + O_2 = 1-amino-4-butylaldehyde + NH_3 + H_2O_2, H_2O_2 + 3-amino-9-ethyl carbazole (+ peroxidase) = $2H_2O$ + *oxidized 3-amino-9-ethyl carbazole* (coloured)

Electrophoresis:

Technique	pH	System	Staining	Source
starch gel, 6 V/cm, 6 h	8.5	S_{11}	FM	hog, human, pig, pea

Recipe:

Dissolve 4 mg putrescine (= 1,4-diaminobutane) or 5 mg cadaverine (= 1,5-diaminopentane) in 7 ml of a 100 mmol l^{-1} phosphate buffer of pH 6.8, containing 100 mmol l^{-1} EDTA and add 2 ml peroxidase solution (25 µg/ml) and 1 ml 3-amino-9-ethyl carbazole (12 mg/ml). Pour on the cut surface of a processed starch gel. Reddish-brown bands indicate zones of enzyme activity [23].

D-Amino-acid oxidase (1.4.3.3)

Reaction scheme:

D-Amino acid + H_2O + O_2 = keto acid + H_2O_2 + NH_3, H_2O_2 + 3-amino-9-ethyl carbazole (+ peroxidase) = $2H_2O$ + *oxidized 3-amino-9-ethyl carbazole* (coloured)

Electrophoresis:

Technique	pH	System	Staining	Source
starch gel, 5 V/cm, 17 h	7.4	S_{12}	FM	human

Recipe:

Dissolve a D-amino acid at a concentration of 30 mmol l^{-1} (248 mg D-phenylalanine) in 50 ml of a 500 mmol l^{-1} Tris-HCl buffer of pH 8 and re-adjust the pH-value to 8.0. Then add 8 mg FAD, 5 mg peroxidase (100 U/mg) and 1 ml 3-amino-9-ethyl carbazole (25 mg/ ml). Drop the solution on the cut surface of a processed starch gel. Alternative staining methods have been given by Harris and Hopkinson [6]. These were found not to be more sensitive than the method given here [24].

L-Amino acid oxidase (1.4.3.2)

Reaction scheme:

An L-amino acid + H_2O + O_2 = a 2-oxoacid + NH_3 + H_2O_2, H_2O_2 + 3-amino-9-ethyl carbazole + H^+ (+ peroxidase) = $2H_2O$ + oxidized *3-amino-9-ethyl carbazole* (brown)

Electrophoresis:

Technique	pH	System	Staining	Source
PAGE, 3 mA/gel	8.3	D_1	FM	human

Recipe:

1 ml L-leucine (13 mg in 10 ml 100 mmol l^{-1} phosphate buffer, pH 7.4) are mixed with 2 ml peroxidase (25 µg/ml), 0.5 ml 3-amino-9-ethyl carbazole (25 mg/ml) and 6.5 ml 0.1 mol l^{-1} phosphate buffer, pH 7.4. Sites of enzyme activity are indicated by brown bands [24].

Aminoacylase (3.5.1.14)

Reaction scheme:

An N-acyl-amino acid + H_2O = a fatty acid anion + an amino-acid, L-amino acid + O_2 (+ snake venome (= amino acid oxidase) = keto acid + NH_3 + H_2O_2, H_2O_2 + o-dianisidine (+ peroxidase) = *oxidized dianisidine* (coloured)

Electrophoresis:

Technique	pH	System	Staining	Source
Cellogel, 10 V/cm, 3 h, 4 °C	8.0	C_4	FM	human, mouse, hamster, Hela cells

Recipe:

Dissolve in 5 ml of a 300 mmol l^{-1} phosphate buffer of pH 7.5 10 mg of N-acetyl-L-methionine or N-formyl-L-methionine, 3 mg peroxidase, 5 mg o-dianisidine hydrochloride and 0.2 ml of a 12.5 mmol l^{-1} $MgCl_2$-$6H_2O$-solution. Drop on the surface of a processed Cellogel[R] sheet and incubate 30 min at 37 °C. Brown bands indicate the location of active enzyme molecules [25]. The enzyme catalyzes the hydrolysis of a large number of acetylated amino acids. But N-acetylated aspartic acid compounds are pour substrates for aminoacylase I [26, 27].

Aminopeptidase (cytosol) (3.4.11.1)

Reaction scheme:

L-Leucyl-4-methoxy-2-naphthylamide + H_2O = L-leucinamide + 4-methoxy-2-naphthol, 4-methoxy-2-naphthol + Fast blue B = a *diazo-dye* (coloured)

Electrophoresis:

Technique	pH	System	Staining	Source
starch gel, 4 V/cm, 18 h, 4 °C	7.4	S_{13}	FM	human, rat

Recipe:

Add to 5 ml of a 100 mmol l^{-1} acetatebuffer of pH 6.5, 1 ml of L-leucyl-4-methoxy-2-naphthylamide-HCl (4 mg/ml), 3.5 ml NaCl (850 mg/100 ml), 0.5 ml NaCN (100 mg/100 ml) and 5 mg Fast blue B. After staining, the gel may be rinsed in saline for serveral minutes and then transferred for a few minutes to a solution of 100 mmol l^{-1} cupric sulphate. Cu^{2+} chelates with the dye formed on coupling 2-naphthyl amine with Fast blue B producing a shift in colour from red to purple [4, 28 – 30].

Aminopeptidase (cytosol) (3.4.11.1)

Reaction scheme:

L-Leucine-4-nitroanilide + H_2O = L-leucine + *4-nitroaniline* (yellow)

Electrophoresis:

Technique	pH	System	Staining	Source
starch gel, 5 V/cm, 18 h, 4 °C	7.4	S_{13}	FM	human, rat

Recipe:

Dissolve in 6 ml of a 50 mmol l^{-1} Tris-HCl buffer of pH 8.0 3 ml L-leucine-4-nitroanilide (1 mg/ml 0.1 mol l^{-1} HCl) and incubate gels at 37 °C. Light yellow bands turning black when the gels are examined under UV-light of a wave-length of 360 nm show the location of active enzyme. To stop the diffusion of 4-nitroanilide the gels are frozen at – 4 °C after the appearance of well visible bands [6, 31].

Aminotransferases (2.6.1. (1 – 6))

Reaction scheme:

A L-amino acid + 2-oxoglutarate = oxaloacetate + L- glutamate, L-glutamate + H_2O + NAD = 2-oxoglutarate + NH_3 + NADH, NADH + PMS = NAD + reduced PMS, reduced PMS + Nitro BT = PMS + *formazane* (blue coloured)

Electrophoresis:

Technique	pH	System	Staining	Source
Cellogel 150 V, h	7.8	C_2	FM	rat

Recipe:

Add to 5 ml of a 200 mmol l^{-1} phosphate buffer of pH 7.6 1 ml aspartic acid (133 mg/ml, pH 7.6) or 1 ml alanine (89 mg/ml), or 1 ml glycine (75 mg/ml) or 1 ml tyrosine (181 mg/ml). To stain for aspartate aminotransferase (2.6.1.1), alanine aminotransferase (2.6.1.2), glycine aminotransferase (2.6.1.4), tyrosine aminotransferase (2.6.1.5) or leucine amiotransferase (2.6.1.6) mix to one of the above mentioned solutions 1 ml of 2-oxoglutaric acid (15 mg/ml), 0.1 ml glutamate dehydrogenase (5 mg/ml), 1 ml NAD (30 mg/ml), 1 ml PMS (2 mg/ml) and 1 ml Nitro BT (10 mg/ml). Impregnate Cellogel[R]-strips with staining solution. Blue bands indicate the sites of enzyme activity [15, 16].

AMP deaminase (3.5.4.6)

Reaction scheme:

AMP + H_2O = IMP + NH_3, NH_3 + 2-oxoglutarate + NADH (+ glutamate dehydrogenase) = L-glutamate + H_2O + *NAD* (non-fluorescent)

Electrophoresis:

Technique	pH	System	Staining	Source
starch gel, 5 V/cm, 12 h, 4 °C	7.6	S_{14}	MOL	human

Recipe:

Add to 5 ml of a 100 mmol l^{-1} Tris-HCl buffer of pH 7.6 25 mg 2-oxoglutarate, 10 mg NADH, 50 µl L-glutamate dehydrogenase (500 U/ml), 10 mg ATP-Na_2-$3H_2O$ and 40 mg KCl. Impregnate a porous membrane with the staining solution and put it on the cut surface of a processed starch gel. The sites of the release of ammonia (production of NAD) are detected as non-fluorescent zones on a fluorescent background [32].

α-Amylase (3.2.1.1)

Reaction scheme:

The enzyme hydrolyzes 1,4-α-glycosidic linkages in polysaccharides containing 3 or more 1,4-α-linked D-glucose units.

Electrophoresis:

Technique	pH	System	Staining	Source
Cellulose acetate, 300 V, 3 h	7.5	C_5	AOL	human

Recipe:

Blue starch tablets are grouned to a fine powder with a pestle and mortar and mixed with 5 ml of boiled 2 % Special Agar-Noble per tablet. The mixture is boiled for a further 30 min and then poured onto a glass plate (10 × 30 cm) to make a smooth-surface. Then the plate is set up in a moist chamber with a transparent cover so that it can be observed during the incubation period. The cellulose acetate membrane or the PAA gel plate is now placed on the blue starch gel plate. Active amylase enzymes will be seen as white bands on a blue background [33].

Anthranilate phosphoribosyltransferase (2.4.2.18)

Reaction scheme:

N(-5'-phospho-D-ribosyl)-anthranilate + pyrophosphate = 5'-phospho-α–D-ribose 1-diphosphate + *anthranilate* (fluourescent)

Electrophoresis:

Technique	pH	System	Staining	Source
PAGE	8.3	D_1	FM	*Salmonella typhimurium*

Recipe:

Add to 6 ml of a 100 mmol l^{-1} triethanolamine HCl buffer of pH 8.3 1 ml chorismate (7.7 mg/ml) 1 ml ammonium sulphate (66 mg/ml), 1 ml $MgCl_2$-$6H_2O$ (10 mg/ml) and 1 ml 2-mercaptoethanol (1.6 µl/ml). Active enzyme bands are recognized by the formation of fluourescent anthranilate [35, 36]. When anthranilate synthetase is not associated with anthranilate 5-phosphoribosyltransferase only NH_3 can be used as the amine donor. Association of both enzymes to an enzyme complex allows either NH_3 or glutamine to be utilized as a substrate [34].

Anthranilate synthase (4.1.3.27)

Reaction scheme:

Chorimate + L-glutamine = pyruvate + L-glutamate + *anthranilate* (fluorescent)

Electrophoresis:

Technique	pH	System	Staining	Source
PAGE	8.3	D_1	FM	*Salmonella typhimurium*

Recipe:

Add to 7 ml of a 75 mmol l^{-1} potassium phosphate buffer of pH 7.4 1 ml of chorismate (7.7 mg/ml), 1 ml glutamine (7.3 mg/ml) and 1 ml $MgCl_2$-$6H_2O$ (10 mg/ml). Active enzyme bands are recognized by the formation of fluorescent anthranilate [34 – 36].

Arginase (3.5.3.1)

Reaction scheme:

L-Arginine + H_2O = L-ornithine + urea, urea + H_2O (+ urease) = $2NH_3$ + CO_2, NH_3 + 2-oxoglutarate + NADH (+ glutamate dehydrogenase) = glutamate + H_2O + *NAD* (non-fluorescent)

Electrophoresis:

Technique	pH	System	Staining	Source
starch gel, 5 V/cm, 14 h	8.3	S_{15}	MOL	human

Recipe:

2 ml L-arginine (1.74 g/10 ml, pH 7.6) and 4 mg urease (20 Units) are added to 10 ml of a 100 mmol l^{-1} Tris-HCL buffer of pH 7.6 containing 50 mg 2-oxoglutarate, 20 mg NADH and 100 µl glutamate dehydrogenase (500 Units/ml). Non fluorescent bands indicate zones of enzyme activity [37, 38].

Arginase (3.5.3.1)

Reaction scheme:

L-Arginine + H_2O = L-ornithine + urea, urea + H_2O (+ urease) = $2NH_3$ + CO_2, (NH_3 causes an increase in pH which dissociates dithiothreitol) dithiothreitol-SH = dithiothreitol-S- + H+, dissociated dithiothreitol + Nitro BT = oxidized dithiothreitol + *formazane* (blue coloured)

Electrophoresis:

Technique	pH	System	Staining	Source
starch gel, 10 V/cm, 4 h, 4 °C	6.7	S_{16}	MOL	human

Recipe:

Mix 2 ml arginine (296 mg/2 ml, pH 8.6) to 15 ml of a 2% agar solution cooled down to 45 °C. Then add 2 ml of urease (200 Units/ml), 0.6 ml dithiothreitol (15 mg/ml) and 1 ml Nitro BT solution (5 mg/ml) and poor on the cut surface of a processed starch gel. The appearence of blue bands indicates the existence of active enzyme molecules [6].

Argininosuccinate lyase (4.3.2.1)

Reaction scheme:

L-Argininosuccinate = fumarate + L-arginine, L-arginine + H_2O (+ arginase) = L-ornithine + urea, urea + H_2O (+ urease) = CO_2 + $2NH_3$, NH_3 + 2-oxoglutarate + NADH (+ glutamate dehydrogenase) = L-glutamate + H_2O + *NAD* (non-fluorescent)

Electrophoresis:

Technique	pH	System	Staining	Source
starch gel, 5 V/cm, 12 h, 4 °C	8.1	S_{17}	MOL	human

Recipe:

Add to 5 ml of a 100 mmol l^{-1} Tris-HCl buffer of pH 7.6 50 mg arginino-succinate-Ba-salt, 1 mg arginase (4.0 U), 2 mg urease (Type IV, 20 U), 5 µl L-glutamate dehydrogenase (500 U/ml), 25 mg 2-oxoglutarate and 10 mg NADH-Na_2-salt. Impregnate a porous membrane with the staining solution and put it on the cut surface of a processed starch gel. The sites of release of ammonia are detected as non-fluorescent zones on a fluorescent background [32].

Arylsulphatase (3.1.6.1)

Reaction scheme:

p-Nitrophenylsulphate + H_2O = sulphate + *p-nitrophenol*

Electrophoresis:

Technique	pH	System	Staining	Source
PAGE, 2,5 mA/gel, 1 h	8.9	D_1	FM	*Pseudomonas perfecto-marinus, E. coli*

Recipe:

Following electrophoresis gel columnes are immersed in 3 ml of a 10 mmol l^{-1} p-nitrophenylsulphate solution. The occurence of yellow bands indicates the location of active enzyme molecules. Incubation of gels containing separated bacterial enzyme extracts with an unbuffered solution of 10 mmol l^{-1} sodium dodecylsulphate will result in the precipitation of long chain alcohols due to the presence of alkylsulphatase [39].

Arylsulphatase (3.1.6.1)

Reaction scheme:

4-Methylumbelliferyl sulphate + H$_2$O = sulphate + *4-methylumbelliferone* (fluorescent)

Electrophoresis:

Technique	pH	System	Staining	Source
Cellogel 9 V/cm, 2.5 h	7.0	C$_6$	FM	human

Recipe:

Add to 2 ml of a 500 mmol l^{-1} sodium acetate-acetic acid buffer of pH 5.6 8.8 mg of 4-methylumbelliferyl sulphate-K-salt. Following electrophoresis the processed Cellogel strips are diped into the staining mixture for 30 – 60 sec. Then they are carefully blotted between filter paper and incubated at 37 °C for 1 h in a moist chamber. Afterwards the Cellogel membranes are fixed in 10 % formalin for 4 min and immersed in a 250 mmol l^{-1} sodium carbonate-glycine buffer of pH 10 for 4 min. When viewed in UV-light enzyme activity is indicated by bright fluorescent bands. In some human tissues the arylsulphatase isozymes C and C$_1$ may be removed by acidification (500 mmol l^{-1} sodium acetate-buffer) [6, 38, 40].

Aspartate aminotransferase (2.6.1.1)

(syn: Glutamic-oxaloacetic transaminase)

Reaction scheme:

L-Asparate + 2-oxoglutarate = oxalocetate + L-glutamate, oxaloacetate + NADH (+ malate dehydrogenase) = malate + *NAD* (non fluorescent)

Electrophoresis:

Technique	pH	System	Staining	Source
starch gel 6 V/cm, 18 h	8	S$_4$	AOL	human

Recipe:

Starch gel electrophoresis: mix to 17.5 ml of a 100 mmol l^{-1} Tris-HCl buffer of pH 8.0: 200 mg L-aspartic acid dissolved in 7.5 ml of buffer (pH 8.0), 110 mg of 2-oxoglutaric buffer (pH 8.0), 10 µl malate dehydrogenase (6 000 Units/ml), 5 mg NADH-Na$_2$ and 17.5 ml of 2 % agar solution cooled down to 45 °C. Pour the mixture on the cut surface of a processed starch gel. Non-fluorescent bands observed under a UV-lamp indicate the location of active enzyme [4 – 6].

Aspartate aminotransferase (2.6.1.1)
(alterative stain)

Reaction scheme:

2-Oxoglutarate + aspartate = glutamate + oxaloacetate, oxaloacetate + Fast blue B = *diazo-dye* (coloured)

Electrophoresis:

Technique	pH	System	Staining	Source
starch gel 6 V/cm, 18 h	8	S_4	FM	human

Recipe:

Add to 35 ml of a 100 mmol l^{-1} Tris-HCl buffer of pH 8.5 7.5 ml L-aspartic acid (200 mg dissolved in 7.5 ml of a 100 mmol l^{-1} Tris-HCl buffer adjusted to pH 8 with KOH), 7.5 ml 2-oxoglutaric acid (110 mg dissolved in 7.5 ml 100 mmol l^{-1} Tris-HCl buffer adjusted to pH 8 with KOH) and 250 mg Fast blue B. Filtrate and drop on the cut surface of a processed starch gel. Enzymatically liberated oxaloacetate binds to the diazonium salt Fast blue B resulting in a blue coloured dye [6].

D-Aspartate oxidase (1.4.3.1)

Reaction scheme:

D-Aspartate + H_2O + O_2 = oxaloacetate + NH_3 + H_2O_2, H_2O_2 + 3-amino-9-ethylcarbazole (+ peroxidase) = 2 H_2O + *oxidized 3-amino-9-ethylcarbazole* (coloured)

Electrophoresis:

Technique	pH	System	Staining	Source
starch gel 5 V/cm, 17 h	7.4	S_{12}	FM	human

Recipe:

Dissolve 200 mg D-aspartic acid in 50 ml of a 500 mmol l^{-1} Tris-HCl buffer of pH 8.0 and readjust the pH-value. Add 8 mg FAD, 5 mg peroxidase (100 U/mg) and 1 ml 3-amino-9-ethylcarbazole (25 mg/ml). Sites of enzyme activity are indicated by the formation of brownish bands. Alternative stains have been described [32]. But in general the sensitivity with the positive stain given here is better [6]. With prolonged staining weak isozymes due to superoxide dismutase activity may be observed [6].

Carbonate dehydratase (4.2.1.1)
(syn: carbonic anhydrase)

Reaction scheme:

4-Methylumbelliferyl acetate + H_2O = acetate + *4-methylumbelliferone* (fluorescent)

Electrophoresis:

Technique	pH	System	Staining	Source
starch gel, 12 V/cm, 5 h	8.6	S_{18}	MOL	human

Recipe:

The staining method is based on the fact that carbonic anhydrase can act as an esterase (EC 3.1.1.1): 10 mg 4-methylumbelliferyl acetate are disolved in a few drops of acetone and then mixed with 100 ml of a 100 mmol l^{-1} phosphate buffer of pH 6.5. A porous membrane is impregnated with the staining solution and overlayed on the cut surface of a processed starch gel. Inspection under long wave UV-light indicates active enzyme zones as white fluorescent bands [5, 6].

Carbonate dehydratase (4.2.1.1)

Reaction scheme:

Fluorescein diacetate + H_2O = acetate + *fluorescein* (fluorescent)

Electrophoresis:

Technique	pH	System	Staining	Source
starch gel, 12 V/cm, 5 h	8.6	S_{18}	MOL	human

Recipe:

The staining is based on the fact that carbonic anhydrase can also act as an esterase (3.1.1.1): 10 mg of fluorescein diacetate are dissolved in a few drops of acetone and then mixed with 100 ml 0.1 mol l^{-1} phosphate buffer of pH 6.5. The solution is droped on the cut surface of a processed starch gel which is incubated at 37 °C. Yellow fluorescent zones are inspected under long-wave UV-light. 4-Methylumbelliferyl acetate is the preferred substrate for human CA_1-isozymes whereas fluorescein diacetate is preferentially hydrolyzed by human CA_2-isozymes. CA_1 and CA_2-isozymes are inhibited by 1 mmol l^{-1} acetazolamide ("diamox") and this specific inhibitor is useful to distinguish the carbonic anhydase isozymes from other esterases of similar electrophoretic mobilities. The enzyme from parsley fails to catalyze the hydrolysis of p- and o-nitrophenyl acetates and is reversibly inhibited by p-chloromercuribenzoate or imidazole buffers [5, 41].

Carboxylesterase (3.1.1.1)

Reaction scheme:

4-Methylumbelliferyl acetate (-butyrate) + H_2O = acetate (butyrate) + *4-methylumbelliferone* (fluorescent)

Electrophoresis:

Technique	pH	System	Staining	Source
starch gel, 5 V/cm, 17 h	7.2	S_4 S_{19}	FM	human

Recipe:

In 0.5 ml of N,N-dimethylformamide 1 mg of 4-methylumbelliferyl acetate (-butyrate) are dissolved. The solution is slowly mixed to 10 ml of a 100 mmol l^{-1} phosphate buffer of pH 6.5 and poured on the surface of a processed starch gel. Fluorescent zones, inspected under UV-light, indicate the location of active enzyme molecules [5, 6].

Catalase (1.11.1.6)

Reaction scheme:

$2H^+ + H_2O_2 + 2 Fe^{3+} = 2 Fe^{2+} + 2H_2O$ (no colour change in the presence of $K_3(Fe^{III}(CN)_6)$ and $FeCl_3$), $3 Fe^{II} + 2(Fe^{III}(CN)_6)^{3-} = Fe_3((Fe^{III}(CN)_6)_2$ (dark green dye)

Electrophoresis:

Technique	pH	System	Staining	Source
starch gel, 6 V/cm, 17 h, 4 °C	8.6	S_{20}	FM	human

Recipe:

Add to 45 ml of a 100 mmol l^{-1} phosphate buffer of pH 7.4 5 ml of a 3 % H_2O_2-solution in water. Drop the solution on the cut surface of a processed starch gel and incubate for 15 min. Then rinse the gel with water and afterwards dip it in a freshly prepared mixture of equal volumes of a 2 % $FeCl_3$-solution and a 2 % $K_3[Fe^{III}(CN)_6]$-solution. Mix by gently agitating the container for a few minutes. Finally remove the stain. Zones of enzyme activity appear as yellow bands on a blue green background [5, 6].

Catechol oxidase (1.10.3.1)

Reaction scheme:

3,4-Dihydroxyphenylalanine + O_2 = *DOPA-quinone* (red to black coloured dye)

Electrophoresis:

Technique	pH	System	Staining	Source
PAGE,	8.9	D_1	FM	plants

Recipe:

Dissolve 10 mg 3,4-dihydroxyphenylalanine in 10 ml of a 0.1 mol l^{-1} phosphate buffer of pH 7.5. Sites of active enzyme molecules appear as red to brown coloured bands [42].

Cathepsin B (3.4.22.1)

Reaction scheme:

α-N-Benzyloxycarbonyl-L-arginyl-L-arginine-4-methyl-β-naphthylamide + H_2O = α-N-benzyloxycarbonyl-L-arginyl-L-arginine + *1-methoxy-3-naphthylamide* (fluorescent)

Electrophoresis:

Technique	pH	System	Staining	Source
IEF	4 – 7	I_1	MOL	human

Recipe:

Add to 5 ml of a 100 mmol l⁻¹ sodium phosphate buffer of pH 6.0 1 ml of α-N-benzyloxycarbonyl-L-arginyl-L-arginine-4-methoxy-β-naphtylamide (9 mg dissolved in 1 ml of 5 % dimethylformamide), 1 ml EDTA (3 mg/ml) and 1 ml of L-cystine-HCl-H_2O (10 mg/ml). Caution: β-naphtylamide derivatives are potential carcinogenes. Handle with extreme care!

A piece of a porous membrane is cut to cover the area of the processed flat gel to be stained. The overlay in then placed on to the gel surface and pressed to make firm contact and avoid the entrapment of air bubbles. Then the substrate solution is poured evenly on the membrane. The gel and overlay are then placed in a humidified box and incubated at 37 °C. The progress of the enzyme reaction is assessed from time to time by viewing the membrane still being on the gel under a long-wave UV-lamp where blue bands of the fluorescent reaction product 1-methoxy-3-naphthylamide will be visible. When the reaction has proceeded adequately the overlay is removed and immediately immersed in colour regent. To prepare the colour reagent two stock solutions are needed: (a) a solution of Fast Garnet GBC and (b) a solution of mersalyl acid (= 2-[(3-hydroxymercuri-2-methoxypropyl-)carbamoyl]-phenoxy acetic acid) containing 50 mmol l⁻¹ EDTA-Na_2. The Fast Garnet GBC-solution is prepared as follows: 225 mg of 4-amino-2′,3-dimethylazobenzene are crushed to a powder und dissolved in 50 ml ethanol. Then 30 ml of 1 mol l⁻¹ HCl are added with stirring and finally the solution is diluted to 100 ml with water and stored at 4 °C. The mersalyl acid solution is prepared in the following way: 2.43 g of mersalyl acid are placed in a baker containing a-stirring bar and dissolved with 60 ml of 0.5 mol l⁻¹ NaOH. Then 9.3 g of EDTA-Na_2 are added and the solution is diluted with water to about 450 ml. The EDTA is allowed to dissolve and then the pH-value is adjusted to 4.0 by adding 1 mol l⁻¹ HCl with stirring. Finally 500 ml of 4 % (w/v) aqueous Brej 35 (polyoxy ethylene lauryl ether) are added and the reagent is stored in a brown bottle at room temperature. The colour reagent is prepared by placing 1 ml of the Fast Garnet GBC stock-solution in a test tube standing in a beaker of ice and adding 100 µl of a 0.2 mol l⁻¹ $NaNO_2$-solution under mixing. The reagent is left at least 5 min und then diluted to 100 ml with the mersalyl-Brej 35-reagent. The reagent is kept cool and used the same day. Bands of the red reaction product appear almost immediately. After 5 – 10 min the overlay can be washed in water. In addition to cathepsin B the lysosomale proteinases cathepsin H and L also cleave CBZ-Arg-Arg-β-nathylamide although at a much reduced rate [43 – 45].

Cellulase (3.2.1.4)

Reaction scheme:

Carboxymethyl cellulose + (KJ + J_2) = blue colour

Electrophoresis:

Technique	pH	System	Staining	Source
PAGE 5 mA/tube, 105 min, 4 °C	8.9	D_1	FM	*Citrus sinensis;* *Trichodermata* *viride*

Recipe:

Processed gels are transferred to 12 ml test tubes containing 0.2 mol l⁻¹ sodium-potas-
sium phosphate buffer (pH 6.0) and then the tubes are incubated for 10 min in a water
bath at 37 °C. Afterwards the buffer is decanted and the tubes containing the gels are
stoppered and further incubated in the water bath. Incubation of the gels for periods
longer than 10 min may cause diffusion of substrate and enzyme from the gels into the
external buffer so that it is impossible to locate enzyme activity. After the incubation
period, the buffer is decanted from the gels and the enzymic reaction is stopped by
adding 60 % H_2SO_4 to the test tubes for 5 to 10 min. The staining solution is poured
into the tubes which are kept for 30 – 60 min at room temperature; the duration of
incubation in the staining solution determines the intensity of background colour in
areas with no enzyme activity. The staining solution consists of 2 % KJ and 0.2 % J_2 in
distilled water. The colour reaction is analogous to the well-known starch-iodine
reaction and does not appear where carboxymethyl cellulose is degrated. The duration of incubation in H_2SO_4 determines the staining efficiency by the iodine solution.
Care should also be taken not to incubate gels too long in the staining solution. Jodine
evaporates from the gel edges very easily; therefore it is important to perform the
scanning as rapidly as possible. When gels are kept in stoppered tubes they can be
stored for a few days with no major changes in colour [46].

Cholinesterase (3.1.1.8)

Reaction scheme:

α-Naphtyl acetate + H_2O = acetate + α-naphtol
α-naphtol + Fast Red TR = *diazo-dye* (coloured)

Electrophoresis:

Technique	pH	System	Staining	Source
starch gel 5 V/cm, 5 g	8.6	S_2	FM	human

Recipe:

(SGE) Mix slowly 0.2 ml α-naphtyl acetate (1 % dissolved in 50 % watery acetone) into
10 ml of a 0.2 mol l⁻¹ phosphate buffer of pH 7.1. Finally add 4 mg Fast Red TR-salt dissolve and filtrate. Drop on the cut surface of a processed starch gel. Red bands indicate
active enzyme [5, 6].

Chymotrypsin (3.4.21.1)

Reaction scheme:

Preferential cleavage of peptides at their Tyr-, Trp-, Phe-, or Leu-carboxyl groups

Electrophoresis:

Technique	pH	System	Staining	Source
PAGE 3 – 4 mA/gel	4.2	D_3	see recipe	human

Recipe:

Prior to electrophoresis 0.1 % casein are polymerized into 7.5 % polyacrylamide gels if electrophoresis is performed at acidic conditions; 0.1 % haemoglobin are polymerized into 7.5 % polyacrylamide gels if electrophoresis performed at alkaline conditions. Following electrophoresis gels are incubated in a 0.1 mol l^{-1} Tris-HCl buffer of pH 7.5 for 2 – 12 h at 37 °C changing the buffer frequently. An inclusion of 0.01 % nigrosin in the incubation buffer intensifies the banding patterns when using haemoglobin. Enzymatic active zones appear as pale bands on a brown background [47, 48].

Chymotrypsin (3.4.21.1)
(alternative method)

Reaction scheme:

Acetyl-L-tyrosine-p-nitroanilide + H_2O = Acetyl-L-tyrosine + p-nitroaniline;
p-nitroaniline + $NaNO_2$ + N-(α-naphthyl) ethylenediamine = *diazo-dye* (coloured)

Electrophoresis:

Technique	pH	System	Staining	Source
PAGE 0.1 mA/gel, 1 h	4.3	D_3	FM	bovine, hornet, honey bee

Recipe:

3.4 mg N-acetyl-L-tyrosine-4-nitroanilide are dissolved in 2 ml of a 200 mol l^{-1} Tris-HCl buffer of pH 8.4 containing 5 % (v/v) dimethylformanide. Before use 1 ml of a solution of 0.4 % $NaNO_2$ and 1 ml of a solution of 2 % N-(α-naphthyl) ethylenediamine are added. Processed disc gels are incubated directly in the staining solution at 35 °C for 10 – 30 min. After incubation the gels are washed with water for 1 min to remove the excess of substrate and then treated with an aqueous solution of 12.5 % trichloroacetic acid (w/v) for 30 sec during which time a purple colour becomes visible. The substrate which may precipitate on the surface of the gel is washed out with 200 mmol l^{-1} Tris-HCl buffer of pH 8.4 until the gels become transparent again. The detection limit is about 120 ng chymotrypsin/gel.

To detect *trypsin* (3.4.21.4) the same procedures can be used but replacing N-acetyl-L-tyrosine-4-nitroanilide by a solution of 4.4 mg N-α-benzoyl-DL-arginine-4-nitroanilide-HCl in 2 ml of a 200 mmol l^{-1} Tris-HCl buffer of pH 8.4 containing 5 % (v/v) dimethylformamide [148].

Citrate (si)-synthase (4.1.3.7)

Reaction scheme:

Oxaloacetate + acetyl-CoA = citrate + HS-CoA; HS-CoA + dichlorophenol indophenol = oxidized CoA + reduced dichlorophenol indophenol; MTT + reduced dicholorophenol indophenol = dicholorophenol indophenol + *reduced MTT* (blue coloured)

Electrophoresis:

Technique	pH	System	Staining	Source
Cellogel 10 V/cm, 3 h	8.0	C_8	FM	human

Recipe:

Prepare the following solutions: (a) dissolve 1.32 mg oxaloacetate in 1 ml of a 100 mmol l^{-1} Tris-HCl buffer of pH 7.6, (b) dissolve 11 mg acetyl CoA-Li$_3$-3H$_2$O in 1 ml of a 100 mmol l^{-1} Tris-HCl buffer of pH 7.6 and (c) prepare a solution of 24 mg 2.6-dichloro-phenol indophenol-Na (DCPIP) in 10 ml of a 100 mmol l^{-1} Tris-HCl buffer of pH 7.6. The staining solution is prepared by mixing of 500 µl oxaloacetate-solution, 250 µl acetyl-CoA-solution and 1.25 ml of DCPIP-solution. Processed cellogel strips are impregnated with 155 µl per strip of staining solution and incubated for 10 – 15 min at room temperature in a moist chamber. Active enzyme zones appear as white zones against a blue background. A counterstain with MTT is possible (0.5 mg MTT per ml of 100 mmol l^{-1} Tris-HCl buffer of pH 7.6). If doing so the remaining DCPIP and MTT is removed by extensive washing of the strips. Enzyme bands now appear as purple bands on a white background [49].

Creatine kinase (2.7.3.2)

Reaction scheme:

ATP + creatine = ADP + phosphocreatine; ADP + phosphoenolpyruvate (+ pyruvate kinase) = pyruvate + ATP; pyruvate + NADH (+ lactate dehydrogenase) = lactate + *NAD* (non-fluorescent)

Electrophoresis:

Technique	pH	System	Staining	Source
starch gel 13 V/cm, 18 h	7	S_{21}	AOL	human

Recipe:

(SG) 30 mg creatine, 20 mg ATP-Na$_2$-3H$_2$O, 40 mg magnesium acetate-4H$_2$O, 40 mg potassium acetate, 15 mg phosphoenolpyruvate, 10 mg NADH-Na$_2$, 5 µl pyruvate kinase (400 Units/ml and 50 µl lactate dehydrogenase (2750 Units /ml) are dissolved in 10 ml 0.5 mol l^{-1} Tris HCl, pH 6.0. Paper overlay. Enzymatic bands appear as non-fluorescent bands at a wavelength of 340 nm [1, 14, 50].

3':5'-Cyclic-nucleotide phosphodiesterase (3.1.4.17)

Reaction scheme:

3':5'-cyclic-AMP + H$_2$O + 5'-AMP, 5'-AMP + ATP + (adenylate kinase) = ADP + ADP, ADP + phosphoenol pyruvate (+ pyruvate kinase) = ATP + pyruvate, pyruvate + NADH$_2$ (+ lactate dehydrogenase) = lactate + *NAD* (non-fluorescent)

Electrophoresis:

Technique	pH	System	Staining	Source
starch gel 4 V/cm, 18 h, 4 °C	7.7	S_{22}	AOL	human

Recipe:

Add to 10 ml of a 0.1 mol l^{-1} Tris-HCl buffer of pH 7.9 : 21 mg cyclic-AMP-H$_2$O, 40 mg MgSO$_4$-7H$_2$O, 60 mg KCl, 15 mg CaCl$_2$-6H$_2$O, 7 mg ATP-Na$_2$-3H$_2$O, 6 mg phosphoenol pyruvate-K, 30 mg NADH-Na$_2$, 5 µl adenylate kinase (360 U/ml), 5 µl pyruvate kinase (400 U/ml) and 5 µl lactate dehydrogenase (2750 U/ml). Mix and add to 5 ml of a 2% agar solution cooled down to 45 °C. Pour on the cut surface of a processed starch gel. Active enzyme zones appear as fluorescent bands under UV-light [1].

Alternative staining methods for polyacrylamide gel electrophoresis are described in [51, 52].

Cystathionine-β-synthase (4.2.1.22)

Reaction scheme:

Cystine + Pb^{2+} + 2-mercaptoethanol = S-β-hydroxyethyl-L-systeine + 2 H$^+$ + *PbS* (brown to black coloured)

Electrophoresis:

Technique	pH	System	Staining	Source
PAGE	8.9	D_1	FM	Yeast

Recipe:

Immediately after electrophoresis the gel rods are placed in a solution of 100 mmol l^{-1} phosphate buffer of pH 7.8 containing L-cystine-HCl-H$_2$O (3.5 mg/ml), 2-mercapto-ethanol (4 µl/ml), pyridoxal-5′-phosphate-H$_2$O (26 mg/ml) and Pb II-acetate-3H$_2$O (76 mg/ml). The reaction is stopped by placing gels into 5% acetic acid. The development of black-brown bands indicates active enzyme zones [53].

Cystyl aminopeptidase (3.4.11.3)

Reaction scheme:

L-Cystinyl-di-β-naphthylamide + H$_2$O = L-cystinyl-β-naphthylamide + naphthol; naphtol + Fast Black K = *diazo-dye* (black coloured)

Electrophoresis:

Technique	pH	System	Staining	Source
PAGE 5 mA, 105 – 125 V, 60 min	8.6	D_4	FM	human (blood serum)

Recipe:

30 mg L-cystinyl-di-β-naphthylamide are dissolved in 10 ml dimethylformamide. To this solution 30 mg of Fast Black K-salt and 90 ml of a 200 mmol l^{-1} Tris-maleate-

NaOH buffer of pH 6.0 are added. The incubation is carried out at room temperature during 90 min. At a final concentration of 20% dimethylformamide completely inhibits enzyme activity. At a concentration of 10% dimethylformamide most of the enzyme remains active [54, 55].

Cytidine deaminase (3.5.4.5)

Reaction scheme:

Cytidine + H_2O = uridine + NH_3; NH_3 + 2-oxoglutarate + NADH (+ glutamate dehydrogenase) = glutamate + H_2O + *NAD* (non-fluorescent)

Electrophoresis:

Technique	pH	System	Staining	Source
starch gel 7 V/cm, 14 h	5.7	S_{23}	POL	human

Recipe:

Add 80 mg cytidine, 20 mg dithiothreithol and 20 mg ADP-Na_2 to 10 ml of a 0.1 mol l^{-1} Tris-HCl buffer of pH 7.6 containing 50 mg 2-oxoglutarate, 20 mg NADH and 100 µl glutamate dehydrogenase (500 U/ml). Non-fluorescent bands indicate active enzyme molecules [56, 57].

Cytidine deaminase (3.5.4.5)
(alternative method)

Reaction scheme:

Cytidine + H_2O = uridine + NH_3, (NH_3 elevates the pH-value and as a cause dithiothreitol dissociates:) dithiothreitol-SH (+ OH^-) = dithiothreitol-$S^{(-)}$ + $H^{(+)}$, dissociated dithiothreitol + Nitro BT = oxidized dithiothreitol + *reduced Nitro BT* (blue)

Electrophoresis:

Technique	pH	System	Staining	Source
starch gel 4 V/cm, 19 h (cooling)	6.7	S_{23}	AOL	human

Recipe:

Wash starch gel following electrophoresis with 20 mmol l^{-1} citric acid adjusted to pH 6.7 with NaOH. Decant buffer and pour the agar overlay on top. Mix to 10 ml of a cytidine solution (15 mg/10 ml H_2O), 10 ml of a 2% ionagar solution cooled down to 55 °C. Then add 10 mg dithiothreitol and 0.3 ml Nitro BT-solution (10 mg/ml), mix and pour over the cut surface of a processed starch gel. The appearance of blue bands indicates the existence of active enzyme molecules [6].

Deoxyribonuclease I (3.1.21.1)

Reaction scheme:

DNA + ethidium bromide = DNA-ethidiumbromide (fluorescent)

DNA + H_2O (+ DNase) = oligonucleotides
(oligonucleotides are non-fluorescent in the presence of ethidium bromide)

Electrophoresis:

Technique	pH	System	Staining	Source
IEF 50 V/cm, 4 – 8 h	4 – 10	I_3	AOL	mammals

Recipe:

The processed IEF gel still sticking to the glass-support is immersed for 20 min into a 100 mmol l^{-1} Tris-HCl buffer of pH 7.0. Then it is taken off and dried. A 2 mm thick Plexiglass frame is finally put on it and fixed with clamps. Afterwards the gel is heated under an infrared-lamp and the staining solution is poured into the frame. DNase active bands appear in the overlay under a wave length of 200 nm as dark bands on a fluorescent background. The substrate solution consists of 2.2 g $CaCl_2$, 2.5 g $MnCl_2$, 0.5 g DNA and 0.05 g ethidium bromide per liter of a 0.2 mol l^{-1} Tris HCl buffer of pH 7.0. Composition of the staining solution : equal volumes of a 2 % agar solution in H_2O cooled down to 60 °C are mixed with substrate solution heated to 60 °C and poured on th gel [58].

Dihydrouracil dehydrogenase (NADP+) (1.3.1.2)

Reaction scheme:

5,6-Dihydrouracil + NADP = uracil + NADPH, NADPH + PMS = $NADP^+$ + reduced PMS, reduced PMS + MTT = *reduced MTT* (blue coloured)

Electrophoresis:

Technique	pH	System	Staining	Source
PAGE 2.5mA/gel, 2 – 4 °C	8.3	D_1	FM	rat

Recipe:

Dissolve in 1.5 ml of a 200 mmol l^{-1} phosphate buffer of pH 7.5 7.2 mg of 5,6-dihydrouracil, 5 mg NADP-Na-$2H_2O$, 38 mg ATP-Na_2-$3H_2O$ and 13 mg Mg Cl_2-$6H_2O$. Mix and add 5.5 ml MTT (1 mg/ml), 0.25 ml PMS (1 mg/ml) 1 ml NaCl (6 mg/ml) and 4.75 ml aqua dest. Acitve enzyme zones appear as blue bands [59, 60].

Dipeptidase (3.4.13.11)

Reaction scheme:

L-Alanyl-β-naphthylamide + H_2O = L-alamine + β-naphthol, β-naphtol + Fast red B = *diazo-dye* (coloured)

Electrophoresis:

Technique	pH	System	Staining	Source
PAGE	8.9	D_1	FM	rat

Recipe:

Dissolve 100 mg L-alanyl-β-naphthylamide in 10 ml of aceton and add 1 ml to 10 ml of 100 mmol l^{-1} Tris-HCl buffer of pH 7.1. Incubate disc gels at 37 °C for 45 min. Then decant the substrate solution and replace by 10 ml of a 100 mmol l^{-1} Tris-HCl buffer of pH 7.1 containing 100 mg Fast red B. After 10 min the coupling solution is decanted and replaced by water [61, 62].

Endo-β-N-acetylglucosaminidase (3.2.1.96)

Reaction scheme:

Endohydrolysis of 1,4-N-acetamidodeoxy-β-D-glucosidic linkages in mannosyl-glycoproteins

Electrophoresis:

Technique	pH	System	Staining	Enzyme
Cellogel 250 V, 20 min	6,5	C$_9$	MOL	bovine *Strephococcus*

Recipe:

Following electrophoresis the moist Cellogel strip is laid flat on a glass plate and rolled firmly with a clean glass test tube to express air bubbles. It is then overlayed with a second Cellogel strip saturated with substrate solution. The substrate solution consists of a 50 mmol l^{-1} phosphate-citrate buffer of pH 5 containing 8 mg/ml NaCl and 1 mg/ml hyaluronic acid prepared from human umbilical cord. The overlay is firmly pressed on the enzyme containing membrane and the "sandwich" is incubated in a moist chamber at 37 °C for 30 min. Following incubation the overlay is hung to dry in air, stained in 1% Alcian Blue 865 in 7% acetic acid for 1 min and finally rinsed in distilled water. Until no additional Alcian Blue is washed out, the overlay is again hung to dry; zones of enzyme activity appear as white bands on a uniformly blue background [63].

Enolase (4.2.1.11)

Reaction scheme:

2-Phospho-D-glycerate = phosphoenolpyruvate + H$_2$O, phosphoenolpyruvate + ADP (+ pyruvate kinase) = pyruvate + ATP, ATP + glucose (+ hexokinase) = glucose-6-phosphate, glucose-6-phosphate + NADP (+ glucose-6-phosphate dehydrogenase) = 6-phosphogluconate + NADPH, NADPH + PMS + MTT = NADP + PMS + *reduced MTT* (blue)

Electrophoresis:

Technique	pH	System	Staining	Enzyme
Cellogel 8 mA, 50 min	8.4	C$_9$	FM	rabbit

Recipe:

Add to 3 ml of a 100 mmol l^{-1} Tris-HCl buffer of pH 7.5 1 ml 2-phosphoglycerate-Na_3-$6H_2O$ (5 mg/ml), 1 ml ADP-Na_3-$2H_2O$ (0.36 mg/ml), 50 µl pyruvate kinase (800 U/ml), 1 ml glucose (1.8 mg/ml), 50 µl hexokinase (600 U/ml), 1 ml NADP-Na_2-$2H_2O$ (4.1 mg/ml), 50 µl glucose-6-phosphate dehydrogenase (2 800 U/ml), 1 ml $MgCl_2$-$6H_2O$ (4 mg/ml), 1 ml AMP-H_2O (78 mg/ml) and 1 ml MTT (0.4 mg/ml). All reagents are dissolved in 100 mmol l^{-1} Tris-HCl, pH 7.5. Immediately before use 100 µl PMS (240/ml) are added. Zones of enzyme activity appear as blue bands. The appearance of adenylate kinase is prevented by the inclusion of excess AMP and by use of catalytic amounts of ADP, the latter being regenerated by glucose and hexokinase. Omission of AMP and use of higher levels of ADP lead to the appearance of extra bands of activity which are not dependent on the presence of 2-phosphoglycerate and which are presumably due to adenylate kinase [64].

Enolase (4.2.1.11)

Reaction scheme:

3-Phospho-D-glycerate + 2,3-bisphospho-D-glycerate (+ phosphoglyceromutase) = 2,3-bisphospho-D-glycerate + 2-phospho-D-glycerate, 2-phospho-D-glycerate (+ enolase) = phosphoenol pyruvate + H_2O, phosophoenl pyruvate + ADP (+ pyruvate kinase) = ATP + pyruvate, pyruvate + NADH (+ lactate dehydrogenase) = lactate + *NAD* (non fluorescent)

Electrophoresis:

Technique	pH	System	Staining	Enzyme
starch gel 5 V/cm, 30 h	6.5	S_{24}	MOL	human

Recipe:

Add to 50 ml of a 50 mmol l^{-1} Tris-HCl buffer of pH 7.6 50 mg 3-phosphoglycerate-Na_3, 35 mg NADH-Na_2-$3H_2O$, 50 mg ADP-Na_3-$2H_2O$, 40 mg $MgCl_2$-$6H_2O$, 50 µl phosphoglyceromutase (2 500 U/ml), 50 µl pyruvate kinase (400 U/ml) and 50 µl lactate dehydrogenase (2 750 U/ml). Mix and saturate a porous membrane bubble free with that solution. Place it on the cut surface of a processed starch gel and view under UV-light. Active zones appear as non-fluorescent bands on a fluorescent background [6].

β-D-Fructofuranosidase (3.2.1.26)

Hydrolysis of terminal, non-reducing β-D-fructofuranoside residues in β-D-fructofuranosides

Electrophoresis:

Technique	pH	System	Staining	Source
PAGE 3 mA/gel,	8.3	D_1	FM	plants

Recipe:

Add to 10 ml of a 85 mmol l⁻¹ citrate-phosphate buffer of pH 6.5: 342 mg sucrose 0.1 ml glucose oxidase (grade I, 0.02 mg/ml), 0.1 ml peroxidase (horse-radish, 12 mg/ml) and 0.1 ml 3-amino-9-ethylcarbazole (50 mg/ml); filtrate. Active enzyme bands appear as brownish zones [65, 66].

Fructokinase (2.7.1.4)

Reaction scheme:

ATP + D-fructose = ADP + D-fructose-6-phosphate, D-fructose-6-phosphate + (glucose phosphate isomerase) = D-glucose-6-phosphate, D-glucose-6-phosphate + NADP (+ glucose-6-phosphate dehydrogenase) = D-glucono-lactone-6-phosphate + NADPH, NADPH + PMS = NADP + reduced PMS, reduced PMS + MTT = PMS + *reduced MTT* (blue coloured)

Electrophoresis:

Technique	pH	System	Staining	Source
starch gel, 2.5 h		S_{22}		*Drosophila*

Recipe:

100 mg fructose, 25 mg ATP-Na₂-3H₂O, 0.5 ml NADP-solution (5 mg/ml), 0.7 U glucose-phosphate isomerase, 0.5 ml MTT (5 mg/ml) and 0.25 ml PMS-solution (5 mg/ml) are added to 25 ml of a 60 mmol l⁻¹ Tris-HCl buffer of pH 8.0. Finally 25 ml of a 2 % agarose-solution in 25 mmol l⁻¹ MgCl₂-6H₂O cooled to 60 °C are added. The mixture is then poured onto the gel which is incubated for 1 h at 37 °C [67, 68].

Fructose-bisphosphatase (3.1.3.11)

Reaction scheme:

D-Fructose-1,6-bisphosphate + H₂O = D-fructose-6-phosphate + ortho-phosphate, ortho-phosphate + Ca²⁺ = Ca₃(PO₄)₂-(H₂O)ₙ (white precipitate)

Electrophoresis:

Technique	pH	System	Staining	Source
PAGE 2 mA/gel, 1.5 h	8.6	D_1	FM	cattle

Recipe:

Following electrophoresis the gels are equilibrated in a 50 mmol l⁻¹ glycine-KOH buffer of pH 10 for 20 – 30 min at 37 °C. Then they are transferred into a solution of 10 ml 50 mmol l⁻¹ glycine-KOH buffer of pH 10, containing 20.3 mg MgCl₂-6H₂O, 75 mg KCl, 26 mg fructose-1,6-bisphosphate-Na₃-8H₂O and 11 mg CaCl₂. The appearance of white bands indicates the excistance of active enzyme [21, 69, 70].

Fructose-bisphosphate aldolase (4.1.2.13)

Reaction scheme:

Fructose-1,6-diphosphate = dihydroxyacetone phosphate + D-glyceraldehyde-3-phosphate, dihydroxyacetone phosphate (+ triose phosphate isomerase) = glyceraldehyde-3-phosphate, glyceraldehyde-3-phosphate + NAD + (glyceraldehyde-3-phosphate-dehydrogenase) = 3-phospho-D-glyceroyl phosphate + NADH, NADH + PMS = NAD + reduced PMS, reduced PMS + MTT = PMS + *reduced MTT* (blue)

Electrophoresis:

Technique	pH	System	Staining	Source
starch gel 12 V/cm, 5 h cooling	7.4	S_{12}	AOL	human

Recipe:

Add to 25 ml of a 0.1 mol l^{-1} Tris-HCl buffer of pH 8 100 mg fructose-1,6-diphosphate-Na$_3$-8H$_2$O, 20 mg NAD, 60 mg sodium arsenate (Na$_2$AsO$_4$-7H$_2$O), 50 µl glyceraldehyde-3-phosphate dehydrogenase (800 U/ml), 1.5 ml MTT (5 mg/ml) and 0.5 ml PMS (5 mg/ml H$_2$O). Mix this solution with 25 ml of 2 % agar cooled down to 45 °C. Enzyme zones appear blue. In human tissues in addition a so called aldolase B exists which catalyzes the break down of fructose-1-phosphate [6].

L-Fucose dehydrogenase (1.1.1.122)

Reaction scheme:

L-Fucose + NAD = L-fucono-1,5-lactone + NADH, NADH + PMS = NAD + reduced PMS, reduced PMS + MTT = PMS + *reduced MTT* (blue coloured)

Electrophoresis:

Technique	pH	System	Staining	Source
PAGE	8.9	D_1	FM	pork (liver)

Recipe:

Dissolve in 10 ml of a 100 mmol l^{-1} Tris-HCl buffer of pH 8.0 49 mg L-Fucose, 43 mg NAD, 1.5 mg PMS and 4 mg MTT. Polyacrylamide gel rods are immersed in the staining solution for approximately 30 min at 37 °C. The pork liver enzyme is active only with NAD whereas the enzyme from *Pseudomonas spe.* can utilize NAD and NADP. With arabinose as substrate it is about half as active as with fucose [171].

α-L-Fucosidase (3.2.1.51)

Reaction scheme:

4-Methylumbelliferyl-α-L-fucoside + H$_2$O = α-L-fucose + *4-methylumbelliferone* (fluoerescent)

Electrophoresis:

Technique	pH	System	Staining	Source
starch gel, 5 V/cm, 20 h, 4 °C	7.0	S_{25}	MOL	human

Recipe:

Dissolve in 10 ml 0.1 mol l^{-1} phosphate-citrate buffer of pH 4.8 4 mg 4-methylumbelli-feryl-α-L-fucoside. Impregnate a porous membrane with the staining solution and position the membrane on a processed starch gel. Incubate and spray the membrane afterwards with a diluted NH_4OH-solution and view fluorescent bands under UV-light [6].

Fumarate hydratase (4.2.1.2)

Reaction scheme:

Fumarate + H_2O = malate, malate + NAD (+ malate dehydrogenase) = oxaloacetate + NADH, NADH + PMS + MTT = NAD + *reduced MTT* (blue coloured)

Electrophoresis:

Technique	pH	System	Staining	Source
Cellogel 10 V/cm, 3 h	7.0	C_{11}	FM	human, hamster

Recipe:

Dissolve in 5 ml of a 25 mmol l^{-1} K, Na-phosphate buffer of pH 7.5 80 mg fumaric acid and readjust the pH with NaOH. Add to 1 ml of this solution 10 µl malate dehydrogenase (6 000 U/ml) 200 µl NAD (10 mg/ml), 400 µl PMS (0.4 mg/ml) and 400 µl MTT (2 mg/ ml). Impregnate a processed Cellogel strip with this solution and observe the appearance of blue bands. Incubate 30 min at 37 °C [5].

Galactokinase (2.7.1.6)

Reaction scheme:

ATP + D-galactose (+ galactokinase) = ADP + α-D-galactose-1-phosphate, UDP glucose + α-D-galactose-1-phosphate (+ UDPglucose-hexose-1-phosphate uridylyl-transferase) = UDPgalactose + α-D-glucose-1-phosphate, α-D-glucose-1,6-bisphosphate + α-D-glucose-1-phosphate (+ phosphoglucomutase) = α-D-glucose-6-phosphate + α-D-glucose-1,6 bisphosphate, D-glucose-6-phosphate + NADP(+ glucose-6-phosphate dehydrogenase) = D-glucono-α-lactone-6-phosphate + NADPH, 6-phospho-D-gluconate + NADP (+ phosphogluconate dehydrogenase) = D-ribulose-5-phosphate + CO_2 + *NADPH* (fluorescent)

Electrophoresis:

Technique	pH	System	Staining	Source
starch gel 5 V/cm, 6 h	8.0	S_{26}	AOL	rat, human

Recipe:

Add to 1.5 ml of a 1 mol l^{-1} Tris-HCl buffer of pH 8.0 100 µl galactose (7.2 mg/ml), 200 µl ATP-Na$_2$-3H$_2$O (182 mg/ml), 750 µl UDPglucose-Na$_2$ (8 mg/ml), 75 µl glucose-1,6-diphosphate (tetra cyclohexyl ammonium salt-4H$_2$O) (1.4 mg/ml), 750 µl NADP-Na$_2$ (16 mg/ml), 0,2 mg UDPglucose-hexose-1-phosphate uridylyltransferase (20 U/mg), 30 µl phosphoglucomutase (2 000 U/ml), 60 µl glucose-6-phosphate dehydrogenase (700 U/ml), 30 µl phosphogluconate dehydrogenase (120 U/ml), 750 µl MgCl$_2$-6H$_2$O (8 mg/ml) and 750 µl 2-mercaptoethanol (10 µl/ml). Mix with 5 ml of 1.5 % agar solution cooled down to 45 °C and pour on the cut surface of a processed starch gel [6].

α-D-Galactosidase (3.2.1.22)

Reaction scheme:

4-Methylumbelliferyl-α-galactoside + H$_2$O = α-D-galactose + *4-methylumbelliferone* (fluorescent)

Electrophoresis:

Technique	pH	System	Staining	Source
starch gel 3 V/cm, 15 h, 4 °C	7.0	S$_{27}$	FM	human

Recipe:

Dissolve in 20 ml 200 mmol l^{-1} citrate-phosphate buffer of pH 4.0 10 mg 4-methylumbelliferyl-α-galactoside. Drop on the cut surface of a processed starch gel. Spray after some time with ammonia to intensify fluorescence. View under UV-light [6].

Glucose dehydrogenase (1.1.1.47)

Reaction scheme:

β-D-Glucose + NAD(P) (+ glucose dehydrogenase) = D-gluconic acid + NAD(P)H, NAD(P)H + PMS + MTT = NAD(P) + PMS + *reduced MTT* (blue coloured)

Electrophoresis:

Technique	pH	System	Staining	Source
starch gel 10 V/cm, 17 h, 4 °C	9.0	S$_{28}$	FM	human

Recipe:

To 50 ml 50 mmol l^{-1} phosphate buffer of pH 7.5 9 mg D-glucose, 20 mg NAD, 1 ml MTT (5 mg/ml H$_2$O) and 1 ml PMS (5 mg/ml H$_2$O) are added. The staining solution is dropped on the cut surface of a processed starch gel. Active enzyme bands appear as blue zones. The enzyme also oxidizes glucose-6-phosphate and galactose-6-phosphate [6].

Glucose oxidase (1.1.3.4)

Reaction scheme:

β-D-Glucose + O$_2$ = D-glucono-δ-lactone + H$_2$O$_2$, H$_2$O$_2$ + aminoethylcarbazole + H$^+$ (+ peroxidase) = H$_2$O + *oxidized aminoethylcarbazole* (coloured)

Electrophoresis:

Technique	pH	System	Staining	Source
PAGE, 3 mA/gel,	8.3	D_1	FM	rat

Recipe:

1 ml D-glucose (180 mg/10 ml) are mixed with 2 ml peroxidase solution (25 µg/ml), 0.4 ml 3-amino-9-ethyl carbazole (25 mg/ml) and 6.6 ml 0.1 mol l^{-1} phosphat buffer of pH 7.0 and used as substrate solution [24].

Glucose-1-phosphate uridylyltransferase (2.7.7.9)

Reaction scheme:

UDP glucose + pyrophosphate (+ glucose-1-phosphate uridylyltransferase) = UTP + α-D-glucose-1-phosphate, α-D-glucose-1-phosphate + α-D-glucose-1,6-bisphosphate (+ phosphoglucomutase) = α-D-glucose-1,6-bisphosphate + α-D-glucose-6-phosphate, α-D-glucose-6-phosphate + NADP (+ glucose-6-phosphate dehydrogenase) = 6-phosphogluconate + NADPH, NADPH + PMS + MTT = NADP + PMS + *reduced MTT* (blue coloured)

Electrophoresis:

Technique	pH	System	Staining	Source
Cellogel 200 V, 2.5 h	7.0	C_{11}	FM	human

Recipe:

Add to 0.8 ml of a 360 mmol l^{-1} Tris-HCl buffer of pH 8.0: 0.2 ml MgCl$_2$-6H$_2$O (20.3 mg/ml), 6 mg UDPglucose-Na$_2$-2H$_2$O, 0.2 ml pyrophosphate-Na$_2$-10H$_2$O (22,5 mg/ml H$_2$O), 0.2 ml glucose-1,6-diphosphate (tetracyclohexylammonium salt-4H$_2$O, 0.175 mg/ml H$_2$O), 0.2 ml NADP-Na$_2$-2H$_2$O (4 mg/ml H$_2$O), 0.2 ml EDTA pH 7.0 (14.6 mg/ml), 5 µl glucose-6-phosphate dehydrogenase (140 U/ml), 5 µl phosphoglucomutase (400 U/ml), 0.2 ml MTT (2 mg/ml H$_2$O) and 0.2 ml PMS (0.4 mg/ml H$_2$O). Impregnate Cellogel strips and observe the appearance of blue bands [5, 6].

Glucose-1-phosphate uridylyltransferase (2.7.7.9)

(alternativ stain)

Reaction scheme:

UTP + α-D-glucose-1-phosphate (+ glucose-1-phosphate uridylyltransferase) = UDPglucose + pyrophosphate, pyrophosphate + Ca^{2+} = *Ca-pyrophosphate* (insoluble)

Electrophoresis:

Technique	pH	System	Staining	Source
PAGE, 3 – 4 mA/gel, 5 °C	8.9	D_1	FM	Yeast

Recipe:

To 5 ml of a 100 mmol l⁻¹ glycine-KOH buffer of pH 9.0 are added: 1 ml glucose-1-phosphate-Na$_2$-4H$_2$O (1.9 mg/ml), 1 ml UTP-Na$_3$-2H$_2$O (10.7 mg/ml), 1 ml MgCl$_2$-6H$_2$O (16,24 mg/ml), 1 ml CaCl$_2$ (2,2 mg/ml) and 1 ml of distilled water. Following electrophoresis the gels were soaked in 50 mmol l⁻¹ glycine KOH buffer of pH 9.0 for 20 – 30 min at 37 °C. They then are transferrred to the staining mixture and incubated at 37 °C. Active enzyme zones appear as milky white bands [21].

Glucose-6-phosphate dehydrogenase (1.1.1.49)

Reaction scheme:

D-Glucose-6-phosphate + NADP$^+$ = D-glucono-δ-lactone-6-phosphate + NADPH, NADPH + PMS = NADP + reduced PMS, reduced PMS + Nitro BT = PMS + *reduced Nitro BT* (blue)

Electrophoresis:

Technique	pH	System	Staining	Source
starch gel 13 V/cm, 5 h	8	S$_{29}$	AOL	mammals plants *Drosophila*

Recipe:

40 mg glucose-6-phosphate-Na$_2$, 3 mg NADP, 3 mg Nitro BT, 0.2 mg PMS and 10 mg MgCl$_2$-6H$_2$O are dissolved in 10 ml 0.04 mol l⁻¹ Tris-HCL buffer of pH 8.0. Blue bands indicate enzyme activity [6, 14, 72].

Glucose-phosphate isomerase (5.3.1.9)

Reaction scheme:

D-Fructose-6-phosphate (+ glucose-phosphate isomerase) = D-glucose-6-phosphate, D-glucose-6-phosphate + NADP (+ glucose-6-phosphate dehydrogenase) = D-glucono-δ-lactone-6-phosphate + NADPH, NADPH + PMS + MTT = NADP + PMS + *reduced MTT* (blue coloured)

Electrophoresis:

Technique	pH	System	Staining	Source
starch gel 6 V/cm, 33 V/cm, 2 h, 4 °C	7.2	S$_{30}$	MOL	mouse

Recipe:

3.2 ml 332 mmol l⁻¹ Tris-citric acid buffer of pH 8.0 are added to 38.4 ml distilled water. Then are added: 1 ml glucose-6-posphate dehydrogenase (50 U/ml), 1 ml D-fructose-6-phosphate (grade I, 75 mg/ml), 1 ml NADP-Na$_2$-2H$_2$O (10 mg/ml), 1 ml PMS (1.8 mg/ml), 1 ml MTT (5 mg/ml) and 6.4 ml MgCl$_2$-6 H$_2$O (50.75 mg/ml). Following electrophoresis a nitrocellulose membrane (ø 47mm, 0.45 or 0.20 µm pore size, Sartorius) attached to an O-ring with silicone lubricant is placed on the micro starch gel, the

O-ring upside. The assembly is left at room temperature for about 10 min while buffer from the gel soakes through the membrane. When the membrane is uniformly wet, the well is filled with stain. After staining, the stain is removed, the membrane is rinsed with water and the O-ring removed. The membrane is then lifted from the gel and washed for 1 h with a mixture of water, methanol and glacial acetic acid (5:5:1 (v/v)). The silicon lubricant is removed and the membrane is dried at 37 °C [73 – 75].

α-D-Glucosidase (3.2.1.20)

Reaction scheme:

4-Methylumbelliferyl-α-D-glucopyranoside + H_2O = glucose + *4-methylumbelliferone* (fluorescent)

Electrophoresis:

Technique	pH	System	Staining	Source
starch gel 13 V/cm, 4 h, 4 °C	5.8	S_3	MOL	human

Recipe:

Dissolve in 20 ml of a 100 mmol l^{-1} citrate-buffer of pH 4.0 10 mg 4-methylumbelliferyl-α-D-glucopyranoside. Apply on a porous membran and place the membrane on the cut surface of a processed starch gel. Spray after some time with ammonia to intensify fluorescence. Monitor fluorescent bands under UV-light [6].

α-D-Glucosidase (3.2.1.20)

Reaction scheme:

α-D-6-Bromo-2-naphtyl-α-D-glucopyranoside + H_2O = 6-bromo-2-naphthol + glucose, 6-bromo-2-naphthol + Fast blue B = *diazo-dye* (coloured)

Electrophoresis:

Technique	pH	System	Staining	Source
IEF 250 – 500V, 4 – 5 h, 4 °C	5 – 7	I_1 but at pH 5 – 7	FM	*Saccharomyces cerevisiae*

Recipe:

After electrophoresis the ampholytes containing polyacrylamide gel is soaked for 5 min in a 50 mmol l^{-1} potassium phosphate buffer of pH 6.9 containing 1 mmol l^{-1} EDTA (29.2 mg/100 ml). While the gel is soaking 5 mg 6-bromo-2-napthyl-α-D-glucopyranoside are dissolved in 0.5 ml dimethylsulphoxide and then added to 10 ml of 50 mmol l^{-1} phosphate buffer pH 6.9 containing 0.292 mg EDTA/ml. The gel is taken out of the buffer and the substrate-solution is spread over the gel surface incubating it for 15 min at room temperature. At the end of the incubation period 20 mg of Fast blue B are dissolved in 2 ml EDTA-containing phosphate buffer of pH 6.9. This solution is immediately added to the substrate solution. The gel is incubated for 3 – 5 min until visible bands appear. The staining procedure may be repeated to intensify the staining [76].

β-D-Glucuronidase (3.2.1.31)

Reaction scheme:

4-Methylumbelliferyl-β-D-glucuronide + H_2O = glucuronate + *4-methylumbelliferone* (fluorescent)

Electrophoresis:

Technique	pH	System	Staining	Enzyme
starch gel 13 V/cm, 5 h	8	S_4	FM	human

Recipe:

20 mg 4-methylumbelliferyl-β-Dglucuronide-3H$_2$O are dissolved in 10 ml of a 0.025 mmol l^{-1} citrate-phosphate buffer of pH 4. The solution is droped on the cut surface of a processed starch gel [6, 72].

L-Glutamate dehydrogenase (NADP) (1.4.1.4)

Reaction scheme:

L-Glutamate + H_2O + NADP$^+$ = L-oxoglutarate + NH_3 + NADPH, NADPH + PMS = NADP + reduced PMS, reduced PMS + MTT = PMS + *reduced MTT* (coloured)

Electrophoresis:

Technique	pH	System	Staining	Source
starch gel 3 V/cm, 17 h	8.1	S_{17}	AOL	human, plants, *Drosophila*

Recipe:

70 mg L-glutamic acid-Na-salt are dissolved in 20 ml 0.5 mol l^{-1} Tris-HCl, pH 7.6. 5 mg NADP-Na$_2$ in 1 ml H_2O, 0.5 mg MTT in 1 ml H_2O and 5 mg PMS in 1 ml H_2O are added. 20 ml of a 2% agar solution cooled to 45 °C are mixed with the solution (AOL) and poured on the cut starch gel surface. In starch gel electrophoresis good resolution of glutamate dehydrogenase appears to depend on cooking the starch gel only to the point where it is just to degase [6, 14].

Glutamine synthetase (6.3.1.2)

Reaction scheme:

ADP + ortho-phosphate + L-glutamine = ATP + L-glutamate + NH_3; NH_3 + FeCl$_3$ = *coloured dye*

Electrophoresis:

Technique	pH	System	Staining	Source
PAGE	9.5	D_1	FM	*E. coli*

Recipe:

Add to 5 ml of a 100 mmol l^{-1} imidazole-HCl buffer of pH 7.2 1 ml of L-glutamine (45.8 mg/ml), 1 ml ADP-Na$_3$-2H$_2$O (0.21 mg/ml), 1 ml disodium hydrogenarsenate

(NaHAsO$_4$-7H$_2$O) (62.4 mg/ml), 1 ml MgCl$_2$-6 H$_2$O (0.61 mg/ml) and 1 ml hydroxyl-amine-HCl (41.4 mg/ml). After incubation of the gels at 37 °C colour is developed by placing the gels in 2 ml of a solution containing 33% FeCl$_3$, 2% trichloroacetic acid and 0.25 N HCl. Parallel incubations in similar reaction mixtures containing no ADP or arsenate serve as controls [77].

Glutaminyl-peptide-glutamyl transferase (2.3.2.13)

Reaction scheme:

N-5-Amino-3-thiapentyl-5-dimethyl-aminonaphthalene-1-sulphonamide + casein = *casein* (fluorescent)

Electrophoresis:

Technique	pH	System	Staining	Source
agarose gel 1 mm thin polymerized on a polyester film 20 V/cm, 1 h, 10 °C	8.6	75 mmol l^{-1} sodium-barbital containing 2 mM EDTA, pH 8.6	MOL	rabbit, human

Recipe:

Add to 5 ml of a 100 mmol l^{-1} Tris-HCl buffer of pH 7.5 1 ml of monodansyl thiocada-verine-1/2-fumarate (2.31 mg/ml), 1 ml of casein (7 mg/ml), 1 ml thrombin (100 000 U/ml), 1 ml dithiothreitol (13.86 mg/ml) and 1 ml CaCl$_2$ (7.8 mg/ml). After electropho-resis the agarose gel is covered with a porous membrane moistened with the staining solution. It is incubated at 37 °C in a humid chamber to 1 h. The overlay is then re-moved and the gel is quickly rinsed in distilled water and immediately immersed in 10% aqueous trichloroacetic acid for 20 min. Afterwards the gel is immersed again for 2 times 5 min in trichloroacetic acid. The excess fluorescent stain is removed by repeated washes (4 × 5 min) in 10% aqueous acetic acid and in 50 mmol l^{-1} Tris-HCl buffer of pH 7.5 until neutral. Fluorescent bands are observed at 254 nm. Mono-dansyl thiocadaverine-1/2-fumarate (N-(5-amino-3-thiopentyl)-5-dimethyl-amino-1-naphtalene-sulphonamide-1/2-fumarate) was synthesized in one step from dansyl chloride and bis(2-aminoethyl) sulphide.

Transamidases (transglutaminases, endo-glutamine: ε-lysine transferases) com-prise a group of Ca^{2+}-dependent thiol enzymes catalyzing the formation of inter-molecular-glutamyl-ε-lysine bridges between some native proteins like fibrin and cold insoluble globulin [78].

Glutathione peroxidase (1.11.1.9)

Reaction scheme:

t-Butyl hydroperoxide + 2 reduced glutathione (+ glutathione peroxidase) = t-butyl-alcohol + oxidized glutathione + 2 H$_2$O, oxidized glutathione + NADPH (+ glutathione reductase) = 2 reduced glutathione + *NADP* (non-fluorescent)

Electrophoresis:

Technique	pH	System	Staining	Source
starch gel, 4 V/cm, 16 h, 4 °C	7.5	S$_{13}$	FD	human

Recipe:

Add to 10 ml 100 mmol l⁻¹ K_2HPO_4/KH_2PO_4 buffer of pH 7.0 30 mg reduced glutathione, 1 mg t-butyl hydrogenperoxide, 8 mg NADPH-Na₄, 20 mg EDTA-Na₂-2H₂O and 0.1 ml glutathione reductase (120 U/ml). Drop on the cut surface of a processed starch gel and observe non-fluorescent bands on a fluorescent background [16].

Glutathione reductase (1.6.4.2)

Reaction scheme:

Oxidized glutathione + NADPH (+ glutathione reductase) = NADP + 2 reduced glutathione, reduced glutathione + 2 nitrobenzoic acid = *coloured dye*

Electrophoresis:

Technique	pH	System	Staining	Source
starch gel 8–10V/cm, 4 h	8.0	S₂₁ to pH 8.0	AOL	human, plants, *Drosophila*

Recipe:

Add to 6 ml of a 200 mmol l⁻¹ Tris-HCl buffer of pH 8.0 1 ml oxidized glutathione (67.44 mg/ml), 1 ml NADPH-Na₄ (5 mg/ml), 1 ml 2-nitrobenzoic acid (2.94 mg/ml) and 1 ml EDTA (192.72 mg/ml). EDTA and 2-nitrobenzoic acid are added to the buffer and heated only as much as necessary to bring the dye into solution. After the mixture reaches a temperature of 47 °C NADPH and glutathione are added. Then the staining solution is mixed with 10 ml of a 2 % agar solution cooled down to 45 °C. The solution is poured on the cut surface of a processed starch gel. Yellowish bands mark the site of enzyme activity. Bands may also appear in the abscence of substrate [14].

Glyceraldehyde-phosphate dehydrogenase (1.2.1.12)

Reaction scheme:

D-Glyceraldehyde 3-phosphate + ortho-phosphate + NAD (+ glyceraldehyde-phosphate dehydrogenase) = 3-phospho-D-glyceroylphosphate + NADH, NADH + PMS + MTT = NAD + PMS + *reduced MTT* (blue coloured)

Electrophoresis:

Technique	pH	System	Staining	Source
starch gel 4 V/cm, 17 h, 4 °C	8.6	S₃₁	AOL	human

Recipe:

Add to 25 ml of a 50 mmol l⁻¹ Tris-HCl buffer of pH 7.5 2.5 µmol D-glyceraldehyde-3-phosphate, 30 mg NAD, 50 mg Na₂HAsO₄-7H₂O, 50 mg pyruvate-Na₂ 1 ml MTT (5mg/ml H₂O) and 0.5 ml PMS (2 mg/ml H₂O). Mix and add to 25 ml of a 2 % agar solution cooled down to 55 °C. Prepare the glyceraldehyde-3-phosphate from the diethylacetal barium salt using DOWEX 50 following the instructions of the manufacturer and assay the product by the method described by [6, 79].

Glyceraldehyde-phosphate dehydrogenase (1.2.1.12)

(alternative stain)

Reaction scheme:

3-Phospho-D-glyceroylphosphate + NADH (+ glyceraldehyde-phosphate dehydrogenase) = D-glyceraldehyde-3-phosphate + NADP + ortho-phosphate + Ca^{2+} = *Ca-phos-
phate* (precipitate)

Electrophoresis:

Technique	pH	System	Staining	Source
PAGE	8.9	D_1	FM	rabbit
3-4 mA/gel, 5 °C				

Recipe:

Add to 6.7 ml of a 200 mmol l^{-1} Tris-HCl buffer of pH 8.8 12.6 mg 3-phosphoglycerate-
Na$_3$, 1 ml phosphoglyceraldehyde kinase (0.1 mg/ml), 0.1 ml NADH-Na$_2$-3H$_2$O (5.5 mg/
ml), 0.1 ml ATP-Na$_2$-3H$_2$O (60.5 mg/ml), 0.1 ml dithiothreitol (15.4 mg/ml), 1 ml Mg Cl$_2$-
6H$_2$O (40,6 mg/ml) and 1 ml CaCl$_2$ (22.2 mg/ml). Processed polyacrylamide gels are
soaked in the buffer at 37 °C for 20 – 30 min and then transferred to the staining mixture. Gels are viewed against a dark background and stored in 50 mmol l^{-1} glycine-
KOH of pH 10 containing 0.5 mg/ml CaCl$_2$ and an antibacterial agent. This stain cannot be used for starch gels [21].

Glycerol-3-phosphate dehydrogenase (NAD) (1.1.1.8)

Reaction scheme:

sn-Glycerol-3-phosphate + NAD$^+$ = dihydroxyacetone phosphate + NADH, NADH +
PMS + Nitro BT = NAD + *reduced Nitro BT* (blue coloured)

Electrophoresis:

Technique	pH	System	Staining	Source
starch gel	8	S_4	AOL	human
13 V/cm, 5 h				

Recipe:

5 ml of substrate solution consisting of 1.0 mol l^{-1} Na-α-glycerolphosphate, pH 7.0
(21.6 g/100 ml H$_2$O adjusted to pH 7 with 1 N HCl) are mixed with 25 mg NAD, 15 mg
Nitro BT, 1 mg PMS, 10 ml 0.2 mol l^{-1} Tris-HCl buffer of pH 8 and 35 ml H$_2$O. The solution is poured on the cut surface of a processed starch gel and incubated at 37 °C in the
dark [72].

Glycerol-3-phosphate dehydrogenase (NAD) (1.1.1.8)

(alternative stain)

Reaction scheme:

sn-Glycerol-3-phosphate + NAD$^+$ = dihydroxyacetone phosphate + NADH, NADH +
PMS + MTT = NAD + PMS + *reduced MTT* (blue coloured)

Electrophoresis:

Technique	pH	System	Staining	Source
starch gel 5 V/cm, 17 h, 4 °C	8.0	S_{32}	AOL	human

Recipe:

Add to 20 ml 60 mmol l⁻¹ Tris-HCl buffer of pH 8.0 650 mg DL-glycerol-3-phosphate-Na_2-$5H_2O$, 200 mg pyruvate-Na, 20 mg NAD, 1 ml MTT (5 mg/ml H_2O) and 1 ml PMS (5 mg/ml H_2O). Heat substrate solution to 45 °C and mix with 25 ml of a 2 % agar solution cooled down to 45 °C. Pour on the cut surface of a processed starch gel and observe the formation of blue bands. Sodium pyruvate is added to minimize the staining of lactate dehydrogenase isozymes, *sn*-glycerol-3-phosphate is synonym to R-glycerol-3-phosphate [6, 35].

Glycollate oxidase (1.1.3.1)

Reaction scheme:

Glycollate + O_2 (+ glycollate oxidase) = glyoxylate + H_2O_2, H_2O_2 + 3,3,5,5′ tetramethylbenzidine (+ peroxidase) = $2H_2O$ + oxidized *tetramethylbenzidine* (brown coloured)

Electrophoresis:

Technique	pH	System	Staining	Source
PAGE 5 mA/tube, 4 °C	8.9	D_1	FM	rat

Recipe:

Add to 6.6 ml of a 100 mmol l⁻¹ phosphate buffer of pH 7.4 1 ml glycollate (3.8 mg/ml phosphate buffer adjusted to pH 7.4), 2 ml peroxidase (25 µg/ml phosphate buffer, pH 7.4) and 0.4 ml 3,3′,5,5′-tetramethylbenzidine (500 mg/100 ml dimethylformamide). The peroxidase solution should be prepared freshly [24].

Guanine deaminase (3.5.4.3)

Reaction scheme:

Guanine + H_2O (+ guanine deaminase) = xanthine + NH_3, xanthine + H_2O + PMS (+ xanthine oxidase) = urate + reduced PMS, reduced PMS + MTT = PMS + *reduced MTT* (blue coloured)

Electrophoresis:

Technique	pH	System	Staining	Source
starch gel 2 V/cm, 17 h, 4 °C	8.6	S_{33}	AOL	human

Recipe:

Add to 20 ml of a 200 mmol l⁻¹ Tris-HCl buffer of pH 7.6 3 ml of guanidine solution (50 mg of guanidine are dissolved in 10 ml of warm 0.1 N NaOH and the solution is finally filled up to 50 ml with distilled water), 25 µl xanthine oxidase (4 U/ml), 1.5 ml MTT

(5 mg/ml H_2O) and 0.5 ml PMS (5 mg/ml H_2O). Warm the substrate solution to 40°C and mix with 25 ml of a 2% agar solution [6].

Guanylate Kinase (2.7.4.8)

Reaction scheme:

GMP + ATP (+ guanylate kinase) = GDP + ADP, ADP + phosphoenolpyruvate (+ pyruvate kinase) = pyruvate + ATP, pyruvate + NADH (+ lactate dehydrogenase) = lactate + *NAD* (non-flourescent)

Electrophoresis:

Technique	pH	System	Staining	Source
starch gel 4 V/cm, 20 h, 4 °C	7.4	S_{22}	MOL	human

Recipe:

Add to 10 ml of a 100 mmol l⁻¹ Tris-HCl buffer of pH 7.5 10 mg ATP-Na_2-$3H_2O$, 25 mg GMP-Na_2-$5H_2O$, 11 mg phosphoenolpyruvate-Na-H_2O, 12 mg NADH-Na_2-$3H_2O$, 1 ml $MgCl_2$-$6H_2O$ (40.6 mg/ml), 2 ml KCl (3.75 mg/100 ml), 0.2 ml $CaCl_2$ (5.55 mg/100 ml), 25 µl pyruvate kinase (400 U/ml) and 50 µl lactate dehydrogenase (2 750 U/ml). Use the staining solution to impregnate a porous membrane. Place the membrane on the cut surface of a processed starch gel and observe the appearance of nonfluorescent bands on a flourescent background [6].

Hexokinase (2.7.1.1)

Reaction scheme:

ATP + α-D-glucose (+ hexokinase) = ADP + glucose-6-phosphate, glucose-6-phosphate + NADP (+ glucose-6-phosphate dehydrogenase) = 6-phosphogluconate + NADPH, NADPH + PMS + MTT − NADP + PMS + *reduced MTT* (blue coloured)

Electrophoresis:

Technique	pH	System	Staining	Source
starch gel 20 V/cm, 4.5 h, 4 °C	7.4	S_{12}	FM	human

Recipe:

Add to 40 ml of a 100 mmol l⁻¹ Tris-HCl buffer of pH 7.5 900 mg α-D-glucose, 40 mg ATP-Na_2-$3H_2O$, 1.5 ml NADP-Na_2-$2H_2O$ (5 mg/ml H_2O), 1 ml MTT (5 mg/ml H_2O), 0.5 ml PMS (5 mg/ml H_2O), 40 µl glucose-6-phosphate dehydrogenase (140 U/ml) and 10 ml $MgCl_2$-$6H_2O$ (4.06 mg/ml). Pour on the cut surface of a processed starch gel and observe the formation of blue bands. In human tissues up to 4 different isozymes have been observed. Isozyme III is inhibited by the glucose concentration given here while the other isozymes are not. Isozyme III is active when 9 mg glucose are used instead of 900 mg glucose. Isozymes I – III can also use fructose as a substrate, while isozyme IV can not [6].

Homoserine dehydrogenase (1.1.1.3)

Reaction scheme:

L-Homserine + NAD(P) = L-aspartate-β-semialdehyde + NAD(P)H, NAD(P)H + PMS + MTT = NAD(P) + PMS + *reduced MTT* (blue coloured)

Electrophoresis:

Technique	pH	System	Staining	Source
PAGE	7.9	D_6	FM	*E. coli*

Recipe:

Add to 5 ml of a 200 mmol l⁻¹ Tris-HCl buffer of pH 8.9 1 ml L-homoserine (59.5 mg/ml), 1 ml NADP-Na$_2$-2H$_2$O (8.23 mg/ml), 1 ml HCl (300 mg/ml), 1 ml PMS (24.48 mg/100 ml H$_2$O) and 1 ml MTT (12 mg/10 ml H$_2$O). Gel rods are immersed in the staining solution and the formation of blue bands is observed [80].

3-Hydroxybutyrate dehydrogenase (1.1.1.30)

Reaction scheme:

D-3-Hydroxybutyrate + NAD = acetoacetate + NADH, NADH + PMS + MTT = NAD + PMS + *red. MTT* (blue coloured)

Electrophoresis:

Technique	pH	System	Staining	Source
Starch gel	7.4	S_{34}	FM	human

Recipe:

Add to 25 ml 500 mmol l⁻¹ phosphate buffer of pH 7.4 10 ml 0.1-hydroxybutyrate, 100 mg NAD, 2.5 mg PMS, 12.5 mg MTT, 10.2 mg MgCl$_2$-6H$_2$O, 575 mg NaCl and 65 ml H$_2$O. Drop the staining solution on the cut surface of a processed starch gel and incubate the gel at 37 °C until blue bands appear. 3-Hydroxybutyrate dehydrogenase also oxidizes other 3-hydroxymonocarboxylic acids [17].

β-Hydroxysteroid dehydrogenase (1.1.1.51)

Reaction scheme:

Testosterone + NAD(P) = 4-androsterone-3,17-dione + NAD(P)H, NAD(P)H + PMS + MTT = NAD(P) + *red. MTT* (blue coloured)

Electrophoresis:

Technique	pH	System	Staining	Source
PAGE	8.9	D_7	FM	human

Recipe:

Add to 1.8 ml of a 25 mmol l⁻¹ Na$_2$HPO$_4$ buffer of pH 9.5 0.2 ml NAD-3H$_2$O (1.43 mg/ml), 1 ml MTT (0.42 mg/ml) 0.1 ml PMS (1 mg/ml) and 0.4 ml pregnenolone (0.32 mg/ml) in dimethylformamide. ß-Hydroxysteroid dehydrogenase practically did not penetrate acrylamide gels in the absence of chloral hydrate. The appearance of "nothing-dehydrogenase" bands is possible. A radiochemical assay has also been described [81].

3-Hydroxyacyl-CoA dehydrogenase (1.1.1.35)

Reaction scheme:

L-3-Hydroxyacyl-CoA + NAD = 3-oxoacyl-CoA + NADH
Counterstain:
NADH + PMS + MTT = NAD + PMS + *red. MTT* (blue coloured)
At the positions were NAD is formed white bands remain on the blue background, formed by reduced MTT.

Electrophoresis:

Technique	pH	System	Staining	Source
Cellogel 150 V, 2 h	6.5	C_{13}	FM	human

Recipe:

Add to 50 µl of a 100 mmol l⁻¹ citrate-phosphate buffer of pH 5.3 0.375 mg acetoacetyl-CoA in 150 µl H_2O and 0.35 mg NADH-Na_2-3H_2O in 50 µl H_2O. Apply to a processed Cellogel strip and incubate in a moist chamber. Enzyme bands are seen under UV-light as dark bands on a fluorescent background. A counterstain is possible by passing the Cellogel through a solution of MTT (1 mg/ml) and PMS (0.5 mg/ml) in 100 mM citrate-phosphate buffer of pH 5.3. Isozymes appear as white bands on a purple background. Finally wash throughly with tap water [6, 82].

3α-Hydroxysteroid dehydrogenase (1.1.1.50)

Reaction scheme:

Androsterone + NAD(P) = 5-androstane-3,17-dione + NAD(P)H, NAD(P)H + PMS + MTT = NAD(P) + PMS + *red. MTT* (blue coloured)

Electrophoresis:

Technique	pH	System	Staining	Source
Cellogel 15 V/cm, 30 – 60 min	9.0	C_{14}	FM	*Pseudomonas testosteroni*

Recipe:

Add to 70 ml of 100 mmol l⁻¹ phosphate buffer of pH 8.0 7 ml of a hydroxysteroid (1.5 mg/ml), 2 ml NAD-3H_2O (10 mg/ml), 1 ml PMS (5 mg/ml) and 4 ml MTT (5 mg/ml). The staining solution is divided into several equal parts and specific steroid substrates are added to each part. As hydroxysteroids, testosterone, dehydroepiandrosterone androsterone or etiocholanolone give positive reactions, as blank a solution without substrate (but including methanol) is used. During the incubation period direct light should be avoided since unspecific staining may then occur [83].

Hypoxanthine phosphoribosyl transferase (2.4.2.8)

Reaction scheme:

Hypoxanthine + 5′-phospho-α-D-ribose-1-diphosphate (+ hypoxanthine phosphoribosyl transferase) = inosine-5′-monophosphate + pyrophosphate, inosine-5′-phosphate + NAD + H$_2$O (+ inosine-5′-phosphate dehydrogenase) = xanthosine-5′-phosphate + *NADH* (fluorescent)

Electrophoresis:

Technique	pH	System	Staining	Source
starch gel 10 V/cm, 45 min and 5 V/cm, 3.5 h	7.2	S$_{29}$	FM	human

Recipe:

Add to 12 ml of a 200 mmol l^{-1} Tris-HCl buffer of pH 8.1 1 ml hypoxanthine (8.16 mg/ml), 1 ml NAD-3H$_2$O (86.04 mg/ml), 1 ml D-ribose-1-diphosphate-5-phosphate-Mg$_2$-2H$_2$O (18.84 mg/ml), 1 ml MgSO$_4$-7H$_2$O (74.1 mg/ml), 1 ml L-glutathione (red.) (36.84 mg/ml), 1 ml HCl (450 mg/ml), 1 ml allopurinol (8.16 mg/ml) and 1 ml inosine-5′-phosphate dehydrogenase (6 U/ml). Reduced glutathione, D-ribose-1,5-diphosphate and NAD should be prepared fresh for each run. The allopurinol is included in the reaction mixture to inhibit xanthin oxidase activities. An autoradiographic method has also been described. Isozymes are detected by the fluorescence of NADH visible under UV-light. The staining system is not coupled to PMS and MTT because PMS inhibits IMP dehydrogenase [6].

L-Iditol Dehydrogenase (1.1.1.14)

Reaction scheme:

L-Iditol + NAD = L-sorbose + NADH, NADH + PMS + MTT = NAD + PMS + *red. MTT* (blue coloured)

Electrophoresis:

Technique	pH	System	Staining	Source
starch-gel 2 V/cm, 17 h, 4 °C	8.6	S$_{35}$	FM	human

Recipe:

Add to 50 ml 50 mmol l^{-1} Tris-HCl buffer of pH 8.0 125 mg sorbitol, 20 mg NAD-3H$_2$O, 1 ml MTT (5 mg/ml) 1 ml PMS (2.5 mg/ml), 50 mg pyruvate-Na and 50 mg pyrazole. Pyrazole and pyruvate are included into the reaction mixture to inhibit lactate dehydrogenase and alcohol dehydrogenase. Drop staining mixture on the cut surface of a processed starch gel and observe the formation of blue bands [6].

Inorganic pyrophosphatase (3.6.1.1)

Reaction scheme:

Pyrophosphate + H_2O = 2 ortho-phosphate (H_3PO_4), H_3PO_4 + ammonium molybdate (+ H_2SO_4 + ascorbic acid) = *yellow coloured dye*

Electrophoresis:

Technique	pH	System	Staining	Source
starch gel 4 V/cm, 17 h, 4 °C	7.4	S_{36}	AOL	human

Recipe:

Add to 50 ml of a 50 mmol l^{-1} Tris, 10 mmol l^{-1} sodium pyrophosphate buffer of pH 7.8 2.5 ml $MgCl_2$-$6H_2O$ (40.6 mg/ml) and 50 ml 2% agar-solution cooled down to 45 °C. Incubate processed starch gels at 37 °C for 45 – 60 min. Then remove agar overlay and stain for ortho-phosphate by using a second agar overlay prepared from 50 ml 2.5% ammonium molybdate in 4N H_2SO_4, 5 g L-ascorbic acid and 50 ml 2% agar-solution cooled down to 45 °C. 2-Mercaptoethanol should be added to the sample solution to prevent disulphide exchange with oxidized glutathione [6].

Isocitrate dehydrogenase (NADP) (1.1.1.42)

Reaction scheme:

threo-D_s-Isocitrate + $NADP^+$ = 2-oxoglutarate + CO_2 + NADPH, NADPH + PMS + MTT = NADP + PMS + red. *Nitro BT* (blue coloured)

Electrophoresis:

Technique	pH	System	Staining	Source
starch gel 13 V/cm, 5 h	8	S_{37}	FM	human, plants, *Drosophila*

Recipe:

8 ml of substrate solution consisting of 0.1 mol l^{-1} isocitrate-Na_3, pH 7.0 (2.58 g/100 ml H_2O, adjusted to pH 7 with 1 N HCl) are mixed with 15 mg NADP, 15 mg Nitro BT, 1 mg PMS, 50 mg $MgCl_2$-$6H_2O$, 10 ml 0.2 mol l^{-1} Tris-HCl buffer of pH 8.0 and 32 ml H_2O. The solution is poured on the cut surface of a processed starch gel. Incubation is performed at 37 °C in the dark [6, 14, 72].

Lactate dehydrogenase (1.1.1.27)

Reaction scheme:

Lactate + NAD = pyruvate + NADH, NADH + PMS + MTT = NAD + PMS + *reduced MTT* (blue coloured)

Electrophoresis:

Technique	pH	System	Staining	Source
starch gel 4 v/cm, 17 h	7.0	S_{38}	AOL	human

Recipe:

Add to 20 ml of a 50 mmol l^{-1} Tris-HCl buffer of pH 8.0 100 mg L-lactate-Ca-5H$_2$O, 10 mg NAD, 1 ml MTT (5 mg/ml), 0.5 ml PMS (5 mg/ml) and 20 ml 2% agar-solution cooled down to 45 °C. Often weak bands also occur in the absence of lactate. This so called "nothing-dehydrogenase" reaction is probably due to substrate bound to enzyme protein. The reverse reaction (pyruvate + NADH = lactate + NAD) can also be used to detect active enzyme bands. The following method can be used to detect this reaction: 20 ml of a 50 mmol l^{-1} Tris-HCl buffer, pH 8.0, 100 mg pyruvate-Na and 10 mg NADH containing are used to impregnate a porous membrane, which is placed on the cut surface of a processed starch gel. Active enzyme bands appear as non-fluorescent zones on a fluorescent background. LDH-X is relatively more active with the substrates α-hydroxy butyrate and α-hydroxy valerate instead of lactate [6].

Lactose synthase (2.4.1.22)

Reaction scheme:

UDPgalactose + D-glucose (+ lactose synthase) = lactose + UDP, UDP + ATP (+ nucleoside-5'-diphosphate kinase) = UTP + ADP, UTP + glucose-1-phosphate (+ UDPglucose pyrophosphorylase) = UDPglucose + pyrophosphate, UDPglucose + 2NAD (+ UDPglucose dehydrogenase) = UDPglucuronic acid + *2NADH* (fluorescent)

Electrophoresis:

Technique	pH	System	Staining	Source
PAGE 5 – 10 V/cm, Gel dimensions 100 × 140 × 1.5 mm	7.8	D_8	AOL	calf bovine

Recipe:

Add to 4 ml of a 50 mmol l^{-1} cacodylate buffer of pH 7.5 1 ml MnSO$_4$-H$_2$O (17 mg/ml), 1 ml UDPgalactose-K$_2$-H$_2$O (10 mg/ml), 1 ml N-acetyl-D-glucosamine (104 mg/ml), 1 ml ATP-Na$_2$-3H$_2$O (47 mg/ml), 100 µl nucleoside-5-diphosphate kinase (935 000 U/ml), 1 ml glucose-1-phosphate-Na$_2$-4H$_2$O (2.26 mg/ml), 100 µl UDPglucose pyrophosphorylase (23 000 U/ml). 1 ml NAD-3H$_2$O (72 mg/ml), and 100 µl UDPglucose dehydrogenase (1 860 U/ml). Mix this solution with 10 ml of a 2% agar solution cooled to 45 °C and pour on the flat side of the polyarylamide gel; view under UV-light the appearance of fluorescent bands.

The enzyme is a complex of two portions. In the absence of the protein α-lactalbumin, the enzyme catalyzes the transfer of galactose from UDP to N-acetyl glucosamine [84].

Lactoyl-glutathione lyase (4.4.1.5)

Reaction scheme:

Reduced glutathione + methylglyoxal (+ glutathione lyase) = *S-D-lactoyl-glutathione* (no colour formation), Counter stain: reduced glutathione + dichlorophenol indophenol = reduced dichlorophenol indophenol + oxidized glutathione, reduced dichlorophenol indophenol + MTT = dichlorophenol indophenol + *red. MTT* (blue coloured)

Electrophoresis:

Technique	pH	System	Staining	Source
starch gel 7.5 V/cm, 14 h, 4 °C	7.8	S_{39}	FM	human

Recipe:

Add to 50 ml 200 mmol l⁻¹ phosphate buffer pH 6.8 0.9 ml methylglyoxal, 250 mg reduced glutathione and 50 mg MTT. Incubate at 37 °C. Then add 60 mmol l⁻¹ dichlorophenol indophenol in 100 mmol l⁻¹ Tris. White patches indicate enzyme zones. Alternative methods proposed by [6, 85].

Leucine dehydrogenase (1.4.1.9)

Reaction scheme:

L-Leucine + H_2O + NAD = 4-methyl-2-oxopentanoate + NH_3 + NADH, NADH + PMS + MTT = NAD + PMS + *reduced MTT* (blue coloured)

Electrophoresis:

Technique	pH	System	Staining	Source
starch gel	8.0	S_{40}	FM	bacteria

Recipe:

Add to 100 ml of a 100 mmol l⁻¹ phosphate buffer of pH 7.0 50 mg L-leucine, 50 mg NAD, 2 mg PMS and 15 mg MTT. Drop the staining solution on the cut surface of a processed starch gel and incubate at 37 °C in the dark until blue bands become visible [17].

Lysine 2-monooxygenase (2.13.12.2)

Reaction scheme:

Lysine + PMS = 5-aminovaleramide + CO_2 + reduced PMS, reduced PMS + p-iodonitrotetrazolium violet = PMS + reduced *p-iodonitrotetrazolium violet* (violet coloured)

Electrophoresis:

Technique	pH	System	Staining	Source
PAGE 5 mA/gel, 20 – 25 min		D_1		*Pseudomonas fluorescens*

Recipe:

Staining solution: 2 ml of a 100 mmol l^{-1} glycine buffer of pH 8.8 containing 1 ml L-lysine-HCl (915 mg/ml), 1 ml PMS (0.15 mg/ml) and 1 ml p-iodonitrotetrazolium violet (1 mg/ml). Incubate processed polyacrylamide gels in the staining solution until violet bands appear [86].

Malate dehydrogenase (1.1.1.37)

Reaction scheme:

L-Malate + NAD = oxalacetate + NADH, NADH + PMS + Nitro BT = NAD + PMS + reduced *Nitro BT* (blue coloured)

Electrophoresis:

Technique	pH	System	Staining	Source
starch gel 13 V/cm, 5 h	8	S$_4$	AOL	plants, *Drosophila*

Recipe:

1.215 g of Na$_2$CO$_3$-H$_2$O are dissolved in 5 ml H$_2$O and cooled in an icebath, then 1.34 g of L-malic acid are added under stirring and the solution is afterwards filled up to 10 ml with H$_2$O. The staining solution consists of 1 ml of this solution and 9 ml of a 0.04 mol l^{-1} Tris-HCl buffer of pH 8.0, containing 5 mg NAD, 3 mg Nitro BT and 0.3 mg PMS [14, 72, 87].

Malate dehydrogenase (oxaloacetate-decarboxylating) (NADP) (1.1.1.40)

Reaction scheme:

L-Malate + NADP = pyruvate + CO$_2$ + NADPH, NADPH + PMS + MTT = NADP + PMS + *red. MTT* (blue coloured)

Electrophoresis:

Technique	pH	System	Staining	Source
starch gel 10 V/cm, 5 h, 4 °C	8.6	S$_{41}$	AOL	human, mouse

Recipe:

Dissolve in 20 ml of a 100 mmol l^{-1} Tris-HCl buffer of pH 7.0 100 mg L-malic acid and readjust pH to 7.0 with NaOH. Then add 2.5 ml Mg Cl$_2$-6H$_2$O (40.6 mg/ml), 1 ml NADP-Na$_2$-2H$_2$O (5 mg/1 ml H$_2$O), 1 ml MTT (5 mg/ml H$_2$O), 0.1 ml PMS (5 mg/ml H$_2$O) and finally mix with 25 ml of a 2 % agar solution cooled down to 45 °C. Pour on the cut surface of a processed starch gel and observe the formation of blue bands. To solubilize the human enzyme an emulsifier is used when preparing tissue homogenates. In man the soluble enzyme form is migrating more anodal than the mitochondrial form. In mouse it is vice versa [6, 72].

Mannitol dehydrogenase (1.1.1.67)

Reaction scheme:

D-Mannitol + NAD = D-fructose + NADH, NADH + PMS + MTT = NAD + PMS + *red. MTT* (blue coloured)

Electrophoresis:

Technique	pH	System	Staining	Source
PAGE	8.9	D_1	FM	*Absidia glauca*

Recipe:

Add to 25 ml of a 300 mmol l⁻¹ glycine-NaOH buffer of pH 9.6 5 ml NAD-3H₂O (43 mg/ml), 5 ml mannitol (218.4 mg/ml), 0.1 ml PMS (5 mg/ml H₂O), 1 ml MTT (5 mg/ml H₂O) and 13.9 ml distilled water. Immerse processed polyacrylamide gel in the staining solution and observe the formation of blue bands [88].

Mannosephosphate isomerase (5.3.1.8)

Reaction scheme:

D-Mannose-6-phosphate (+mannosephosphate isomerase) = D-fructose-6-phosphate, D-fructose-6-phosphate (+ glucosephosphate isomerase) = D-glucose-6-phosphate, D-glucose-6-phosphate + NADP (+ glucose-6-phosphate dehydrogenase) = D-glucono-δ-lactone-6-phosphate + NADPH, NADPH + PMS + MTT = NADP + PMS + *red. MTT* (blue coloured)

Electrophoresis:

Technique	pH	System	Staining	Source
Cellogel 3.5 h	6.2	C_{17}	FM	human, hamster

Recipe:

Add to 1 ml of a 360 mmol l⁻¹ Tris-HCL buffer of pH 8.0 0.1 ml mannose-6-phosphate-Na₂ (20 mg/ml), 0.5 ml NADP-Na₂-2H₂O (4 mg/ml). 0.02 ml glucose-phosphate isomerase (2 mg/ml), 0.01 ml glucose-6-phosphate dehydrogenase (grade II) (1 mg/ml), 0.05 ml CoCl₂ - 6 H₂O (59.5 mg/ml), 0.1 ml PMS (0.8 mg/ml) and 0.1 ml MTT (4 mg/ml). Inpregnate a processed Cellogel strip and observe the occurrence of blue bands after approximately 20 min [5].

α-D-Mannosidase (3.2.1.24)

Reaction scheme:

4-Methylumbelliferyl-α-D-mannopyranoside + H₂O = α-D-mannose + *4-methylumbelliferone* (fluorescent)

Electrophoresis:

Technique	pH	System	Staining	Source
Cellogel 10 V/cm, 2 h	6.5	S_4	MOL	human

Recipe:

Reaction mixture: 4 mg 4-methylumbelliferyl-α-D-mannopyranoside are dissolved in 25 ml of 0.1 mol l^{-1} citrate-phosphate buffer, pH 3.5, 4.5 or 6.5. A cellulose acetate membrane is saturated with reaction mixture and applied on to the processed Cellogel membrane at room temperature for 10 – 30 min. Then the membrane is taken off and replaced by a cellulose acetate membrane saturated with 1 mmol l^{-1} glycine-NaOH, pH 10. The appearance of fluorescent zones is viewed under UV-light. In human tissues the main "acidic" enzyme form is visualized with the staining system at pH 3.5 and the main "neutral" isozyme(s) with the staining system at pH 6.5. At pH 4.5 both isozyme systems are seen [72, 89, 90].

α-D-Mannosidase (3.2.1.24)

(alternative stain)

4-Nitrophenyl-α-D-mannopyranoside + H_2O = α-D-mannose + *4-nitrophenol* (yellow coloured)

Electrophoresis:

Technique	pH	System	Staining	Source
PAGE	8.9	D_1	FM	Jack bean

Recipe:

Substrate solution: dissolve in 5 ml of a 25 mmol l^{-1} citrate buffer of pH 4.5 7.55 mg 4-nitrophenyl-α-D-mannopyranoside. Following electrophoresis gels are rinsed with water and incubated for 30 min at room temperature in a 200 mmol l^{-1} acetate buffer of pH 4.9. Subsequently the gels are immersed for 30 min in substrate solution. During the course of the incubation period, the enzymatic liberation of 4-nitrophenol is observed as a developing yellow band [3].

Melilotate 3-monooxygenase (1.14.13.4)

Reaction scheme:

3-(2-Hydroxyphenyl) propionate + NADH + FAD = 3-(2,3-dihydroxyphenyl) propionate + NAD + $FADH_2$, $FADH_2$ + MTT = FAD + *red. MTT* (blue coloured)

Electrophoresis:

Technique	pH	System	Staining	Source
PAGE	8.9	D_1	FM	*Arthrobacter spe.*

Recipe:

Add to 6 ml of a 10 mmol l^{-1} phosphate buffer of pH 7.3 1 ml melilotic acid (o-hydroxyphenyl propionic acid) (2.24 mg/10 ml), 1 ml NADH-Na$_2$-3H$_2$O (1.4 mg/ml), 1 ml FAD-Na$_2$-2H$_2$O (3.6 mg/10 ml) and 1 ml MTT (10 mg/ml). Immerse processed gels until blue bands of enzyme activity can be seen [91].

Monophenol monooxygenase (1.14.18.1)

Reaction scheme:

4-Methylphenol + O_2 (+ L-proline) = 1,2-dihydroxy-4-methyl benzene + H_2O + *1,2-dioxy-4-methylchinone* (dark purple coloured)

Electrophoresis:

Technique	pH	System	Staining	Source
PAGE	8.9	D_1	FM	mushroom, broadbean, potato, *Neurospora*

Recipe:

Mix equal volumes of 4-methylphenol (2 mg/10 ml of 100 mmol l^{-1} phosphate buffer of pH 7.0) and L-proline (2 mg/10 ml of 100 mmol l^{-1} phosphate buffer of pH 7.0) and immerse processed polyacrylamide gels in this solution until dark purple bands appear [92].

NADH dehydrogenase (1.6.99.3)

Reaction scheme:

NADH + 2,6-dichlorophenol indophenol = NAD + reduced 2,6-dichlorophenol indophenol, reduced 2,6-dichlorophenol indophenol + MTT = 2,6-dichlorophenol indophenol + *reduced MTT* (blue coloured)

Electrophoresis:

Technique	pH	System	Staining	Source
starch gel	8.0	S_{42}	AOL	human, plants, *Drosophila*

Recipe:

To 50 ml of a Tris-HCl buffer of pH 8.5 10 mg NADH-Na_2-$3H_2O$ and 2.5 ml 2,6-dichlorophenol indophenol (2 mg/ml) are added. This solution is mixed with 50 ml of a 2% agar solution cooled down to 45 °C.

Alternative methods have been described [14]. But the method given here is to be preferred. After enzyme preparations have been subjected to certain treatments cytochrome c may act as electron acceptor [6].

NADPH dehydrogenase (1.6.99.1)

Reaction scheme:

NADPH + neotetrazolium chloride = NADP + reduced *neotetrazolium chloride* (red coloured)

Electrophoresis:

Technique	pH	System	Staining	Source
PAGE 3 mA/gel	8.9	D_9	FM	porc

Recipe:

Add to 3 ml of a 5 mmol l^{-1} phosphate buffer of pH 7.5 1 ml NADPH-Na$_4$ (1.66 mg/ml) and 1 ml neotetrazolium chloride (2.51 mg/ml). Immerse processed gels into the staining solution and incubate for 1 min at 37 °C. The enzymic reaction is terminated by the addition of a solution consisting of 40 ml distilled water, 3.6 ml 10% Triton X-100, 5 ml 40% formalin and 10 ml of a 1 mol l^{-1} formate buffer of pH 3.5 [93].

NAD(P) nucleosidase (3.2.2.6)

Reaction scheme:

NAD(P) + H_2O = nicotinamide + ADPribose(P)

Counter stain:
The entire gel is stained blue by the action of a NAD(P) specific dehydrogenase on NAD(P)H in the presence of a suitable substrate and Nitro BT or MTT. At the sites where NAD(P) is destroyed no colour formation occurs [6].

Electrophoresis:

Technique	pH	System	Staining	Source
Cellogel prerun at 200 V for 10 min, then 200 V, 90 min	7.6	C_{18}	FM	porc, human

Recipe:

A processed Cellogel strip is soaked in 50 mmol l^{-1} phosphate buffer of pH 6.5 containing NAD or NADP (approximately 5 mg/ml) and incubated in a moist chamber for 1–2 h. Counterstaining to detect NAD is performed by placing the gel in a solution containing 0.3 ml absolute ethanol, 25 U alcohol dehydrogenase, 10 mg Nitro BT and 1 mg PMS in 50 ml 100 mmol l^{-1} pyrophosphate buffer of pH 8.8. Counterstaining for NADP is performed by using the following solution: 10 mg glucose-6-phosphate-Na$_2$, 25 U glucose-6-phosphate dehydrogenase , 10 mg Nitro BT and 1 mg PMS dissolved in 50 ml 100 mmol l^{-1} Tris-HCl buffer of pH 8.0. Enzymatically active bands of NAD(P) nucleosidase appear as white bands on a dark blue background. The enzyme is membrane bound and must be solubilized by a detergent such as Triton X-100 [6].

Nitrate reductase (NADH) (1.6.6.1)

Reaction scheme:

Nitrate reductase: NADH + nitrate = NAD + nitrite + H_2O, nitrite + N(1-naphthyl)-ethylenediamine-HCl + sulphanilamide = *a red diazo-dye*

Electrophoresis:

Technique	pH	System	Staining	Source
PAGE		D_{10}	FM	*E. coli*, radish

Recipe:

Add to 1.5 ml of a 10 mmol l^{-1} phosphate buffer of pH 7.5 0.2 ml KNO_3 (10.1 mg/ml), 0.5 mg NADH-Na$_2$-3H$_2$O, 0.5 ml N(1-naphthyl)ethylenediamine-HCl (0.02 %) and 1 ml sulphanilamide (0.5 % adjusted to pH 7.0). The gels are incubated for 20 min at 30 °C in the staining solution and then transferred into 0.1 N HCl under which conditions a red diazo-dye is formed. The gels may be scanned at 550 nm within 30 min of the development of the colour. Later on the dye slowly diffuses out of the gel [95]. A staining method for nitrite reductase (1.7.99.3) has been described by Huklesby and Hageman [94, 96].

Nitrogenase (1.18.2.1)

Reaction scheme:

Non-haem iron + α,α-diphyridyl + mercaptoacetic acid = *a coloured dye*

Electrophoresis:

Technique	pH	System	Staining	Source
PAGE 17 V/cm	9.0	D_{11}	FM	*Azotobacter vinelandii*

Recipe:

Gels were submersed in a solution of 0.7 % α,α-dipyridyl containing 8 % mercaptoacetic acid which makes the Fe^{2+} of the non-haem iron enzyme accessible to α,α-dipyridyl. Within 10 min pink bands appear which correspond to the complex of α,α-dipyridyl with Fe^{2+} [100]. It is important to observe the gels within 15 min after adding the α,α-dipyridyl because the pinc colour diffuses. Nitrogenase components are extremely O_2-labile and therefore the buffer tank is enriched with 0.3 mg/ml sodium dithionite. Samples of crude extracts were stored under N_2 in bottles with rubber serum caps and approximately 120 µl of sample is mixed with an equivalent volume of 50 % sucrose in Tris-HCl (0,075 mol l^{-1}, pH 9.0) thoroughly degassed with N before use. Samples were layered anaerobically in a sample well in the gel.

Remarkes: cytochrom c, ferritin and other haeme proteins do not yield a positive reaction with α,α-dipyridyl [97]. Haeme-containing proteins are readily detected by the benzidine reagent. Also, the colour produced with haeme is blue, while the reaction product with nonhaeme iron proteins is red or brown [98 – 100].

Nucleoside triphosphatase (3.6.1.15)

Reaction scheme:

Nucleoside triphosphate + H_2O = nucleoside diphosphate + ortho-phosphate, ortho-phosphate + ammonium molybdate = *blue coloured dye*

Electrophoresis:

Technique	pH	System	Staining	Source
PAGE 17 V/cm	8.5	D_{12}	FM	*Streptococcus fecalis,* mitochondria from animals

Recipe:

Substrate solution: Add to 3 ml of 100 mmol l^{-1} Tris-HCl buffer, of pH 7.5 1 ml ATP-Na$_2$-3H$_2$O (15.13 mg/ml) and 1 ml MgCl$_2$-6H$_2$O (5.08 mg/ml). Incubate processed polyacrylamide gels for 30 min in substrate solution. Transfer the gels afterwards in a solution containing 9.4% perchloric acid, 1% ammonium molybdate and 0.25% reducing mixture (sodium sulphite, sodium bisulphite, 1-amino-2-naphthol-4-sulphonic acid, 6:6:1). In a few minutes blue bands appear against an almost colourless background denoting the position of active enzymes. Eventually the entire background becomes blue. In order to delete GTPase and ITPase activity GTP and ITP may be used as the substrate [101].

Nucleosidetriphosphate-adenylate kinase (2.7.4.10)

Reaction scheme:

GTP + AMP = GDP + ADP, ADP + phosphoenolpyruvate (+ pyruvate kinase) = pyruvate + ATP, ATP + glucose (+ hexokinase) = ADP + glucose-6-phosphate, glucose-6-phosphate + NADP (+ glucose-6-phosphate dehydrogenase) = 6-phosphogluconate + NADPH, NADPH + PMS + MTT = NADP + PMS + *reduced MTT* (blue coloured)

Electrophoresis:

Technique	pH	System	Staining	Source
starch gel 10 V/cm, 4,5 h, 4 °C	5.9	S_{43}	AOL	human

Recipe:

Add to 10 ml of a 300 mmol l^{-1} Tris-HCl buffer of pH 8 15 mg GTP-Na$_2$-H$_2$O, 25 mg AMP-Na$_2$-6H$_2$O, 20 mg phosphoenol-pyruvate-Na-H$_2$O, 1 ml NADP-Na$_2$-2H$_2$O (5 mg/ml H$_2$O), 150 mg KCL, 1 ml glucose (40 mg/1 ml H$_2$O) 25 µl glucose-6-phosphate dehydrogenase (140 U/ml), 25 µl hexokinase (280 U/ml), 50 µl pyruvate kinase (400 U/ml), 1 ml MTT (5 mg/ml H$_2$O), 0.1 ml PMS (5 mg/ml H$_2$O) and 1 ml MgCl$_2$-6H$_2$O (81.2 mg/ml H$_2$O). Mix the reaction mixture with 10 ml of a 2% agar solution cooled down to 45 °C and pour on the cut surface of a processed starch gel. Enzymatically active bands appear as blue zones.

Remarks: Adenylate kinase isozymes AK$_1$ and AK$_2$ from human tissues are also stained. But nucleoside triphosphate-adenylate kinase is not stained in the adenylate kinase assay. ITP, but not ATP or CTP or UTP can replace GTP in the stain. The enzyme is not inhibited by AgNO$_3$. The most cathodal isozyme of human AK$_1$ overlaps the less cathodal isozyme of nucleoside triphosphate-adenylate kinase in tissues were they occur together [6].

Nucleosidetriphosphate-pyrophosphatase (3.6.1.19)

Reaction scheme:

Inosine triphosphate + H_2O = inosine monophosphate + pyrophosphate, pyrophosphate + ammonium molybdate + (H_2SO_4 + asorbic acid) = *blue coloured dye*

Electrophoresis:

Technique	pH	System	Staining	Source
starch gel 1.5 V/cm, 17 h, 4 °C	7.2	S_{29}	MOL	human

Recipe:

Add to 10 ml of a 200 mmol l^{-1} Tris-HCl buffer of pH 7.6 20 mg inosine triphosphate-Na_2-H_2O, 10 ml $MgCl_2$-$6H_2O$ (20.3 mg/ml) and 0.2 ml 2-mercaptoethanol. Use the staining mixture to impregnate a porous membrane, put the membrane on the cut surface of a processed starch gel and incubate at 37 °C for 2 h. Finally remove the overlay and replace by an agar overlay consisting of 20 ml of 25% ammonium molybdate in 4 N H_2SO_4, 1 g ascorbic acid and 20 ml of a 2% agar solution. Active bands appear as blue zones.

Remarks: A change in enzyme activity and an increase in the net negative charge of isozymes may occur due to the oxidation of reactive sulphydryl groups by interaction with oxidized glutathione. It may be reversed or prevented by the addition of 2-mercaptoethanol to a concentration of 10 mM [6].

5'-Nucleotidase (3.1.3.5)

Reaction scheme:

A 5'-ribonucleotide + H_2O = a ribonucleoside + ortho-phosphat, ortho-phosphate + Ca^{2+} = $Ca_3(PO_4)_2$

Electrophoresis:

Technique	pH	System	Staining	Source
PAGE 2 mA/gel, 1.5 h	8.6	D_1	FM	plants

Recipe:

Dissolve 11 mg adenosine-5'-phosphate in 8.5 ml of a 0.1 mol l^{-1} Tris-HCl buffer of pH 8.3, add 0.5 ml 2-mercaptoethanol (0.16 ml/100 ml H_2O) and 1 ml $CaCl_2$ (111 mg/10 ml); filtrate. The appearance of white bands indicates the existance of active enzyme(s) [42]

Oestradiol-17β-dehydrogenase (1.1.1.62)

Reaction scheme:

Oestradiol-17β + NAD = oestrone + NADH, NADH + PMS + MTT = NAD + PMS + *reduced MTT* (blue coloured)

Electrophoresis:

Technique	pH	System	Staining	Source
PAGE 1–3 mA/tube	8.3	D_1	FM	human

Recipe:

Add to 9 ml of a 100 mmol l^{-1} Na$_2$CO$_3$/NaHCO$_3$ buffer of pH 9.2 3 ml oestradiol-17β (0.82 mg/ml), 6 ml NAD-3H$_2$0 (4.31 mg/ml), 5 ml PMS (1.6 mg/ml), 5 ml MTT (0.5 mg/ml) and 62 ml distilled water. Immerse processed polyacrylamid gels in the solution and observe the formation of blue coloured bands [102].

Penicillinase (3.5.2.6)

(syn.: β-Lactamase, Cephalosporinase)

Reaction scheme:

Penicillinase hydrolyzes a number of compounds containing the β-lactam structure

Electrophoresis:

Technique	pH	System	Staining	Source
starch gel 110 min	8.45	S_{44}	FM	*Streptomyces Actinomy-cetales*, yeast, Bluegreen algae

Recipe:

The location of cephalosporinase-activity is detected by spraying the gel with a solution containing 1 g of cephalosporidine (7-[(2-thienyl)acetamido]-3-(1-pyridyl-methyl)cephalosporanic acid) dissolved in 10 ml of a 10 mmol l^{-1} J$_2$ 30 mmol l^{-1}-KJ-solution. The presence of enzyme activity is shown by the occurrence of a white band upon a dark blue background caused by the interaction of the iodine with the starch gel.

In polyacrylamide gels β-lactamase may be detected by using as substrate a highly conjugated cephalosporin which upon hydrolysis results in nitrocefin which has a blue colour [103, 104].

Pepsin (3.4.23.3)

Reaction scheme:

Negative stain: a protein + H$_2$O = peptides, protein + nigrosin = *dark blue coloured protein complex*

Electrophoresis:

Technique	pH	System	Staining	Source
1% agarose 11 V/cm, 3–4 h, 4 °C	8.3	Gel and electrode buffer: 50 mmol l⁻¹ barbital barbituric acid, pH 8.3	FM	human

Recipe:

Immerse the processed gel in a solution containing 0.65% bovine haemoglobin in 60 mmol l⁻¹ HCl pH 1.4 for 10 min and incubate the gel at 37 °C for 1 h in a humid chamber. Then fix the undigested haemoglobin by immersing the gel in 10% acetic acid in 50% methanol for about 18 h. Afterwards stain with a solution containing 10% acetic acid in 50% methanol and 200 mg/l nigrosin. The acid methanolic solution can be used to wash the gel. Active enzyme bands occur as transparent bands on a dark-blue background [6].

Peptidases (A, B, C, E, F and S) (3.4.11 or 13.*) (* = not further specified) and peptidase D (3.4.13.9)

(syn.: dipeptidases, tripeptidases, aminopeptidases; syn.: proline dipeptidase for peptidase D)

Reaction scheme:

A dipeptide + H_2O = L-amino acids, (a tripeptide + H_2O = L-amino acid + dipeptide) L-amino acid + O_2 (+ L-amino acid-oxidase) = keto acid(s) + NH_3 + H_2O_2, H_2O_2 + 9-amino-ethylcarbazole (+ peroxidase) = H_2O + *oxidized 9-amino-ethylcarbazole* (brown coloured)

Electrophoresis:

Technique	pH	System	Staining	Source
starch gel, 5 V/cm, 18 h, 4 °C	7.4	S_{45}	AOL	human

Recipe:

Reaction mixture: 35 ml of 20 mmol l⁻¹ Na_2HPO_4 adjusted to pH 7.5 with 1 N HCl, 20 mg dipeptide (tripeptide), 50 µl snake venom L-amino-acid oxidase (approx. 15 U/ml), 100 µl peroxidase (2500 U/ml), 0.5 ml 100 mmol l⁻¹ $MgCl_2$-6H_2O and 0.5 ml 3-amino-9-ethylcarbazole (25 mg/ml). Mix with 30 ml of a 2% Agar solution [6].

Phosphodiesterase I (3.1.4.1)

Reaction scheme:

α-Naphthyluridine-5′-phosphate + H_2O = uridine-5′-phosphate + α-naphthol, α-naphthol + Fast red TR = *diazo dye* (red coloured)

Electrophoresis:

Technique	pH	System	Staining	Source
PAGE	8.9	D_1	FM	rye, barley, potato

Recipe:

Processed gels are immersed in a solution containing in a 40 mmol l^{-1} phosphate buffer of pH 7.0 4 mg α-naphthyluridine-5′-phosphate and 30 mg Fast red TR. Active enzyme zones appear as red bands. The substrate is not hydrolyzed by nucleotide phosphatases. EDTA of concentrations of 10 mmol l^{-1} inhibit the hydrolysis of the naphthylester [105].

Phosphodiesterase I (3.1.4.1)

(alternative stain)

Reaction scheme:

4-Methylumbelliferyl-thymidine-5′-phosphate + H_2O = thymidine-5′-phosphate + *4-methylumbelliferone* (fluorescent)

Electrophoresis:

Technique	pH	System	Staining	Source
Cellogel 60 min	10.5	C_{19}	FM	Snake venom, human

Recipe:

Following electrophoresis the gels are washed in distilled water containing 1.2 mg/ml 4-methylumbelliferyl-thymidine-5′-phosphate. The release of 4-methylumbelliferone results in the appearence of fluorescent bands. The substrate is almost not attacked by alkaline phosphatase [106].

6-Phosphofructokinase (2.7.1.11)

Reaction scheme:

Fructose-6-phosphate + ATP (+ 6-phosphofructokinase) = ADP + fructose-1,6-diphosphate, fructose-1,6-diphosphate (+ aldolase) = dihydroxyacetone phosphate + glyceraldehyde-3-phosphate, dihydroxyacetone phosphate (+ triose-phosphate isomerase) = glyceraldehyde-3-phosphate, glyceraldehyde-3-phosphate + arsenate + NAD (+ glyceraldehyde-3-phosphate dehydrogenase) = 3-phosphoglyceroylarsenate + *NADH* (fluorescent)

Electrophoresis:

Technique	pH	System	Staining	Source
starch gel 8 V/cm, 17 h, 4 °C	7.75	S_{46}	MOL	human

Recipe:

Add to 20 ml of a 100 mmol l^{-1} Tris-HCl buffer of pH 8.0 12 mg fructose-6-phosphate-Na$_2$-H$_2$O, 12 mg ATP-Na$_2$-3H$_2$O, 7 mg NAD-3H$_2$O, 40 mg MgCl$_2$-6H$_2$O, 200 mg Na$_2$HAsO$_4$-7H$_2$O, 20 µl 2-mercaptoethanol, 400 µl aldolase (90 U/ml), 50 µl triose-phosphate isomerase (10 000 U/ml) and 50 µl glyceraldehydephosphate dehydrogenase (800 U/ml). The reaction mixture is used to impregnate a porous membrane which is placed on the cut surface of a processed starch gel. Active enzyme zones appear as fluorescent bands [6].

Phosphoglucomutase (2.7.5.1)

Reaction scheme:

α-D-Glucose-1,6-biphosphate + α-D-glucose-1-phosphate + (phosphoglucomutase) = α-D-glucose-6-phosphate + α-D-glucose-1,6-bisphosphate, glucose-6-phosphate + NADP (+ glucose-6-phosphate dehydrogenase) = 6-phosphogluconate + NADPH, NADPH + PMS + MTT = NADP + PMS + *reduced MTT* (blue coloured)

Electrophoresis:

Technique	pH	System	Staining	Source
starch gel 13 V/cm, 5 h	8	S$_4$	OL	human

Recipe:

Dissolve 60 mg glucose-1-phosphate-Na$_2$-4H$_2$O (containing at least 1% glucose-1,6-bisphosphate) in 2.0 ml of a 0.2 mol l^{-1} Tris-HCl buffer of pH 7.0. Add 10 mg MgCl$_2$-6H$_2$O, 1 ml glucose-6-phosphate dehydrogenase (160 U/ml), 3 mg NADP, 4 mg MTT and 0.2 mg PMS to 8 ml H$_2$O. Mix both solutions and pour over the cut surface of a processed starch gel. Active enzyme is indicated by the appearance of blue bands [72, 107].

Phosphogluconate dehydrogenase (decarboxylating) (1.1.1.44)

Reaction scheme:

6-Phospho-D-gluconate + NADP$^+$ = D-ribulose-5-phosphate + CO$_2$ + NADPH, NADPH + PMS + Nitro BT = NADP + PMS + reduced *Nitro BT* (blue coloured)

Electrophoresis:

Technique	pH	System	Staining	Source
starch gel 13 V/cm, 5 h	8	S$_{47}$	FM	human, hamster

Recipe:

Dissolve 100 mg 6-phosphogluconate-Na$_3$, 15 mg NADP, 15 mg Nitro BT, 1 mg PMS and 50 mg MgCl$_2$-6H$_2$O in 10 ml of a 0.2 mol l^{-1} Tris-HCl buffer of pH 8.0. Then 40 ml H$_2$O are added. The solution is poured on the cut surface of a processed starch gel which is incubated at 37 °C [72].

Phosphoglycerate kinase (2.7.2.3)

Reaction scheme:

ATP + 3-phosphogycerate (+ phosphoglycerate kinase) = ADP + 3-phospho-D-glyceroylphosphate, 3-phospho-D-glyceroylphosphate + NADH (+ D-glyceraldehyde-phosphate dehydrogenase) = D-glyceraldehyde-3-phosphate + *NAD* (non-fluorescent), D-glyceraldehyde-3-phosphate (+ triose-phosphate-isomerase) = 3-phospho-D-glyceroylphosphate

Electrophoresis:

Technique	pH	System	Staining	Source
starch gel 5 V/cm, 17 h, 4 °C	7.5	S$_7$	MOL	human

Recipe:

Add to 5 ml of a 500 mmol l^{-1} Tris-HCl buffer of pH 7.8 15 mg 3-phosphoglycerate-Na$_3$, 30 mg ATP-Na$_2$-3H$_2$O, 40 mg MgCl$_2$-6H$_2$O, 50 µl glyceraldehyde-3-phosphate dehydrogenase, 10 mg NADH, 10 µl triosephosphate isomerase (10 000 U/ml) and 50 µl glycerol-3-phosphate dehydrogenase (80 U/ml). Impregnate a porous membrane with the staining solution and put the membrane on the cut surface of a processed starch gel. Non-fluorescent bands indicate active enzyme zones.

The reverse reaction is not recommended to stain for phosphoglycerate kinase since the adenylate kinase isozymes also give a positive reaction [6].

Phosphoglyceromutase (2.7.5.3)

Reaction scheme:

2,3-Bisphospho-D-glycerate + 2-phospho-D-glycerate (+ phosphoglyceromutase) = 3-phospho-D-glycerate + 2,3-bisphospho-D-glycerate, 3-phospho-D-glycerate + ATP (+ phosphoglycerate kinase) = ADP + 3-phospho-D-glyceroylphosphate, 3-phospho-D-glyceroylphosphate + NADH (+ glyceraldehyde phosphate dehydrogenase) = D-glyceraldehyde-3-phosphate + *NAD* (non-fluorescent)

Electrophoresis:

Technique	pH	System	Staining	Source
starch gel 13 V/cm, 5 h	8	S$_4$	FM	human

Recipe:

Dissolve 25 mg 2-phospho-D-glycerate-Na$_3$-6H$_2$O, 30 mg NADH-Na$_2$, 20 mg ATP-Na$_2$-3H$_2$O, 40 mg MgCl$_2$-6H$_2$O and 2 mg EDTA-Na$_2$-2H$_2$O in 10 ml 0.1 mol l^{-1} Tris-HCl, pH 8.0 and add 640 Units phosphoglycerate kinase and 200 Units glyceraldehyde-phosphate dehydrogenase. The solution is dropped on the cut surface of a processed starch gel. Upon illumination with UV-light non-fluorescent bands appearing on a fluorescent background indicate active enzyme molecules [6].

Phosphorylase (2.4.1.1)

Reaction scheme:

Glucose-1-phosphate + $(1,4\text{-}\alpha\text{-}D\text{-glucosyl})_{n-1}$ = ortho-phosphate + $(1,4\text{-}\alpha\text{-}D\text{-gluco-}$ $syl)_n$, counter stain: $(1,4\text{-}\alpha\text{-}D\text{-glucosyl})_n + I_2$ = blue colour

Electrophoresis:

Technique	pH	System	Staining	Source
PAGE 150 V ,1.5 mA/gel rod	8.9	D_{13}	FM	human

Recipe:

Following electrophoresis glycogen-containing PAA gels are incubated at 37 °C in a $2\ \text{mmol l}^{-1}$ acetate buffer of pH 5.9 containing 0.5 mg AMP and 2.5 mg glucose-1-phosphate per ml. After the incubation time of 30 min, the gels are removed from the substrate solution and immersed in a 7 % acetic acid solution containing 1 ml of 6 % I_2-4 % KI per liter. A blue-violet band indicates the position of active phosphorylase [108].

Polyribonucleotide nucleotidyltransferase (2.7.7.8)

Reaction scheme:

RNA_n + a nucleoside diphosphate = RNA_{n+1} + ortho-phosphate, RNA_{n+1} + acridine orange (or methylene blue) = coloured complex

Electrophoresis:

Technique	pH	System	Staining	Source
PAGE	8.9	D_1	FM	*Microccus lysodeikticus*

Recipe:

Add to 1 ml of a $250\ \text{mmol l}^{-1}$ Tris-HCl buffer of pH 9.0 1 ml $ADP\text{-}Na_3\text{-}2H_2O$ (53 mg/ml), 1 ml adenylyl-(3′-5′)-adenosine (1 mg/ml), 1 ml $MgCl_2\text{-}6H_2O$ (5.1 mg/ml) and 1 ml EDTA (0.29 mg/ml). Immerse the processed polyacrylamide gels into the incubation mixture for 1 h at 37 °C and wash the gels afterwards with 7 % acetic acid. The polynucleotides formed during incubation are stained either (a) with a solution of 15 % (v/v) acetic acid consisting 2 % (w/v) acridine orange and 1 % (w/v) $La(NO_3)_6\text{-}6H_2O$ [109, 110] or (b) 0.2 % (w/v) methylene blue in 0.4 M sodium acetate, pH 4.7 [111]. In procedure (a) the gels were destained electrophoretically, in method (b) they were destained with running water. Stain (b) is more sensitive but stain (a) gives more stable colours [112].

Purine-nucleoside phosphorylase (2.4.2.1)

Reaction scheme:

Inosine + ortho-phosphate (+ purine-nucleoside phosphorylase) = ribose-1-phosphate + hypoxanthine, hypoxanthine + PMS (+ xanthine oxidase) = xanthine + reduced PMS, reduced PMS + MTT = PMS + *reduced MTT* (blue coloured)

Electrophoresis:

Technique	pH	System	Staining	Source
starch gel 10 V/cm, 45 min and 5 V/cm, 3.5 h.	7.2	S_{29}	AOL	human

Recipe:

Add to 25 ml of a 50 mmol l^{-1} phosphate buffer of pH 7.5 5 mg inosine-H$_2$O, 10 µl xanthine oxidase (4 U/ml) 1 ml MTT (5 mg/ml) and 1 ml PMS (5 mg/ml). Mix the staining solution with 25 ml of a 2 % agar solution cooled down to 45 °C and pour over the cut surface of a processed starch gel. Active enzyme zones appear as blue bands. In human tissues the least anodal zone is the primary isozyme, while the more anodal zones are secondary isozymes derived from the primary [6].

Pyridoxal kinase (2.7.1.35)

Reaction scheme:

Pyridoxal + ATP (+ pyridoxal kinase) = pyridoxal-5′-phosphate + ADP, ADP + phosphoenol pyruvate (+ pyruvate kinase) = ATP + pyruvate, pyruvate + NADH (+ lactate dehydrogenase) = lactate + *NAD* (non fluorescent)

Electrophoresis:

Technique	pH	System	Staining	Source
starch gel 3 – 6 V/cm, 16 h, 4 °C	7.0	S_{48}	MOL	human

Recipe:

Add to 25 ml of a 100 mmol l^{-1} Tris-HCl buffer of pH 8.0 1 mg pyridoxal, 7 mg ATP-Na$_2$-3H$_2$O, 25 mg phosphoenolpyruvate-K, 3.5 mg NADH-Na$_2$-3H$_2$O, 2 mg KCl, 5 mg MgCl$_2$-6H$_2$O and 200 µl lactate dehydrogenase (2 750 U /ml). A porous membrane is impregnated with the reaction mixture and placed on the surface of a cut starch gel. The appearance of non-fluorescent bands is observed under the UV-light. In addition faint bands may also appear due to other enzymes utilizing ATP. It is therefore necessary to carry out a control stain without pyridoxal [6].

Pyruvate kinase (2.7.1.40)

Reaction scheme:

ADP + phosphoenol pyruvate (+ pyruvate kinase) = ATP + pyruvate, pyruvate + NADH (+ lactate dehydrogenase) = lactate + *NAD* (non-fluorescent)

Electrophoresis:

Technique	pH	System	Staining	Source
starch gel 13 V/cm, 5 h	8	S_{49}	FM	human

Recipe:

Dissolve in 10 ml of 0.1 mol l^{-1} Tris-HCl, pH 8.0: 50 mg phosphoenolpyruvate-Na$_3$, 5 mg ADP, 10 mg MgCl$_2$-6H$_2$O, 30 mg NADH, 2 mg fructose-1,6-diphosphate, 2 mg EDTA, 50 mg KCl and 300 Units lactate dehydrogenase. The solution is dropped on the cut surface of the processed starch gel. Active enzyme zones appear as non-fluorescent bands upon illumination with UV-light of 375 nm [6, 72].

Retinol dehydrogenase (1.1.1.105)

Reaction scheme:

Retinol + NAD = reduced retinol + NADH, NADH + PMS + MTT = NAD + PMS+ *reduced MTT* (blue coloured)

Electrophoresis:

Technique	pH	System	Staining	Source
starch gel	8	S$_{50}$	FM	human

Recipe:

Dissolve in 93 ml of a 10 mmol l^{-1} phosphate buffer of pH 7.5 66 mg NAD-3H$_2$O, 2 mg PMS and 17 mg MTT. Finally add 7 ml acetone to which immediately before use 100 mg of all-trans retinol (vitamin A alcohol) were added. Impregnate the cut surface of a processed starch gel with the staining solution at 37 °C in the dark until blue bands appear [17].

Ribonuclease (pancreatic) (3.1.27.5)

Reaction scheme:

A ribonucleic acid + H$_2$O = ribonucleotides.

Counter stain:
ribonucleic acids + Pyronin Y = *dye RNA-complex* (yellow coloured)

Electrophoresis:

Technique	pH	System	Staining	Source
PAGE 8 mA/gel, 4 h, 4 °C	4.5	D$_{14}$	FM	bovine

Recipe:

Following electrophoresis the gels are immersed in a 100 mmol l^{-1} Tris-HCl buffer of pH 7.6 containing 10 mmol l^{-1} NaCl and shaken over night in a gel diffusion destainer. Afterwards they are stained in a solution containing 1% Pyronin Y in 7% acetic acid for 1.5 h at 37 °C and finally destained over night in 7% acetic acid using a diffusion destainer. (Pyronin Y = Tetramethyldiamino-xanthenylchloride). Active enzyme zones appear as clear bands in a yellow coloured gel [113, 114].

Ribosephosphate pyrophosphokinase (2.7.6.1)

Reaction scheme:

AMP + 5-phospho-α-D-ribose-1-diphosphate (+ ribosephosphate pyrophospho-kinase) = ribose-5-phosphate + ATP, ATP + glucose (+ hexokinase) = ADP + gluco-se-6-phosphate, ADP + phosphoenolpyruvate (+ pyruvate kinase) = pyruvate + ATP, glucose-6-phosphate + NADP (+ glucose-6-phosphate dehydrogenase) = 6-phospho-gluconate + NADPH, NADPH + PMS + MTT = NADP + PMS + *reduced MTT* (blue coloured)

Electrophoresis:

Technique	pH	System	Staining	Source
Cellogel prerun: 15 min at 5 h at 300 V, 2 C, after sample application	6.8	C_{22}	MOL	human

Recipe:

Add to 10 ml of a 150 mmol l^{-1} phosphate buffer of pH 7.4 1 ml AMP-Na$_2$ (3.91 mg/ml), 1 ml 5-phospho-α-D-ribose-1-diphosphate-Mg$_2$-2H$_2$O (9.4 mg/ml), 1 ml D-glucose (3.6 mg/ml), 1 ml NADP-Na$_2$-2H$_2$O (6.41 mg/ml), 1 ml MgCl$_2$-6H$_2$O (40.6 mg/ml), 2 ml gly-cerol, 0.5 ml PMS (3.06 mg/ml), 0.5 ml MTT (14.57 mg/ml), 1 ml phosphoenolpyruvate-Na-H$_2$O (4.16 mg/ml), 0.3 ml hexokinase (150 U/ml), 0.3 ml glucose-6-phosphate dehy-drogenase (66.67 U/ml) and 0.3 ml pyruvate kinase (666.7 U/ml). Following electro-phoresis a porous membrane (glass supported 3 MM Whatman, Chromatography paper) is saturated with the histochemical stain and placed on the central 20 cm por-tions of the Cellogel strip. A glass plate is layed on the Cellogel strip and the entire arrangement is wrapped in aluminum foil to avoid non specific light induced pro-duction of the stain. After an incubation time of 1 – 2 h at 37 °C the Cellogel strip is dipped in 10 % acetic acid to stopp the formation of reduced MTT [115].

RNA nucleotidyltransferase (2.7.7.6)

Reaction scheme:

n-Nucleoside triphosphate (+ DNA) = n-pyrophosphate + RNA, pyrophosphate + Ca^{2+} = *Ca-pyrophosphate* (insoluble precipitate)

Electrophoresis:

Technique	pH	System	Staining	Source
PAGE 3 – 4 mA/tube 5 °C	8.9	D_{15}	FM	*E. coli*

Recipe:

Following electrophoresis the gels are soaked for 20 – 30 min in a 100 mmol l^{-1} Tris-HCl buffer of pH 8.0 at 37 °C. Then they are transferred to a staining solution consisting of: 82 ml of a 100 mmol l^{-1} Tris-HCl buffer of pH 8.0, 1 ml GTP-Na_2-H_2O (46.8 mg/ml), 1 ml CTP-Na_2-$2H_2O$ (45.0 mg/ml), 1 ml UTP-Na_3-$2H_2O$ (46.9 mg/ml), 1 ml ATP-Na_2-$3H_2O$ (48.4 mg/ml), 1 ml $MgCl_2$-$6H_2O$ (243.6 mg/ml), 1 ml EDTA (3 mg/ml), 10 ml HCl (150 mg/ml) and 1 ml $CaCl_2$ (222 mg/ml). The formation of white bands indicating a precipitation of calciumpyrophosphate is attended [21].

Sucrose phosphorylase (2.4.1.7)

Reaction scheme:

Sucrose + ortho-phosphate (+ sucrose phosphorylase) = D-fructose + α-D-glucose-1-phosphate, D-fructose + triphenyltetrazolium chloride = oxidised fructose + *reduced triphenyltetrazolium chloride* (pink coloured)

Electrophoresis:

Technique	pH	System	Staining	Source
PAGE 3 – 4 mA/gel 5 °C	8.9	D_1	FM	Jack bean, *E. coli*

Recipe:

Following electrophoresis PAA gels are thoroughly rinsed with distilled water and incubated for 20 min at 30 °C in a solution containing 68.4 mg/ml sucrose in 3 mmol l^{-1} phosphate buffer of pH 6.9. Then the gels are rinsed thoroughly with distilled water and immersed in a freshly prepared solution of 0.1 % triphenyltetrazolium chloride in 1 N NaOH. Care must be taken to protect the gel from undue exposure to light. Intermittent inspection reveals the formation of descret bands and a slow appearance of a diffuse pinc background. At this point the staining is interrupted by washing the gels with 5 % acetic acid [3].

Superoxide dismutase (1.15.1.1)

Reaction scheme:

MTT + PMS + day light = PMS + *reduced MTT* (blue coloured), reduced MTT + O_2^- (+ superoxide dismutase) = *MTT* (colourless)

Electrophoresis:

Technique	pH	System	Staining	Source
starch gel 12 V/cm, 4 h, 4 °C	7.0	S_{48}	AOL	human

Recipe:

Add to 25 ml of a 50 mmol l^{-1} Tris-HCl buffer of pH 8.0 1 ml MTT (5 mg/ml) and 1 ml PMS (5 mg/ml). Mix the staining solution with 25 ml of a 2 % agar solution and apply on the cut surface of a processed starch gel. Expose the agar overlayered gel for several minutes to daylight, then incubate at 37 °C. Enzyme zones appear as white bands on a blue background [6].

Liver mitochondria of rat, mouse and chicken contain a fast migrating CN⁻-sensitive form of superoxide dismutase. Liver mitochondria, as well as whole homogenates of every tissue of mouse and chicken, have two additional slow-migrating CN⁻-insensitive dismutase enzymes. The electrophoresis nitro blue tetrazolium technique according to [116] is unsiutable for the detection of the fast CN⁻-insensitive form of superoxide dismutase [117].

Testosterone 17β-dehydrogenase (NADP⁺) (1.1.1.64)

Reaction scheme:

Testosterone + NADP = 4-androstene-3,17-dione + NADPH, NADPH + PMS + MTT = NADP + PMS + *reduced MTT* (blue coloured)

Electrophoresis:

Technique	pH	System	Staining	Source
IEF	5 – 8	I$_4$	FM	pig
200 V/cm, 3 h at 4 °C				

Recipe:

Add to 35.5 ml of a 10 mmol l⁻¹ Tris-HCl buffer of pH 7.4 1.5 ml testosterone (8.64 mg/10 ml propylene-glycol), 3 ml NADP-Na$_2$-2H$_2$O (4.94 mg/ml), 1 ml 2-mercaptoethanol (1.38 mg/ml), 1 ml EDTA (0.518 mg/ml), 0.2 ml PMS (1.6 mg/ml) and 2.5 ml MTT (0.5 mg/ml). Blue bands develop at the position of active enzyme zones [118].

Tetrahydrofolate Dehydrogenase (1.5.1.3)

Reaction scheme:

5,6,7,8-Tetrahydrofolate + NADP = 7,8-dihydrofolate + NADPH, NADPH + PMS + MTT = NADP + PMS + *reduced MTT* (blue coloured)

Electrophoresis:

Technique	pH	System	Staining	Source
Cellulose acetate	8.6	C$_{23}$	AOL	*E. coli*
30 V/cm, 75 – 120 min, 4 °C				

Recipe:

To 5 ml of a 1.6 % agar solution in a 50 mmol l⁻¹ Tris-HCl buffer of pH 7.5 containing 100 mmol l⁻¹ KCl cooled down to 50 °C, the following solutions are mixed: 2 ml MTT (2 mg/5 ml), 1 ml NADP-Na$_2$-H$_2$O (8.23 mg/ml) and 0.4 ml 5,6,7,8-tetrahydrofolate-2HCl-H$_2$O (9.11 mg/ml). The agar substrate solution is poured evenly on a glass plate and the cellulose acetate strip containing the separated enzymes is placed upside down on the gel surface.

The two *E. coli* enzymes are specifically inhibited to 50 % by 1.5-5 mmol l⁻¹ trimethoprime while the two enzymes from human liver are inhibited at a concentration of 0.3 mmol l⁻¹. Methotrexate inhibits the *E. coli* enzyme in a concentration of 0.1 μmol l⁻¹ [119].

Threonine dehydratase (4.2.1.16)

Reaction scheme:

L-Threonine + H$_2$O = 2-oxobutyrate + NH$_3$ + H$_2$O, 2-oxobutyrate + 2,4-dinitrophenyl-hydrazine = coloured dye

Electrophoresis:

Technique	pH	System	Staining	Source
PAGE	8.9	D$_1$	FM	*Bacillus subtilis, Salmonella typhimurium, E. coli*

Recipe:

Add to 8 ml of 100 mmol l^{-1} phosphate buffer of 8.2 1 ml of L-threonine (47.6 mg/ml) and 1 ml of pyridoxal-5-phosphate-H$_2$O (26.5 mg/ml). The gels are incubated for 30 min in the substrate solution. Subsequently they are treated with a 0.5 N HCl-solution containing 0.025% (w/v) 2,4-dinitrophenylhydrazine. 15 min later they are immersed in a 10% (w/v) KOH-solution. Tris may inactivate one of the 2 isozymes occurring in bacteria [120].

Thrombin (3.4.21.5)

Reaction scheme:

Benzoyl-L-leucyl-L-alanyl-L-arginine-α-naphthyl-ester + H$_2$O = benzoyl-L-leucyl-L-alanyl-L-arginine + α-naphthol, 2-naphtol + Fast Violet B = *coloured diazo dye*

Electrophoresis:

Technique	pH	System	Staining	Source
IEF	3.5	I$_5$	FM	human
200 V, 4 h, 4 °C	− 10			

Recipe:

Staining solution: 9 ml of a 250 mmol l^{-1} phosphate buffer of pH 7.5 containing 10 mg Fast Violet B and 1 ml of benzoyl-L-leucyl-L-alanyl-L-arginine-α-naphthylester (0.123 mg/ml). Blue bands indicate the presence of active enzyme molecules. To stain for prothrombin gels are first placed at 37 °C for 10 min in 10 ml of a 250 mmol l^{-1} phosphate buffer of pH 7.0 containing 1 µg per ml of *Echis carinatus* venom [121].

Transaldolase (2.2.1.2)

Reaction scheme:

D-Erythrose-4-phosphate + D-fructose-6-phosphate (+ transaldolase) = sedoheptulose-7-phosphate + D-glyceraldehyde-3-phosphate, D-glyceraldehyde-3-phosphate + NADP + H$_2$O (+ D-glyceraldehyde-3-phosphate dehydrogenase) = 3-phospho-D-glycerate + NADPH, NADPH + PMS + MTT = NADP + PMS + *reduced MTT* (blue coloured)

Electrophoresis:

Technique	pH	System	Staining	Source
starch gel 6 V/cm, 18 h, 4 °C	7.0	S_{51}	FM	Bifido- bacteria

Recipe:

Add to 13 ml of a 50 mmol l⁻¹ glycine-NaOH buffer of pH 7.6, 1 ml of fructose-6-phosphate-Na_2-H_2O (128 mg/ml), 1 ml erythrose-4-phosphate (4.3 mg/ml), 1 ml sodium arsenate-$7H_2O$ (125 mg/ml), 1 ml D-glyceraldehyde-3-phosphate dehydrogenase (56 U/ml) 1 ml NADP-Na_2-$2H_2O$ (16.5 mg/ml) PMS (0.5 mg/ml) and 1 ml MTT (0.75 mg/ml). Pour the reaction mixture on the cut surface of a processed starch gel and incubate the gel at 4–5 °C for 30 min. Observe the formation of the bands within the next 30 – 60 min while incubating the gels at 37 °C [122].

Transketolase (2.2.1.1)

Reaction scheme:

D-Ribose-5-phosphate + D-xylose-5-phosphate (+ transketolase) = sedoheptulose-7-phosphate + D-glyceraldehyde-3-phosphate, D-glyceraldehyde-3-phosphate + NADP + H_2O (+ D-glyceraldehyde-3-phosphate dehydrogenase) = 3-phospho-D-glycerate + NADPH, NADPH + PMS + MTT = NADP + PMS + *reduced MTT* (blue coloured)

Electrophoresis:

Technique	pH	System	Staining	Source
starch gel 6 V/cm, 18 h, 4 °C	7.0	S_{51}	FM	Bifido- bacteria

Recipe:

Add to 13 ml of a 50 mmol l⁻¹ glycine-NaOH buffer of pH 7.6, 1 ml ribose-5-phosphate-Na_2-$2H_2O$ (117 mg/ml), 1 ml D-xylose-5-phosphate (5 mg/ml), 1 ml sodium arsenate-$7H_2O$ (125 mg/ml), 1 ml D-glyceraldehyde-3-phosphate dehydrogenase (56 U/ml) 1 ml NADP-Na_2-$2H_2O$ (16.5 mg/ml) 1 ml PMS (0.5 mg/ml) and 1 ml MTT (0.75 mg/ml). Pour the reaction mixture on the cut surface of a processed starch gel and incubate the gel at 4–5 °C for 30 min. Observe the formation of the bands within the next 30 – 60 min while incubating the gels at 37 °C [122].

Triacyl glycerol lipase (3.1.1.3)

Reaction scheme:

4-Methylumbelliferyl oleate + H_2O = oleic acid + *4-methylumbelliferone* (fluorescent)

Electrophoresis:

Technique	pH	System	Staining	Source
Cellogel 12 V/cm, 3 h	6.3	C_{24}	FM	human

Recipe:

1 ml of 10 mmol l^{-1} 4-methylumbelliferyl oleate in hexane and 1 ml of 16 mmol l^{-1} L-α-phosphatidylcholine (egg yolk lecithin) in hexane are evaporated together to dryness under a stream of nitrogen, resuspended in 25 ml of 2.4 mmol l^{-1} sodium taurocholate and sonicated in an ice-bath at 50 W for 1 min. After electrophoresis, gels are pre-rinsed for 1 min in a 200 mmol l^{-1} acetate buffer of pH 4 and then stained for 5 min with substrate solution. Excess stain is decanted and the gels are incubated at 37 °C and viewed periodically under long wave UV-light [123].

Triosephosphate isomerase (5.3.1.1)

Reaction scheme:

D-glyceraldehyde-3-phosphate (+ triosephosphate isomerase) = dihydroxyacetone-phosphate, dihydroxyacetone phosphate + NADH (+ α-glycerophosphate dehydrogenase) = α-glycerophosphate + *NAD* (non fluorescent)

Electrophoresis:

Technique	pH	System	Staining	Source
starch gel 8 V/cm, 18 h, 4 °C	9.3	S$_{52}$	AOL	human

Recipe:

Add to 20 ml of a 100 mmol l^{-1} triethanolamine-HCl buffer of pH 8.0, containg 5 mmol l^{-1} EDTA: 2 ml 30 mmol l^{-1} glyceraldehyde-3-phosphate (prepared from the diethyl-acetal-Ba-salt according to the suppliers method), 20 mg NADH-Na$_2$-3H$_2$O and 20 µl α-glycerophosphate dehydrogenase (80 U/ml). Mix with 20 ml of a 2 % agar solution cooled down to 45 °C and pour on the cut surface of a processed starch gel. Observe the formation of non-fluorescent bands on a fluorescent background. An alternative method has been described by [6].

Tripeptide aminopeptidase (3.4.11.4)

Reaction scheme:

A tripeptide + H$_2$O = L-aminoacid + a dipeptide, L-aminoacid + PMS (+ L-amino acid oxidase) = α-ketoacid + NH$_3$ + reduced PMS, reduced PMS + Nitro BT = PMS + *reduced Nitro BT* (blue coloured)

Electrophoresis:

Technique	pH	System	Staining	Source
Cellogel 4 °C, 0,8 mA/cm, 40 min	8.6	C$_{25}$	FM	human

Recipe:

Add to 10 ml of a 50 mmol l^{-1} Tris-HCl buffer of pH 8.0 10 mg L-leucyl-L-glycyl-L-glycine, 2 Units L-amino-acid oxidase, 1 mg PMS and 5 mg Nitro BT. Following electrophoresis the Cellogel strip is impregnated with the staining solution and the formation of coloured bands is observed. The reaction can be stopped using a 7% acetic acid solution [124].

Trypsin (3.4.21.4)
(alternative stain)

Reaction scheme:

denatured haemoglobin + H_2O = peptides (colourless)

Counter stain:
haemoglobin + Nigrosin = *haemoglobin-dye complex* (brown coloured)

Electrophoresis:

Technique	pH	System	Staining	Source
PAGE 3 – 4 mA/tube 2 – 3 h	8.9	D_{16}	FM	human, mouse

Recipe:

Following electrophoresis the gels are incubated for 2 – 12 h at 37 °C in 10 ml of a 100 mmol l^{-1} Tris HCl buffer of pH 7.5 containing 0.01% Nigrosin. The medium is changed several fold until transparent zones appear on an opaque background caused by the unhydrolyzed protein substrate. Stained gels can be stored in methanol/acetic acid [125].

UDPglucose-hexose-1-phosphate uridylyltransferase (2.7.7.12)
(syn: glactose-1-phosphate uridylyltransferase)

Reaction scheme:

UDPglucose + α-D-galactose-1-phosphate (+ UDPglucose-hexose-1-phosphate uridylyltransferase) = α-D-glucose-1-phosphate + UDPgalactose, glucose-1-phosphate (+ phosphoglucomutase + glucose-1,6-diphosphate) = glucose-6-phosphate, glucose-6-phosphate + NADP (+ glucose-6-phosphate dehydrogenase) = 6-phosphogluconate + *NADPH* (fluorescent)

Electrophoresis:

Technique	pH	System	Staining	Source
starch gel 11 V/cm, 18 h	6	S_{53}	AOL	human

Recipe:

Add to 1.5 ml of 0.5 mol l^{-1} glycine-NaOH buffer of pH 8.6: 0.75 ml galactose-1-phosphate-K_2-5H_2O (12 mg/ml), 0.75 ml UDPglucose-Na_2 (8 mg/ml), 75 µl glucose-1,6-diphosphate tetra(cyclohexylammonium)salt-4H_2O (1.4 mg/ml), 0.75 ml $MgCl_2$-6H_2O (8 mg/ml)), 0.75 ml NADP-Na_2 (116 mg/ml), 60 µl glucose-6-phosphate dehydrogenase (700 U/ml), 30 µl phosphoglucomutase (2 000 U/ml), 30 µl phosphogluconate dehydrogenase (120 U/ml) and 0.75 ml 2-mercaptoethanol (10 µl/ml). The appearance of fluorescent bands is inspected under UV-light [6].

Urease (3.5.1.5)

Reaction scheme:

Urea + H_2O (+ urease) = CO_2 + $2NH_3$, NH_3 elevates the pH, causing the formation of dark bands on a yellow background

Electrophoresis:

Technique	pH	System	Staining	Source
PAGE	8.9	D_1	FM	Jack bean

Recipe:

After electrophoresis PAA gels are preequilibrated for 30 min on a shaker bath in 50 mM sodium acetate buffer, pH 5.0, containing 1 mmol l⁻¹ EDTA. The gel is then transferred to a buffer containing 20 mM sodium acetate, 1 mmol l⁻¹ EDTA, pH 5.0, and incubated for an additional 30 min. The gels are further equilibrated in 1 mM EDTA and 0.5% (w/v) cresol red for 30 min. The enzymatic reaction is initiated by subsequently transferring the gels into a solution containing 1 mmol l⁻¹ EDTA, 0.5% cresol red, and 1.5% (w/v) urea. The areas on the gel containing urease activity are visible as increasingly darkening crimson bands on a yellow background. The enzymic reaction is terminated by transferring the gel into a solution of 0.1 mol l⁻¹ lead acetate. After 3–5 min of incubation the crimson colored activity bands appear as opaque white bands over a light yellow transparent background. The yellow background of the gels can be removed by washing the gel twice with 20 mmol l⁻¹ lead acetate solution [126].

Remarks: Urease exhibits dissociation and association properties dependent on ionic strength, pH, and storage time [127, 128].

The rate of color development appears to be fairly linear with increasing enzyme concentrations between 0.6 and 3.0 units.

Xanthine oxidase (1.2.3.2)

Reaction scheme:

Xanthine + H_2O + O_2 = urate + H_2O_2, H_2O_2 + amino-ethylcarbazole + H (+ peroxidase) = H_2O + *oxidized amino-ethylcarbazole* (brown coloured)

Electrophoresis:

Technique	pH	System	Staining	Source
PAGE 3 mA/gel	8.3	D_1	FM	rat

Recipe:

68 mg of hypoxanthine are dissolved under heating in 10 ml of a 0.1 mol l⁻¹ phosphate-buffer of pH 7.4. To 1 ml of this solution 2 ml of a peroxidase solution (25 µg/ml), 0.4 ml 3-amino-9-ethylcarbazole (25 mg/ml) and 6.6 ml 0.1 mol l⁻¹ phosphate buffer of pH 7.4 are added. The appearance of brown bands is inspected [24].

Tab. 6.2.1. Buffer system used in Cellogel[R] electrophoresis (remarks) [Ref.]

C_1: Add B to A until pH equals 5.6;
A: 25 mmol l^{-1} citrate-Na$_3$-H$_2$O; B: saturated solution of citric acid. Source: man, hamster.
5 mm (1); 4 h (2); (3) [5].

C_2: Add B to A until pH equals 6.5;
A: 40 mmol l^{-1} K$_2$HPO$_4$; B: 0.1 N KOH. Source: bacteria. 5 mm (1); 2.5 h (200 V) (2); (3) [89, 129].

C_3: Add B to A until pH equals 7.8;
A: 20 mmol l^{-1} Tris; B: saturated solution of citric acid. Source: man, hamster.
5 mm (1); 2 h; (3) [5].

C_4: Add B to A until pH equals 7.8;
A: 50 mmol l^{-1} Tris; B: 1 mmol l^{-1} H$_3$PO$_4$ (0.057 ml H$_3$PO$_4$ 85 % (w/v) in 1 l H$_2$O). Source: rat; 5 mm
(1); 4 h (150 V); (4) [15, 16].

C_5: Add B to A until pH equals 8.8;
A: 500 mmol l^{-1} Tris; B: 1 N HCl (31.43 ml HCl 32% in 1 l H$_2$O). Source: man (serum); 5 mm (1);
3.5 h (200 V); [130].

C_6: Add B to A until pH equals 8.0;
A: 20 mmol l^{-1} Tris, 1 mmol l^{-1} MgCl$_2$-6H$_2$O; B: 40 mmol l^{-1} barbituric acid. Source: mammalia,
chicken; 5 mm (1); 3 h (200 V), 4 °C; (5) [25].

C_7: 20 mmol l^{-1} phosphate, 0.02% NaCl, 0.1% bovine serum albumine, pH 7.5. Source: man; 5 mm
(1) [5, 8].

C_8: Electrode buffer: 36 mmol l^{-1} sodium barbital, 36.6 mmol l^{-1} sodium acetate, acetic acid to
pH 7.0 and addition of 1 mmol l^{-1} 2-mercaptoethanol.
Buffer: Cellogel strips are equilibrated over night in electrode buffer and then immersed in
electrode buffer containing 0.05% bovine serum albumine [39, 40].

C_9: 24 mmol l^{-1} veronal, adjusted with 100 mmol l^{-1} HCl to pH 8.6 [5, 8, 47, 48].

C_{10}: 100 mmol l^{-1} Tris, adjusted with 100 mmol l^{-1} NaH$_2$PO$_4$ to pH 8.0 [5, 8, 49].

C_{11}: 10 mmol l^{-1} NaH$_2$PO$_4$ adjusted with 100 mmol l^{-1} H$_3$PO$_4$ to pH 6.5 [5, 8, 63].

C_{12}: 8 mmol l^{-1} Na$_2$HPO$_4$-2H$_2$O, 1 mmol l^{-1} MgCl$_2$-6H$_2$O, 4 mmol l^{-1} EDTA, 0.2 mmol l^{-1} 2-
mercaptoethanol, adjusted with saturated citric acid solution to pH 6.8. Source: man, hamster;
20 mm (1); 3 h; (3) [5, 8, 64].

C_{13}: Add B to A until pH equals 7.0;
A: 10 mmol l^{-1} Na$_2$HPO$_4$-2H$_2$O; B: saturated citric acid solution. Source: man, hamster; 5 mm (1);
3 h; (3)[5].

C_{14}: Add B to A until pH equals 7.0;
A: 10 mmol l^{-1} Na$_2$HPO$_4$-2H$_2$O; B: 10 mmol l^{-1} NaH$_2$PO$_4$-H$_2$O. Source: man, hamster. 5 mm (1);
2 h; (3) [5].

C_{15}: 61.4 mmol l^{-1} Tris, 4 mmol l^{-1} EDTA, 13.6 mmol l^{-1} citric acid, pH 7.5. Source: man, mouse. 5 mm
(1); 2.5 h; (3) [8, 73 – 76].

C_{16}: Add B to A until pH equals 7.5.
A: 40 mmol l^{-1} Tris, 4 mmol l^{-1} EDTA; B: 40 mmol l^{-1} citric acid, 4 mmol l^{-1} EDTA-Na$_2$. Source:
man, hamster. 5 mm (1); 3 h; (3) [5].

Tab. 6.2.1 (continued)

C_{17}: 20 mmol l⁻¹ Tris, 20 mmol l⁻¹ Veronal, 1 mmol l⁻¹ MgCl₂-6H₂O, pH 8.0, add 0.2 ml 1 mol l⁻¹ 2-mercaptoethanol to 1 l of buffer just before use. Source: man, hamster; middle of separation distance (1); 3 h; (3) [5].

C_{18}: 100 mmol l⁻¹ Tris, adjusted to pH 6.5 with saturated citric acid solution. Source: man. 2 h (150 V), 25 °C [82].

C_{19}: 50 mmol l⁻¹ Glycine, adjusted with 0.1 N NaOH to pH 9.0. Source: *Pseudomonas testosteroni*. 0.5 – 1 h (15 V cm⁻¹) [83].

C_{20}: Add A to B until pH equals 7.0;
A: 20 mmol l⁻¹ Na₂HPO₄-2H₂O; B: 20 mmol l⁻¹ NaH₂PO₄-H₂O. Source: man, hamster; 5 mm (1); 3 h; (3) [5, 6, 8].

C_{21}: Add A to B until pH equals 7.0;
A: 10 mmol l⁻¹ Na₂HPO₄-2H₂O; B: 1.54 mmol l⁻¹ citric acid. Source: man, hamster; 5 mm (1); 3 h; (3), [5].

C_{22}: Add A to B until pH equals 7.5;
A: 33.67 mmol l⁻¹ Tris, 4 mmol l⁻¹ EDTA; B: 6.3 mmol l⁻¹ citric acid, 1.3 mmol l⁻¹ EDTA-Na₂. Source: man, hamster; 5 mm (1); 2.5 h; (3) [8].

C_{23}: Add A to B until pH equals 7.5;
A: 100 mmol l⁻¹ boric acid; B: saturated Tris solution. Source: man, hamster; 5 mm (1); 3.5 h; (3), [5].

C_{24}: 50 mmol l⁻¹ triethanolamine, pH 10.5;
Source: man, snake; 5 mm (1); 1 h (15 V cm⁻¹); (3) [5].

C_{25}: Add A to B until pH equals 7.5;
A: 18.33 mmol l⁻¹ Tris, 4 mmol l⁻¹ EDTA; B: 1.67 mmol l⁻¹ citric acid, 4 mmol l⁻¹ EDTA-Na₂. Source: man, mouse; 5 mm (1); 4 h; (3), [8].

C_{26}: 44.80 g Tricine, 203 mg MgCl₂-H₂O, 0.057 ml H₃PO₄ (85 % w/v), 100 ml Triton X-100 (30 %), 14.61 mg Ribose-5-phosphate at 1 l distilled water; adjust pH to 6.8 with 1 mol l⁻¹ KOH. Source: man; 5 mm (1); 5 h (300 V) 2 °C [115].

C_{27}: Adjust 250 mmol l⁻¹ barbital with 1 N HCl to pH 8.6; Source: man, mammalia; 5 mm (1); 3 h (250 V) 4 °C; sample in water, 5 to 50 µl (1 to 150 µg protein) applied per gel strip [123].

C_{28}: 2 mmol l⁻¹ Tris, 2 mmol l⁻¹ veronal, 0.1 mmol l⁻¹ MgCl₂-6H₂O, pH 8.0, add 0.2 ml 1 mmol l⁻¹ 2-mercaptoethanol per liter of buffer, just before use [5, 8].

Remarks to buffer systems used in Cellogel^R electrophoresis:

(1) Point of application apart from the cathodic end of separation area.
(2) Running time.
(3) The sample buffer consists of a 5 mmol l⁻¹ phosphate buffer of pH 6.4, containing 1 mmol l⁻¹ EDTA-Na₂, 1 mmol l⁻¹ 2-mercaptoethanol, 0.1 mmol l⁻¹ diisopropyl fluorophosphate and 0.02 mmol l⁻¹ NADP. 2 to 3 µl sample are applied per gel strip.
(4) The sample buffer consists of 5 mmol l⁻¹ Tris-phosphate, pH 7.8, containing 10 mmol l⁻¹ α-keto-glutarate, 1 mmol l⁻¹ pyridoxal-5-phosphate, and 0.1 % Triton X-100.
(5) The sample buffer consists of 10 mmol l⁻¹ Tris-HCl, pH 7.5, containing 1 mmol l⁻¹ EDTA, 1 mmol l⁻¹ 2-mercaptoethanol and 50 mmol l⁻¹ ε-aminocapronic acid. 2 to 5 µl sample were applied per gel strip.

Tab. 6.2.2. Disc electrophoresis, buffer systems[a]

System

D_1: Electrode buffer: 0.6 g Tris, 2.88 g glycine ad 1 l of bidistilled water, pH 8.3
Large pore gel: 1 part of solution A, 2 parts of solution B, 1 part of solution C,
4 parts of solution D.
A: 48 ml 1 N HCl, 5.98 g Tris, 0.46 ml Temed to 100 ml with bidistilled water, pH 6.7.
B: 10 g acrylamide, 2.5 g BIS to 100 ml with bidistilled water.
C: 4 mg riboflavin in 100 ml of bidistilled water.
D: 40 g sucrose to 100 ml with distilled water.

Small pore gel: 1 part of solution E, 2 parts of solution F, 1 part bidistilled water,
4 parts of solution G, E: 48 ml 1 N HCl, 36.3 g Tris, 0.23 ml Temed to 100 ml with bidistilled water,
pH 8.9. F: 28 g acrylamide, 0.735 g BIS to 100 ml of distilled water,
G: 140 mg ammoniumperoxodisulphate in 100 ml of distilled water [3, 4].

D_2: Electrode buffer: 140 mmol l⁻¹ β-alanine, 350 mmol l⁻¹ acetic acid, pH 4.5
Large pore gel: 60 mmol l⁻¹ KOH, 63 mmol l⁻¹ acetic acid, pH 6.8 (acrylamide + BIS = 3.125 g/100
ml (acrylamide: BIS = 10:2.5)
Small pore gel: 60 mmol l⁻¹ KOH, 376 mM acetic acid, pH 4.3 (acrylamide + BIS = 7.7 g/100 ml
(acrylamide: BIS = 30:0.8) [7].

D_3: As for system D_5 but including 0.5% Triton X-100 into the stacking gel and 1% Triton X-100 into
the separation gel. The concentration of ammoniumperoxydisulphate is reduced to 90 mg/
100 ml of distilled water. The gels are kept in a moist chamber for 12 h before use [47, 48].

D_4: Electrode buffer: 300 mmol l⁻¹ borate, 60 mmol l⁻¹ NaOH, pH 8.6; large pore gel: 60 mmol l⁻¹
Tris, phosphate, pH 6.9 (3.125% T, 0.625% BIS); separation gel: 52 mmol l⁻¹ borate, adjusted with
NaOH to pH 8.6 (8.75% T) [54, 55].

D_5: Electrode buffer: 5.52 g diethylbarbituric acid, 1 g Tris in 1 l distilled water, pH 7.0.
Large pore gel: 1 vol A, 2 vol B, 1 vol C and 4 vol D.
A: 39 ml 1 mmol l⁻¹ H₃PO₄, 4.95 g Tris, 0.46 ml Temed in 100 ml distilled water,
B: 10 g acrylamide, 2.5 g BIS in 100 ml distilled water,
C: 4 mg riboflavine in 100 ml distilled water,
D: 40 g sucrose in 100 ml distilled water.
Separation gel: 1 vol E, 2 vol F, 1 vol distilled water, 4 vol G.
E: 48 ml 1 N HCl, 6.85 g Tris, 0.46 ml Temed in 100 ml distilled water, pH 7.5.
F: 30 g acrylamide, 0.8 g BIS in 100 ml distilled water,
G: 140 mg ammonium persulphate in 100 ml distilled water (freshly prepared) [77, 131].

D_6: Sample gel buffer: 27 mmol l⁻¹ imidazole-HCl, pH 6, including 10 mmol l⁻¹ L-threonine;
Separation gel buffer: 88 mmol l⁻¹ Tris-HCl, pH 7.9, containing 10 mmol l⁻¹ L-threonine;
Electrode buffer: 30 mmol l⁻¹ asparagine, adjusted to pH 7.3 with Tris, containing 5 mmol l⁻¹
L-threonine, 50 µl dithiothreitol and 1 ml 0.02% bromophenol blue.
The sample was applied in 0.3 ml of 100 mmol l⁻¹ imidazole-HCl buffer of pH 6, containing
5 mmol l⁻¹ L-threonine, 5 µM dithiothreitol and 30% sucrose [80].

D_7: Large and small pore gel contain 10% glycerol and 1 mmol l⁻¹ 2-mercaptoethanol and 10 mg/ml
chloralhydrate.
Chloralhydrate was dissolved in the gel mixture immediately prior to addition of catalyst. The
persulphate concentration was six times the concentration given by Davis [4], i.e. 8.4 mg/ml [81].

D_8: Separation gel buffer (7.5% T): to a mixture of 10 ml of 1 N HCl and 10 ml of distilled water,
imidazole is added until the pH-value reaches 7.8; then the buffer is diluted to 50 ml.
Large pore gel buffer: same as separation gel buffer except that imidazole is added to bring the
pH to 5.8.

Tab. 6.2.2 (continued)

System

D_8: Electrode buffer: 5.52 g diethyl barbituric acid are dissolved in 500 ml of distilled water, then the pH-value is adjusted to pH 7.0 with imidazole and finally the total volume is brought to 1 l with distilled water [84].

D_9: As system D_1 but including 0.1% Triton X-100 into the gels and electrode buffer [93].

D_{10}: Electrophoresis is performed in 6% acrylamide, 0.2% BIS (N,N′-methylenebisacrylamide) and 10% glycerol. After thorough mixing of the above components in a buffer containing 15.5 g of Tris and 7 ml of HCl per liter, pH 7.8, polymerization was carried out by the addition of 5 mg of ammonium persulphate and 20 µl of N, N, N′, N′-tetramethylethylenediamine to 50 ml of gel solution. Gels were prerun for 4 h at 4 °C and 1.5 mA/gel in the same buffer.
Afterwards the cathodic buffer compartment was filled with a buffer consisting of 3.62 g Tris and 6.02 g N-Tris(hydroxymethyl)-methylglycine per liter, pH 8.1 while the anodal buffer compartment was filled with a buffer containing 12.1 g Tris and 4.15 ml HCl per liter, pH 8.1. The sample was layered on the gel in 10% glycerol and run at 1.5 mA/gel and 4 °C [94 – 96].

D_{11}: Electrode buffer: 65 mmol l^{-1} Tris-borate, pH 9.0, containing 0.3 mg/ml $Na_2S_2O_4$ [97 – 100].
Gel buffer: according to reference [4] (system D_1).

D_{12}: Gel and electrode buffer: 15 mmol l^{-1} Tris, 7 mmol l^{-1} glycine, pH 8.5 (5 – 20 µg protein/gel (ø 6 mm)) [101].

D_{13}: 7.5% T gels, prepared according to reference [4] (system D_1) but containing 0.01% glycogen. 100 µl tissue samples are mixed with 200 µl of distilled water, 300 µl 60 mmol l^{-1} thioglycollate solution and 300 µl 40% sucrose. 100 µl aliquots are applied to a round gel of usual size [108].

D_{14}: Small pore gel solution: To 20 ml of either aques 0.06% (w/v) poly (U) or aques 0.01% highly polymerized yeast RNA are added 5 ml of a solution consisting of 0.48 mol l^{-1} KOH containing 17.2% (v/v) acetic acid and 4% (v/v) N,N,N′,N′-tetramethylethylenediamine, pH 4.3 and 10 ml of a solution containing 30% (w/v) acrylamide and 0.8% (w/v) N, N′-methylenebisacrylamide. The solution is degassed.
Large pore gel solution: A 1 ml portion of a solution containing 2.9% (v/v) acetic acid and 0.46% (v/v) N,N,N′,N′-tetramethylethylenediamine in 0.48 mol l^{-1} KOH pH 5.8 is added to 2 ml of a solution containing 12.5% (w/v) acrylamide and 2% (w/v) N,N′-methylenebisacrylamide. Finally 4 ml of distilled water are added and the resulting mixture is degased at reduced pressure. To prepare the small gel 5 ml of a 0.6% freshly prepared ammonium persulphate solution are added to 35 ml of the small pore gel solution, containing 348 mg of spermine tetrahydrochloride. This mixture is dispersed into glass tubes to a height of 7.5 cm and overlayered with a small quantity of water. After 1 h the polymerization is complete and the water layer is removed. 1 ml of a 0.6% ammonium persulphate solution is then added to 7 ml of large pore gel solution containing 69.6 mg of spermine tetrahydrochloride. The tops of the small pore gels are washed with this mixture and then 0.25 ml of the mix ture are added to each of the small pore gels. Water is used to overlay the large pore gels.
Sample preparation: 50 µl of RNase and 50 µl of 50 mmol l^{-1} spermine tetrahydrochloride and 0.05% Methyl Green in 50% glycerol are layered on top of the gels.
Electrode buffer: 0.35 mol l^{-1} β-alanine containing 0.8% (v/v) acetic acid, pH 4.5.
The poor migration of highly charged polycathionic ribonucleases in the presence of negatively charged synthetic polynucleotides (or yeast RNA) is compensated by the inclusion of a high concentration of spermin in the entire electrophoretic system [113, 114].

D_{15}: System D_1 but polymerizing the small pore gel in the presence of 1.6 mg/ml calf thymus DNA [21].

Tab. 6.2.2 (continued)

System

D_{16}: System D_1, but using a solution of 50 mmol l^{-1} Tris-HCl, pH 8.9 as electrode buffer.
Preparation of 15% PAA gels, containing heat-denaturated haemoglobin: 2.5% (w/v) haemo-globin in 100 mmol l^{-1} Tris-HCl, pH 7.5 is heated to 70 °C for 5 min. After sonification at 105 W/cm^2 and 5 °C for 5 min, a sufficient amount of this solution is added to the apropriate gel mixture to yield a final concentration of 0.1% (w/v) [125].

---

a) Remarks

(1) 1 mA/gel (ø 5 mm, length 90 mm) for 150 min; *Drosophilia*, silk moth; 2 mU/gel.
(2) 3 mA/gel (ø 5 mm, length 75 mm) for 70 min; sweet potatos tubers, wheat germ;
     250 mg protein/gel.
(3) 5.0% T, 0.12% BIS; 5 mA/gel (ø 5 mm, length 75 mm) for 150 min; *Micrococcus luteus*; purified enzyme.
(4) 1 – 2 mA/gel (ø 6mm, length 75 mm) for 4 – 6 h under cooling; man; 25 – 50 µl purified enzyme solution containing 20% (w/v) sucrose.
(5) 2 mA/gel (ø 6 mm, length 75 mm) for 110 min; *Pseudomonas aeruginosa*; purified enzyme.
(6) 10% T, 0.25% BIS, 3 mA/gel (ø 5 mm, length 43 mm plus 2.5 mm sample gel) for 2 h;
     calf intestine; 2 mU/sample gel.
(7) *Bacillus sphaericus, Pseudomonas spe.*, bovine, pea; enzyme preparation of various purity.
(8) Pea seedlings.
(9) 7.7% T, 0.184% BIS; 150 V for 2 h; mammalia.
(10) 11.7% T, 5% BIS; 3 mA/gel (ø 6 mm, length 65 mm) for 150 min; *Aspergillus oryzae*; purified enzyme.
(11) 5% T, 0.21% BIS; pre-electrophoresis at 2.5 mA/gel (ø 5 mm, length 65 mm) for 30 min, followed by 1.25 mA/gel for 20 h: *Salmonella typhimurium*; purified enzyme in 100 mmol l^{-1} phosphoric acid, pH 7.4 containing 0.1 mmol l^{-1} EDTA, 0.4 mmol l^{-1} 2-mercaptoethanol, 12.5% glycerol.
(12) 3 mA/gel (ø 6mm, length 75 mm) for 150 min; *E. coli*, bacteria, cerials; crude extract in 50 mmol l^{-1} phosphate, pH 7.8 containing 0.1 mmol l^{-1} EDTA, 20% sucrose (ø 1 U/gel).
(13) 2 mA/gel (ø 6mm, length 75 mm) for 2 h; cathodic buffer: 49.8 mmol l^{-1} glycine, 51.9 mmol l^{-1} Tris, pH 9.45; anodic buffer: 50 mmol l^{-1} HCl, 62.5 mmol l^{-1} Tris: large pore gel: 32 mmol l^{-1} H$_3$PO$_4$, 58.8 mmol l^{-1} Tris; separation gel: 60 mmol l^{-1} HCl, 375 mmol l^{-1} Tris; cereals, mice.
(14) 8% T, 0.33% BIS; 5 mA/gel (ø 6 mm, length 75 mm) for 150 min: citrus leaf; sample buffer: 200 mmol l^{-1} phosphoric acid, pH 7.0, containing 1 mol l^{-1} NaCl and 0.05% L-cysteine; ≈ 2 U/gel.
(15) 3 mA/gel (ø 5 mm, length 75 mm) for 2 h; electrode buffer: 50 mmol l^{-1} citrate-phosphate, pH 4.2 or 40 mmol l^{-1} Tris glycine, pH 4.2; separation gel: before gelatination 0.1% caseine or haemoglobine included; mouse; purified enzyme.
(16) 3 mA/gel (ø 6 mm, length 75 mm) for 80 min; electrode buffer: 8.26 mmol l^{-1} Tris adjusted with diethylbarbituric acid to pH 7.0; large pore gel: 48.75 mmol l^{-1} H$_3$PO$_4$, 51.09 mmol l^{-1} Tris, pH 5.8; separation gel: 70.71 mmol l^{-1} Tris, 60 mmol l^{-1} HCl, pH 7.5; yeast.
(17) 5 mA/gel (ø 6 mm, length 65 mm plus 6 mm sample gel) for 1 h at 4 °C; human pregnancy serum, 10 µl/gel.
(18) Mammalia (liver).
(19) Radish, yeast.
(20) Mammalia, plants.
(21) *Pseudomonas fluorescens, P. aeruginosa, P. putida*
(22) *E. coli*
(23) Human placenta
(24) 100 V/gel (100 × 140 × 1.5 mm) for 2 h with cooling; bovine milk; sample buffer: separation gel buffer plus 50% glycerol.
(25) *Pseudomonas fluorescens*
(26) *Absidia glauca*; 80 µg protein/gel.

(27) 10% T, 0.2% BIS; 100 mmol l⁻¹ A/gel for 3 h under cooling; jack beans; 3 g yeast mannan per 100
     ml of separation gelsolution were added as substrate for α-mannosidase prior to polymeriz-
     ation.
(28) *Arthrobacter* spec.
(29) Mushroom
(30) 5% T, 0.061% BIS; rat liver.
(31) *E. coli*; radish cotyledons; *Curcubita pepo; Zea mays*
(32) *Azotobacter vinelandi*
(33) Mammalia urine
(34) *E. coli, Micrococcus lysodeikticus*
(35) Jack bean meal
(36) *Salmonella thyphimurium*

Tab. 6.2.3. Isoelectric focusing (IEF)

System

I_1:   5% acrylamide, 3% BIS
       Gel dimensions: 250 × 115 × 2 mm
       Separation distance: 100 mm
       Carrier ampholytes: 2% (Ampholine, Pharmacia, Freiburg, Germany), pH 3.5 – 10.
       Focusing: 400 V to 1000 V (90 min).

I_2:   ultrathin IEF with gels polymerized to a reactive polyester foil.
       Gel dimensions: 50 × 50 × 0.05 mm.
       Separation distance: 30 mm
       Carrier ampholytes: 3% (Servalyte), pH 3 – 10, 5% Glycerin
       Prefocusing: 400 V (5 min)
       Focusing: 1200 V (4 min), 1500 – 1800 V (1 – 2 min).

I_3:   4.7% acrylamide (0.14% BIS) gels containing 5% glycerol
       Gel dimensions: 125 × 50 × 0.8 mm
       Separation distance: 120 mm
       1.5% ampholyte ("Bio-Lyte"), pH 4 – 10.
       Focusing: 500 V (4 – 8 h) at 4 °C.

I_4:   IEF gels were prepared by mixing 1 ml of 8% (w/v) ampholytes of pH 5 – 8, 1 ml of 0.004% (w/v)
       riboflavin, 2 ml of 30% (w/v) acrylamide 0.8% (w/v) N,N'-methylenebisacrylamide and 4 ml
       of water. Glass tubes 9 cm long (ø 2.7 mm) were filled to a height of 7 cm with gel solution and
       overlayed with distilled water. After polymerisation the water was removed and a mixture of
       8% (w/v) ampholytes (pH 5 – 8), 0.004% (w/v) riboflavin, 30% (w/v) acrylamide 0.8% (w/v)
       N, N'-methylenebisacrylamide and enzyme solution (100 – 200 µg of protein) (1:1:2:4 by
       volumne) was layered on the gels and then overlayered with distilled water. After photo-
       polymerization the water layer was removed and the glass tubes were inserted into the electro-
       phoretic apparatus. The cathodic compartment contained 0.2 mol l⁻¹ ethylenediamine, while
       the anodic compartment was filled with 0.2 mol l⁻¹ acetic acid.

I_5:   2% ampholyte pH 3.5 – 10, catholyte: 0.1 mol l⁻¹ NaOH, anolyte: 0.1 mol l⁻¹ H_3PO_4.

Tab. 6.2.4. Starch gel electrophoresis, buffer systems

System

S_1: Electrode buffer: 410 mmol l^{-1} citric acid adjusted to pH 5.0 with 10 N NaOH.
   Gel buffer: 16 mmol l^{-1} succinic acid, 18.4 mmol l^{-1} Tris, pH 5.0.
   8 – 10 Vcm^{-1} for 4 h (cooling); man.

S_2: Electrode buffer: 300 mmol l^{-1} boric acid adjusted to pH 8.0 with 10 N NaOH.
   Gel buffer: 76 mmol l^{-1} Tris, 7 mmol l^{-1} citric acid, pH 8.6.
   11 Vcm^{-1} for 5 h (cooling); man.

S_3: Electrode buffer: 220 mmol l^{-1} Tris, 86 mmol l^{-1} citric acid, pH 5.7.
   Gel buffer: 8 mmol l^{-1} Tris, 4 mmol l^{-1} citric acid, pH 5.7.
   11 Vcm^{-1} for 5 h (cooling); man.

S_4: Electrode buffer: 500 mmol l^{-1} Tris, 16 mmol l^{-1} EDTA-Na$_2$, 650 mmol l^{-1} boric acid, pH 8.0.
   Gel buffer: 1 in 10 diluted electrode buffer.
   11 Vcm^{-1} for 5 h (cooling); man.

S_5: Electrode buffer: 130 mmol l^{-1} Tris, 43 mmol l^{-1} citrate, pH 7.0.
   Gel buffer: 9 mmol l^{-1} Tris, 3 mmol l^{-1} citric acid, pH 7.0.
   11 Vcm^{-1} for 5 h (cooling); man.

S_6: Electrode buffer: 100 mmol l^{-1} Tris, 28 mmol l^{-1} citric acid, pH 7.5.
   Gel buffer: 1 in 10 diluted electrode buffer.
   11 Vcm^{-1} for 5 h (cooling); man.

S_7: Electrode buffer: 100 mmol l^{-1} phosphate pH 6.5.
   Gel buffer: 1 in 10 diluted electrode buffer.
   11 Vcm^{-1} for 5 h (cooling); man.

S_8: Electrode buffer: 155 mmol l^{-1} Tris, 43 mmol l^{-1} citric acid, pH 7.0.
   Gel buffer: dilute 66.7 ml of electrode buffer to one liter.
   11 Vcm^{-1} for 5 h (cooling); man.

S_9: Electrode buffer: 30 mmol l^{-1} Tris borate, pH 7.6.
   Gel buffer: 1 in 10 diluted gel buffer.
   2 mA for 2 h per 200 mm long, 30 mm wide, 3 mm thick 12 % starch gel; 25 °C.

S_{10}: Electrode buffer: 300 mmol l^{-1} boric acid adjusted with 1 N NaOH to pH 8.0.
   Gel buffer: 76 mmol l^{-1} Tris, 7 mmol l^{-1} citric acid, pH 8.6.
   8 – 10 Vcm^{-1} for 4 h (cooling); man.

S_{11}: Electrode buffer: 300 mmol l^{-1} boric acid, 60 mmol l^{-1} NaOH, pH 8.5.
   Gel buffer: 30 mmol l^{-1} boric acid, 13 mmol l^{-1} NaOH, pH 8.5.
   15 Vcm^{-1} for 4 h (cooling); man.

S_{12}: Electrode buffer: 100 mmol l^{-1} Tris, 100 mmol l^{-1} maleic anhydride, 10 mmol l^{-1} EDTA,
   10 mmol l^{-1} MgCl$_2$, pH 7.4.
   Gel buffer: 1 in 10 diluted electrode buffer.
   15 Vcm^{-1} for 4 h (cooling); man.

S_{13}: Electrode buffer: 100 mmol l^{-1} Tris, 100 mmol l^{-1} NaH$_2$PO$_4$ adjusted with 1 N NaOH to pH 7.4.
   Gel buffer: 1 in 20 diluted electrode buffer.
   15 Vcm^{-1} for 4 h (cooling); man.

Tab. 6.2.4 (continued)

System

S_{14}: Electrode buffer: 250 mmol l^{-1} Tris, 70 mmol l^{-1} citric acid, pH 7.6.
Gel buffer: 1 in 15 diluted electrode buffer.
5 Vcm^{-1} overnight (cooling); man.

S_{15}: Electrode buffer: 100 mmol l^{-1} Tris adjusted with acetic acid to pH 8.3.
Gel buffer: 1 in 10 diluted electrode buffer.
5 Vcm^{-1} overnight (cooling); man.

S_{16}: Electrode buffer: 20 mmol l^{-1} citric acid adjusted with 1 N NaOH to pH 6.2 (cathode) and 6.9 (anode).
Gel buffer: 5 mM histidine-HCl, pH 6.7.
5 Vcm^{-1} overnight (cooling); man.

S_{17}: Electrode buffer: 100 mmol l^{-1} Tris, 100 mmol l^{-1} NaH$_2$PO$_4$ adjusted to pH 8.1 with 1 N NaOH.
Gel buffer: 1 in 10 diluted electrode buffer.
3 Vcm^{-1} overnight (cooling).

S_{18}: Electrode buffer: 900 mmol l^{-1} Tris, 500 mmol l^{-1} boric acid, 20 mmol l^{-1} EDTA, pH 8.6, diluted 1 in 14 before use.
Gel buffer: 1 in 40 diluted stock solution.
5 Vcm$^-$ for 17 h (cooling); man.

S_{19}: Electrode buffer: 100 mmol l^{-1} Tris and 100 mmol l^{-1} maleic anhydride adjusted to pH 7.2 with 10 N NaOH.
Gel buffer: 1 in 10 diluted electrode buffer.
17 Vcm^{-1} for 4 h (cooling); man.

S_{20}: Electrode buffer: 900 mmol l^{-1} Tris, 500 mmol l^{-1} boric acid, 20 mmol l^{-1} EDTA, pH 8.6, diluted 1 in 7 before use.
Gel buffer: 1 in 10 diluted stock solution.
5 Vcm^{-1} overnight (cooling); man.

S_{21}: Electrode buffer: 410 mmol l^{-1} citric acid adjusted to pH 8.0 with NaOH.
Gel buffer: 5 mmol l^{-1} DL-histidine-HCl, pH 8.0 adjusted with 2 N NaOH.
5 Vcm^{-1} overnight (cooling); man.

S_{22}: Electrode buffer: 100 mmol l^{-1} Tris, 100 mmol l^{-1} maleic acid, 10 mmol l^{-1} MgCl$_2$-6H$_2$O adjusted to pH 7.4 with NaOH.
Gel buffer: 1 in 10 diluted electrode buffer.
5 Vcm^{-1} overnight (cooling); man.

S_{23}: Electrode buffer: 245 mmol l^{-1} NaH$_2$PO$_4$ and 110 mmol l^{-1} citrate-Na$_3$ to pH 5.7.
Gel buffer: 1 in 50 diluted electrode buffer.
4 Vcm^{-1} for 19 h (cooling); man.

S_{24}: Electrode buffer: 100 mmol l^{-1} Tris, 100 mmol l^{-1} maleic anhydride, pH 6.5.
Gel buffer: 1 in 10 diluted electrode buffer.
5 Vcm^{-1} for 30 h (cooling); man.

S_{25}: Electrode buffer: 40 mmol l^{-1} phosphate, pH 7.0.
Gel buffer: 1 in 10 diluted electrode buffer.
5 Vcm^{-1} overnight (cooling); man.

Tab. 6.2.4 (continued)

System

S_{26}: Electrode buffer: 10 mmol l⁻¹ Tris-HCl, pH 8.0, containing 4 mmol l⁻¹ MgCl₂-6H₂O, 4 mmol l⁻¹ EDTA, 4 mmol l⁻¹ N-acetylcysteine.
Gel buffer: same as electrode buffer.
5 Vcm⁻¹ overnight (cooling).

S_{27}: Electrode buffer: 200 mmol l⁻¹ phosphate, pH 7.0.
Gel buffer: electrode buffer 1 in 20 diluted.
5 Vcm⁻¹ for 16 h (cooling); man.

S_{28}: Electrode buffer: 100 mmol l⁻¹ Tris-HCl, pH 9.0.
Gel buffer: electrode buffer diluted 1 in 5.
10 Vcm⁻¹ for 17 h (cooling); man.

S_{29}: Electrode buffer: 40 mmol l⁻¹ LiOH, 440 mmol l⁻¹ boric acid, pH 7.2.
Gel buffer: 1 volume electrode buffer, 9 volume distilled water and 90 volume of a 15 mmol l⁻¹ Tris, 4 mmol l⁻¹ citric acid buffer of pH 7.2.
10 Vcm⁻¹ overnight (cooling); man.

S_{30}: Electrode buffer: 47 mmol l⁻¹ citric acid adjusted with Tris to pH 7.2.
Gel buffer: 7 ml of electrode buffer diluted in 250 ml of water.
10 Vcm⁻¹ overnight (cooling); man.

S_{31}: Cathodal buffer: 661 mmol l⁻¹ Tris, 83 mmol l⁻¹ citric acid, pH 8.6 containing 60 mg NAD in 100 ml of buffer.
Anodal buffer: cathodal buffer without NAD.
Gel buffer: 10 ml of anodal buffer diluted to a final volume of 275 ml and addition of 25 mg EDTA-Na₂. When preparing the starch gel 30 mg NAD in 2 ml H₂O are added to 200 ml of cooked starch suspension just prior to degasing fully.
10 Vcm⁻¹ overnight (cooling); man.

S_{32}: Electrode buffer: 687 mmol l⁻¹ Tris, 157 mmol l⁻¹ citric acid, pH 8.0.
Gel buffer: electrode buffer, diluted 1 in 30.
10 Vcm⁻¹ overnight (cooling); man.

S_{33}: Electrode buffer: 300 mmol l⁻¹ Tris, adjusted with HCl to pH 8.6.
Gel buffer: electrode buffer diluted 1 in 15.

S_{34}: Electrode buffer: 100 mmol l⁻¹ Tris, 100 mmol l⁻¹ maleic acid, 10 mmol l⁻¹ EDTA-Na₂-2H₂O, 10 mmol l⁻¹ MgCl₂-6H₂O, pH 7.4.
Gel buffer: 1 in 10 diluted electrode buffer.

S_{35}: Electrode buffer: 150 mmol l⁻¹ triethanolamine, adjusted to pH 8.6 with concentrated HCl.
Gel buffer: 1 in 5 diluted gel buffer.
10 Vcm⁻¹ overnight (cooling); man.

S_{36}: Electrode buffer: 100 mmol l⁻¹ Tris, 100 mmol l⁻¹ maleate, 10 mmol l⁻¹ EDTA, pH 7.4.
Gel buffer: 1 in 10 diluted electrode buffer.

S_{37}: Electrode buffer: 500 mmol l⁻¹ Tris, 16 mmol l⁻¹ EDTA-Na₂, 650 mmol l⁻¹ borate, pH 8.0.
Gel buffer: 1 in 10 diluted electrode buffer including 0.035 mg NADP per ml of gel.
3 Vcm⁻¹ for 17 h (cooling); man.

Tab. 6.2.4 (continued)

System

S_{38}: Electrode buffer: 200 mmol l^{-1} sodium phosphate, pH 7.0.
Gel buffer: 1 in 20 diluted electrode buffer.
7.5 Vcm^{-1} for 14 h (cooling); man.

S_{39}: Electrode buffer: 200 mmol l^{-1} Tris-histidine, adjusted to pH 7.8 with HCl.
Gel buffer: 1 in 10 diluted electrode buffer.
7.5 Vcm^{-1} for 14 h (cooling); man.

S_{40}: Electrode buffer: 500 mmol l^{-1} Tris, 645 mmol l^{-1} boric acid, 16 mmol l^{-1} EDTA-Na$_2$-2H$_2$O, pH 8.0.
Gel buffer: 50 mmol l^{-1} Tris, 100 mmol l^{-1} boric acid, 1.6 mmol l^{-1} EDTA-Na$_2$-2H$_2$O, pH 8.0.
5 Vcm^{-1} overnight; bacteria.

S_{41}: Electrode buffer: 54 mmol l^{-1} Tris, 23.5 mmol l^{-1} citrate, pH 8.6.
Gel buffer: 1 in 10 diluted electrode buffer.
5 Vcm^{-1} overnight; man.

S_{42}: Electrode buffer: 410 mmol l^{-1} sodium citrate, 410 mmol l^{-1} citric acid, adjusted to pH 8.0.
Gel buffer: 5 mmol l^{-1} histidine, adjusted to pH 8.0 with 2 N NaOH.
5 Vcm^{-1} overnight; man.

S_{43}: Electrode buffer: 110 mmol l^{-1} citrate-Na$_3$, 245 mmol l^{-1} NaH$_2$PO$_4$, adjusted to pH 5.9 with NaOH.
Gel buffer: 10 ml electrode buffer mixed with 800 ml distilled water, pH adjusted to 5.9 with
0.2 mol l^{-1} citric acid and final volume made up to 1 l with distilled water.
10 Vcm^{-1} for 4.5 h (cooling); mammalia.

S_{44}: Electrode buffer: 330 mmol l^{-1} borate, pH 8.45.
Gel buffer: 50 mmol l^{-1} glycine-NaOH, pH 8.9.
20 Vcm^{-1} for 2 h (cooling); bacteria.

S_{45}: Electrode buffer: 100 mmol l^{-1} Tris, 100 mmol l^{-1} NaH$_2$PO$_4$, pH 7.4.
Gel buffer: 1 in 20 diluted electrode buffer.
5 Vcm^{-1} for 18 h (cooling); man.

S_{46}: Electrode buffer: 100 mmol l^{-1} Tris-phosphate, pH 7.75.
Gel buffer: 1 in 10 diluted electrode buffer; before degasing the starch gel, 2-mercaptoethanol
at a final concentration of 10 mmol l^{-1} and ATP at a final concentration of 0.2 mmol l^{-1} are
added.
8 Vcm^{-1} for 17 h (cooling); man.

S_{47}: Electrode buffer: 500 mmol l^{-1} Tris, 16 mM EDTA-Na$_2$, 650 mmol l^{-1} borate, pH 8.0.
Gel buffer: 1 in 10 diluted electrode buffer, containing 20 mg NADP per ml of gel.
8 Vcm^{-1} for 18 h (cooling); man.

S_{48}: Electrode buffer: 100 mmol l^{-1} phosphate buffer, pH 7.0.
Gel buffer: 1 in 10 diluted electrode buffer.
3 – 6 Vcm^{-1} for 16 h (cooling); man.

S_{49}: Electrode buffer: 500 mmol l^{-1} Tris, 16 mmol l^{-1} EDTA-Na$_2$, 650 mmol l^{-1} borate, pH 8.0, containing 66 mg ATP per 500 ml of buffer at the cathode.
Gel buffer: 1 in 10 diluted electrode buffer, containing 66 mg ATP per 500 ml of gel.
3 – 6 Vcm^{-1} for 17 h (cooling); man.

Tab. 6.2.4 (continued)

System

S_{50}: Electrode buffer: 41.4 g KH_2AsO_4, 250 ml 1 M KOH, distilled water at 1 l, pH 8.0.
Gel buffer: 0.69 g $KH_2A_sO_4$, 4 ml 1 M KOH, distilled water at 1 l, pH 8.0.
8 – 10 Vcm⁻¹ for 4 h (cooling); rat.

S_{51}: Electrode buffer: 16.3 g Tris, 9 g citric acid per liter distilled water, pH 7.0.
Gel buffer: 1 in 15 diluted electrode buffer.
5 Vcm⁻¹ for 16 h (cooling); man.

S_{52}: Electrode buffer: 110 mmol l⁻¹ Tris, 4 mmol l⁻¹ EDTA, adjusted to pH 9.3 with HCl.
Gel buffer: 1 in 10 diluted.
5 Vcm⁻¹ for 16 h (cooling); man.

S_{53}: Electrode buffer: 50 mmol l⁻¹ Tris, 50 mmol l⁻¹ NaCl, 3 mmol l⁻¹ EDTA, 3 mmol l⁻¹ 2-mercaptoethanol, adjusted to pH 8.0 with HCl.
Gel buffer: 8 mmol l⁻¹ histidine, 3 mmol l⁻¹ EDTA, 3 mmol l⁻¹ 2-mercaptoethanol, adjusted to pH 6.0 with HCl.
11 Vcm⁻¹ for 18 h (cooling); mammalia.

6.3 References

1. Harris H, Hopkinson DA, Robson EB (1962) Nature 196: 1296 – 1298
2. Robson EB, Harris H (1966) Ann Hum Genet 29: 403 – 408
3. Gabriel O, Wang SF (1969) Anal Biochem 27: 545 – 554
4. Davis BJ (1964) Ann NY Acad Sci 121: 404 – 427
5. Van Someren H, van Henegouwen HB, Los W, Wurzer-Figurelli E, Doppert B, Veroloet, M. and Meera Khan P (1974) Humangenetik 25: 189-201
6. Harris H, Hopkinson DA (1976) Handbook of enzyme electrophoresis in human genetics, North Holland Publ Comp, Amsterdam Oxford Amer Elsevier Publ Comp Inc, New York
7. Tallman JF, Brady RO, Quirk JM, Villalba M, Gal AE (1974) J Biol Chem 249: 3489 – 3499
8. Meera Khan P (1971) Arch Biochem Biophys 145: 470 – 483
9. Weinbaum G, Markman R (1966) Biochim Biophys Acta 124: 207 – 209
10. Ramponi G (1975) 1.3-Diphosphoglycerate phosphatase. In: Wood WA (ed) Methods in Enzymology 42. Academic Press, New York San Francisco London, pp 409 – 426
11. Schrader WP, Bryer DJ (1982) Arch Biochem Biophys 215: 107 – 115
12. Schäfer HJ, Scheurich P, Rathgeber G (1978) Hoppe-Seyler's Z Physiol Chem 359: 1441 – 1442
13. Trewyn RW, Kerr JJ (1981) J Biochem Biophys Methods 4: 299 – 307
14. Brewer GJ, Sing CF (1970) An introduction to isozyme techniques, Academic Press, New York
15. Dikov AL, Lolova IS (1974) Acta Histochem 51: 102 – 108
16. Lolova I, Dikov A (1975) Acta Histochem 53: 12 – 27
17. Shaw CR, Prasad R (1970) Biochem Genet 4: 297 – 320
18. Tigerstrom von RG, Razell WT (1968) J Biol Chem 243: 2691 – 2702
19. Fisher ZA, Turner BM, Dorkin HL, Harris H (1974) Ann Hum Genet London 3: 341 – 353
20. Klebe RJ, Schloss J, Mock L, Link CR (1981) Biochem Genet 19: 921 – 927
21. Nimmo HG, Nimmo GA (1982) Anal Biochem 121: 17 – 22
22. Eady RR, Large PJ (1968) Biochem J 106: 245 – 255
23. Holmstedt B, Tham R (1959) Acta Physiol Scand 45: 152 – 163

24. Tsuge H, Nakamishi Y (1980) Activity staining for flavoprotein oxidases. In: Mc Cormick DB, Wright LD (eds) Methods in Enzymology 1980, 66E (Vitamins and coenzymes) Academic Press New York London San Francisco, pp 344–350
25. Qavi H, Kit S (1980) Biochem Genet 18: 669–679
26. Birnbaum SM, Levintow L, Kingsley RB, Greenstein JP (1952) J Biol Chem 194: 455–470
27. Kördel W, Schneider F (1976) Biochim Biophys Acta 445: 446–457
28. Nachlas MM, Moris B, Rosenblatt D, Seligman AM (1960) J Biophys Biochem Cytol 7: 261–264
29. Lewis WHP, Harris H (1967) Nature 215: 351–355
30. Baker IP (1974) Biochem Genet 12: 199–201
31. Strongin AYA, Azavenkova NM, Vaganova TI, Levin ED, Stepanov VM (1976) Anal Biochem 74: 597–599
32. Nelson RL, Povey J, Hopkinson DA, Harris H (1977) Biochem Genet 15: 1023–1035
33. Takeuchi T, Matsushima T, Sugimura T, Kozu T, Takeuchi T, Takemoto T (1974) Clin Chim Acta 54: 137–144
34. Grove TH, Levy HR (1975) Anal Biochem 65: 458–465
35. Zalkin H, Kling D (1968) Biochemistry 7: 3566–3573
36. Henderson EJ, Nagano H, Zalkin H, Hwang LH (1970) J Biol Chem 245: 1416–1423
37. Nelson RL, Povey MS, Hopkinson DA, Harris H (1977) Biochem Genet 15: 1023–1035
38. Farron F (1973) Anal Biochem 53: 264–268
39. Payne WJ, Fitzgerald JW, Dogson KJ (1974) Appl Microbiol 27: 154–158
40. Rattazzi MC, Marks JS, Davidson RG (1973) Amer J Hum Genet 25: 310–316
41. Tobin AJ (1970) J Biol Chem 245: 2656–2666
42. Rothe GM (1972) Beitr Biol Pflanzen 48: 433–444
43. Mort JI, Leduc M (1982) Anal Biochem 119: 148–152
44. Mac Gregor RR, Hamilton JW, Shofstall RE, Cohn DV (1979) J Biol Chem 254: 4423–4427
45. Barrett AJ (1976) Anal Biochem 76: 374–376
46. Goren R, Huberman M (1976) Anal Biochem 75: 1–8
47. Reisfeld RA, Lewis UJ, Williams DE (1962) Nature 201: 281–283
48. Dahlmann B, Jany KD (1975) J Chromatogr 110: 174–177
49. Craig I (1973) Biochem Genet 9: 351–358
50. Eppenberger ME, Eppenberger HM, Kaplan NO (1967) Nature 214: 239–241
51. Tsou KC, Lo KW, Yip KF (1974) FEBS Letters 45: 47–49
52. Solomon SS, Palazzola M, King LT (1977) Diabetes 26: 967–972
53. Wilhardt I, Wiederanders B (1975) Anal Biochem 63: 263–266
54. Kleiner H, Schram E (1966) Clin Chim Acta 14: 377–385
55. Kleiner H, Brouet-Yager M (1972) Clin Chim Acta 40: 177–180
56. Nelson RL, Povey MS, Hopkinson DA, Harris H (1977) Biochem Genet 15: 1023–1035
57. Teng Y-S, Anderson JE, Giblett ER (1975) Amer J Hum Genet 27: 492–497
58. Kim HS, Liao TH (1982) Anal Biochem 119: 96–101
59. Hallock RO, Yamada EW (1973) Anal Biochem 56: 84–90
60. Smith AE, Yamada EW (1971) J Biol Chem 246: 3610–3617
61. Beck CS, Hasinoff CW, Smith ME (1968) J Neurochem 15: 1297–1301
62. Adams CWM, Glenner GG (1962) J Neurochem 9: 233–239
63. Herd JK, Tschida J, Motycka L (1974) Anal Biochem 61: 133–143
64. Hullin DA, Thompson RJ (1977) Anal Biochem 82: 240–242
65. Faye L (1981) Anal Biochem 112: 90–95
66. März L, Barna J, Ebermann R (1976) J Chromatogr 123: 495–496
67. Jelnes JE (1971) Hereditas 67: 291–293
68. Hjorth JP (1970) Hereditas 64: 146–148
69. Colombo G, Marcus F (1973) Biol Chem 248: 2743–2745
70. Hubert E, Marcus F (1974) FEBS Letters 40: 37-40
71. Schachter H, Sareny J, Mc Guire EJ, Roseman S (1969) J Biol Chem 244: 4785–4792
72. Siciliano MJ, Shaw CR (1976) Separation and visualization of enzymes on gels. In: Smith I (ed) Chromatographic and electrophoretic techniques vol 2 W, Heinemann, London, pp 185–209
73. Tsuyuki H, Roberts E, Kerr RH, Ronald AP (1966) J Fish Res Bd Can 23: 929–933
74. Peterson AC, Frair PM, Wong GG (1978) Biochem Genet 16: 681–690

75. Melrose TR, Brown CGD, Sharma RD (1980) Res Vet Sci 29: 298 – 304
76. Lowenstein A, Spielman L, Mowshowitz DB (1982) Anal Biochem 120: 66 – 70
77. Kimura K, Miyakawa A, Imai T, Sasakawa T (1977) J Biochem 81: 46 – 476
78. Samuelsson B, Stenberg P, Pandolfi M (1982) Graefe's Arch Clin Exp Ophthalmol 218: 233 – 236
79. Schneider AS (1969) Trisoephosphate isomerase deficiency. In: Yunis YY (ed) Biochemical Methods in Red Cell Genetics Bd XIII Academic Press, New York San Francisco London, pp 189 – 200
80. Oglivie JW, Sightler JH, Clark RB (1969) Biochemistry 8: 3557 – 3567
81. O'Conner JL, Edwards DP, Bransome ED (1977) Anal Biochem 78: 205 – 212
82. Craig I, Tolley E, Bobrow M (1975) Baltimore Conference: Third international workshop on human gene mapping. Birth defects: Original articles series XII.7. The National Foundation New York 1976, pp 114 – 126
83. Skalhegg BA (1974) Eur J Biochem 46: 117 – 125
84. Pierce M, Cummings RD, Roth S (1980) Anal Biochem 102: 441 – 449
85. Parr CW, Bagster IA, Welch SG (1977) Biochem Genet 15: 109 – 113
86. Flashner MI, Massey V (1974) J Biol Chem 249: 2579 – 2586
87. Brewbaker JL, Upadhya MD, Mäkinen Y, Mc Donald T (1968) Physiol Plant 21: 930 – 940
88. Ueng J, Hartanowicz T-H, Lewandoski C, Keller J Holick M, Mc Guinness ET (1976) Biochemistry 15: 1743 – 1749
89. Poenaru L, Dreyfus JE (1973) Clin Chim Acta 43: 439 – 442
90. Eppenberger ME, Eppenberger HM, Kaplan NO (1967) Nature 214: 239 – 241
91. Levy CC (1967) J Biol Chem 242: 747 – 753
92. Jolley RL, Nelson RM, Robb DA (1969) J Biol Chem 244: 3251 – 3257
93. Ichihara K, Kusunose E, Kusunose M (1973) Eur J Biochem 38: 463 – 472
94. Mc Gregor CH, Schnaitman CA, Normansell DE, Hodgins MG (1974) J Biol Chem 249: 5321 – 5327
95. Ingle J (1968) Biochem J 108: 715 – 724
96. Hucklesby DP, Hagemann RM (1973) Anal Biochem 56: 591 – 592
97. Hill, R. (1931) Proc Roy Soc Series B 106: 205 – 214
98. Briel WJ, Westphal J, Stieghorst M, Davis L, Shah VK (1974) Anal Biochem 60: 237 – 241
99. Shah VK, Davis LC, Brill WJ (1972) Biochim Biophys Acta 256: 498 – 511
100. Shah VK, Brill WJ (1973) Biochim Biophys Acta 305: 445 – 454
101. Abrams A, Baron C (1967) Biochemistry 6: 225 – 220
102. Karavolas HJ, Baedecker ML, Engel LL (1970) J Biol Chem 245: 4948 – 4952
103. O'Callaghan CH, Morris A, Kirby SM, Shingles AH (1972) Antimicrob Agents Chemother 1: 283 – 288
104. Mattew M, Harris AH (1976) J Gen Microbiol 94: 55 – 67
105. Lerch B (1968) Experientia 24: 889 – 890
106. Hawley DM, Tsou KC, Hodes ME (1981) Anal Biochem 117: 18 – 23
107. Spencer N, Hopkins DA, Harris H (1968) Ann Hum Genet 32: 9 – 14
108. Takeo K, Nitta K, Nakamura S (1974) Clin Chin Acta 57: 45 – 54
109. Klee CB (1969) J Biol Chem 244: 2558 – 2566
110. Richards EG, Coll JA, Gratzer WB (1965) Anal Biochem 12: 452 – 471
111. Peacock AC, Dingman CW (1967) Biochemistry 6: 1818 – 1827
112. Fitt PS, Fitt EA, Wille H (1968) Biochem J 110: 475 – 479
113. Karpetsky TP, Davies GE, Shriver KK, Levy CD (1980) Biochem J 189: 277 – 284
114. Grossbach W, Weinstein IB (1968) Anal Biochem 22: 311 – 320
115. Lebo RV, Martin DW (1978) Biochem Genet 16: 905 – 916
116. Bauchamp C, Fridovitch I (1971) Biochim Biophys Acta 44: 276 – 287
117. De Rosa G, Duncan DJ, Keen CL, Hurley LS (1979) Biochim Biophys Acta 566: 32 – 39
118. Inano H, Ohba H, Tamaoki BI (1981) J Steroid Biochem 14: 1347 – 1355
119. Schalhorn A, Wilmanns W (1977) Res Exp Med 169: 213 – 219
120. Hatfield GW, Umbarger HE (1970) J Biol Chem 245: 1736 – 1741
121. Hitomi Y, Kanda T, Niinobe M, Fujii S (1981) Clin Chim Acta 119: 157 – 164
122. Scardovi V, Sgorbati B, van Leeuwenhoek A (1974) J Micribiol Serol 40: 427 – 440
123. Cortner JA, Coates PM, Swoboda E, Schnatz JD (1976) Pediat Res 10: 927 – 932
124. Sugiara M, Ito Y, Hirano K (1977) Anal Biochem 81: 481 – 484

125. Andary TD, Dabich D (1974) Anal Biochim 57: 457 – 466
126. Shaik MB, Guy AL, Pancholy SK (1980) Anal Biochem 103: 140 – 143
127. Blattler DP, Reithel FJ (1970) Enzymologia 39: 193 – 199
128. Fishbein WW, Nagarajan K, Scurzi K (1975) Structural classes of jackbean urease variants and their relation to a structural classification of isozymes. In: Markert C (ed) Isozymes Vol I Academic Press New York, pp 403 – 417
129. Poenaru C, Weber A, Dreyfus JC, Overdijk B, Hooghwinkel GJM (1974) FEBS Letters 41: 181 – 184
130. Posen S, Neale FC, Path MC, Birkett DJ, Brudenellwoods S (1967) Am J Clin Pathol 48: 81 – 86

Isozymes may be generated by different enzyme loci (a) (isoenzymes), (b) alleles of a locus (allozymes) or (c) post-translational modifications (secondary isozymes). Differences in isozyme numbers and isoenzyme properties can be used for evolutionary studies. But quantitations of genetic variation among or within populations are obtainable only from allozyme frequencies.

7.1 Allozymes as Gene Markers

Diploid, sexually reproducing organisms receive one complete set of chromosomes from each parent. Accordingly, they have each gene in duplicate; the homologous genes are called alleles. If more than one allele occurs at the locus of a structural gene, the corresponding allozymes can either appear in a single diploid (heterozygotic) organism, or in different (homozygotic) members of a species. The number of alleles at a given locus depends on the species and on the locus itself. Genes are characterized as polymorphic if they comprise two or more alleles with one allele having a frequency of ≤ 99 (≤ 95)%.

The genes coding for allozymes are identified by an abbreviation of the enzyme name, e.g., "AAT" for aspartate amino transferase. If a gene comprises more than one locus (precisely defined site on the DNA molecule), then these loci are indicated by a capital letter (or a number) and linked to the gene name as a suffix, e.g., AAT-A or AAT-B (or AAT-1 or AAT-2). The corresponding alleles may be indicated by consecutive numbers such as AAT-A1, AAT-A2 (or AAT-1_1, AAT-1_2) etc.

Allozymes can be used in genetic studies if the following conditions are fulfilled:

(a) the genotype is phenotypically expressed through given enzyme patterns, which can be observed in the laboratory. A prerequisite for this is that the enzyme forms of interest segregate in a Mendelian pattern and that at least one locus per enzyme system is polymorphic.

Specific genotypes can be identified using:

(a) examination of progeny resulting from controlled crossings;
(b) comparative studies of haploid and diploid tissues of an individual (e.g., haploid endosperm and diploid embryo of one seed of a gymnosperm, haploid pollen and diploid tissue of angiosperms);

(c) examination of the progeny resulting from a female parent of known genotype or
(d) by applying Hardy-Weinberg's Law in theoretical models [1].

7.1.1 Enzyme Structure and Compartmentalization

Genetic interpretation of isozyme patterns is considerably aided by the somewhat predictable subunit composition of many enzymes and their localization within particular cell compartments. Most enzymes assayed for electrophoretic mobility are either monomers (made of one polypeptide chain), dimers (composed of two subunits) or tetramers (made of four subunits) (Chap. 1). Usually allozymes of a single gene are codominantly inherited and therefore visible on an electrophoretic separation medium as phenotypes (colored bands). In a homozygote each enzyme locus codes for one allozyme only. But if more than one locus codes for enzymes of the same substrate specificity, interlocus hybrid forms may occur with dimeric or tetrameric enzymes. In heterozygotes, monomeric enzymes display two bands, one from each of the contributing parental alleles. But if the enzyme is a dimer then the two different subunits coded by the contributing parental alleles may form one heterodimer in addition (the proportions of the homomeric and heteromeric forms would be: $AA:AB:BB = 1:2:1$). In case of a tetrameric enzyme in heterozygotes, two homotetramers and three heterotetramers may appear ($A_4:A_3B:A_2B_2:AB_3:B_4 = 1:4:6:4:1$). Dimeric or tetrameric intralocus allozymes usually migrate to different positions on the gel, resulting in three or five banded phenotypes (Fig. 7.1).

Allozyme patterns become more complex when they are coded by more than one isozyme locus. Many enzymes for example of the glycolysis and the pentose phosphate pathway appear in several subcellular compartments such as cytosol, plastids, mitochondria and microbodies [2] (Chap. 1). These enzymes do not form interlocus hybrids.

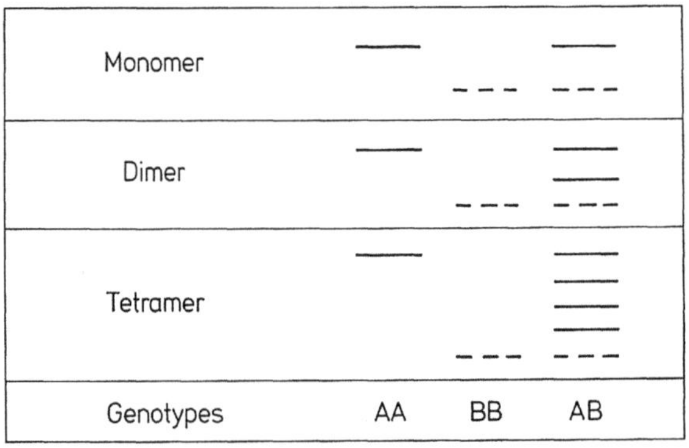

Fig. 7.1. Hypothetical phenotypes (enzyme bands) occuring in homozygotes (*AA, BB*) and heterozygotes (*AB*) for monomeric (one subunit), dimeric (two subunits) and tetrameric (four subunits) enzymes, if subunits associate at random in heterozygotes. If *B* represents a null allele, *dashed lines* indicate the absence of a band

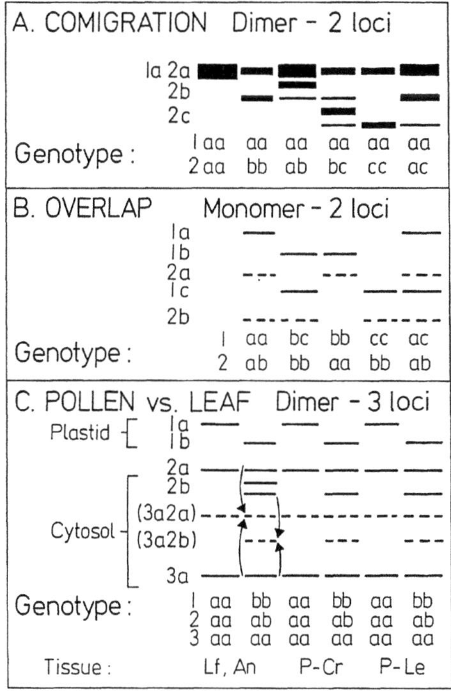

Fig. 7.2 A – C. A: Hypothetical phenotypes for two loci if the alleles "a" of locus 1 (allele 1a) and locus 2 (allele 2a) migrate to the same position upon electrophoresis but do not form interlocus hybrid forms. Interlocus hybrids do not occur if two loci code for (iso)enzymes of different cell compartments. Band width denotes relative amounts of enzyme activity. **B:** Phenotypes for two non-interacting loci which overlap in migration. *Dashed lines* denote products of a second locus. **C:** Hypothetical banding patterns for diploid tissues (e.g., leaf (*Lf*) or anther (*An*)) as compared to haploid pollen (*P*) which has been crushed (*Cr*) or leached (*Le*) after removal from anthers. The same two individuals have been loaded three times on the gel. *Curved arrows* show the origin of interlocus bands. *Dashed lines* denote interlocus hybrid bands. Figure taken with permission from [2]

Sometimes the products of two or more enzyme loci may overlap on an electrophoretic support medium, or may co-migrate to the same position (Fig. 7.2) [2]. Such patterns can be interpreted by changing the separation conditions or by isolating enzymes from a single compartment. In plants haploid tissues are helpful in solving that problem. Pollen leachates, for example, contain only cytoplasmic isozymes [3]. In some taxa, soaking pollen from the anthers of several flowers in 0.5 – 1 ml of a simple extraction buffer (50 mmol l⁻¹ Tris-HCl, pH 8.0, 0.2 – 1 mmol l⁻¹ EDTA, 14 mmol l⁻¹ 2-mercaptoethanol) for 4 – 12 h [4]) releases only the cytosolic isozyme of glucose phosphate isomerase. Crushed extracts of pollen or leaves, however, reveal plastid and cytosolic isozymes. Besides, pollen extracts of diploid heterozygotes lack intralocus oligomeric isozyme forms (bands) (Fig. 7.2). But, if more than one isozyme locus exists for a given cell compartment, then interlocus oligomeric isozyme forms (coloured bands) may appear in pollen leachates. The number of loci coding for a given set of isozymes appearing in more than one intracellular compartment is highly conserved [5]. Deviations from the predicted number of isozyme loci suggest possible duplication events, polyploidy, null alleles or technical faults. On the other hand, enzymes assayed with artificial substrates which do not occur within organisms (peroxidases, esterases, peptidases, phosphatases) often vary with respect to isozyme number and localization [2]. Though frequently used to assess genetic variability, such enzymes are of limited value for phylogenetic comparisons because of uncertain homologies.

If isozymes are not well resolved on gels, heterozygotes may display a single broad zone of activity instead of the predicted two or more distinct bands. In organisms with null alleles which lack enzymatic activity, homozygotes and heterozygotes become visually indistinguishable for monomeric enzymes; heterozygotes show unexpected two- or four- banded phenotypes in dimeric and tetrameric enzymes (Fig. 7.1). Because of these and other constraints, the genetic basis of observed patterns should be verified by isozyme analysis of progeny arrays from controlled crosses or by comparing isozyme patterns of haploid and diploid tissues. Predictions not supported by full genetic analyses should be reported as putative loci and genotypes.

7.1.2 Duplicated Loci and Polyploidy

In plants enzyme loci may often have been duplicated. When two or more loci encode isozymes which are occurring in the same subcellular compartment, interlocus heteromers may appear, giving rise to very complex banding patterns (Fig. 7.3). In the plant species *Clarkia* a duplication of cytosolic phosphoglucose isomerase (PGI) resulted in a three-banded phenotype for double homozygotes (A_1A_1, A_1B_2, B_2B_2), and a six-banded pattern for plants heterozygous for one of the duplicated cytosolic loci (A_1A_1, A_1A_2, A_2A_2; A_1B_1; A_2B_2; B_2B_2 (A,B = loci; 1,2 = alleles)). Ten cytosolic bands occur in double heterozygotes (A_1A_1, A_1A_2, A_2A_2; A_1B_1; A_1B_2; A_2B_1; A_2B_2; B_1B_1, B_1B_2, B_2B_2)[6]. Depending on the distances between bands, overlap and co-migration may result in phenotypes with less than the expected number of six or ten bands [2].

Diploid plants often show independent segregation for each unlinked copy of a duplicated enzyme locus [2]. However, heterozygous phenotypes in related tetraploids may be fixed. Relative to expected values for Hardy-Weinberg equilibrium, allotetraploid populations may show an excess of heterozygotes [2], although such excesses can also occur in diploid populations (e.g., via selection against homozygotes). "Fixed heterozygosity" in allopolyploids occurs when the contributing parental genomes are homozygous for different alleles.

Polyploids of autotetraploid origin may show greater heterozygosity and enzyme multiplicity than related diploids. Besides, due to greater similarity in the duplicated genomes compared to alloploids, chromosomes will pair with more than one homologue of each set. Multivalent formations more or less influence electrophoretic phenotypes depending on whether an autopolyploid is homozygous for a given locus,

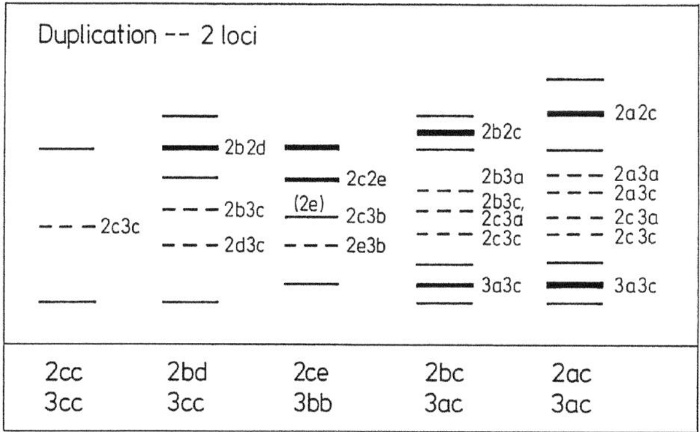

Fig. 7.3. Hypothetical banding patterns for two enzyme loci if subunits of enzymes associate to hybrid forms within the same subcellular compartment; the two loci presumably evolved by duplication. The subunit compositions of the dimeric enzyme are given (*at right*) for interlocus (*dashed lines*) and intralocus (*thick lines*) enzyme forms. Various band numbers result in heterozygotes depending in part on the migration distances of the heteromers. Heterodimers are assumed to migrate to positions intermediate to homodimers encoded by the same alleles. In lanes 3 and 4 (from left to right), different isozymes would have migrated to the same position on the separation medium and would consequently not have been resolved. Figure taken with permission from [2]

arising by duplication of a homozygous genotype in one individual (A_1A_1 leading to $A_1A_1A_1A_1$) or heterozygous, as in cases where the genomes arise from two or more genotypes within the same population [2, 7]. Consequences of the formation of such multivalent associations at meiosis include the following: (a) more than two alleles at a locus may be expressed in a single individual (e.g. $A_1A_1A_2A_3$), (b) both balanced ($A_1A_1A_2A_2$) and unbalanced ($A_1A_1A_1A_2$, $A_1A_2A_2A_2$) heterozygotes with different staining intensities may appear in the progeny of a given autotetraploid, and (c) progeny arrays will show tetrasomic segregation ratios. In general, the combination of differing band intensities and ratios generates autotetraploid patterns which are distinct from those expected in allotetraploids showing disomic inheritance [2, 8]. However, multivalents are not always formed, and pairing factors can generate disomic segregation ratios [2, 7, 9].

7.2 Population Genetic Measures

A population may be defined as a group of individuals of the same species. The species may be asexual but usually its members also cross and are interfertile. Therefore, the model of an ideal population is based on (a) an infinite population size and (b) an equal chance for each individual to participate in sexual reproduction (random mating or panmixia).

The genetic structure of populations can be quantitated by (a) allelic measures, (b) genotypic measures and (c) gene pool measures. Each of these measures can be taken to estimate genetic variation *within* populations or *between* populations.

7.2.1 Calculation of Allelic Frequencies for Codominant Autosomal Genes

Genetic studies using allozymes are always based on discontinuous variation (absence or presence of enzyme bands). The number of alleles (allelic frequencies) occurring within a population (taxon) results from:

$$p = \frac{(2\,H_o + H_e)}{2\,N} \tag{1}$$

with the following definitions:

H_o = number of homozygous individuals carrying that allele,
H_e = number of heterozygotes for that allele, and
N = number of individuals examined [10].

It is conventional to label alleles in order of their decreasing electrophoretic mobility. The faster migrating allele of two alleles may be named AA, A_1, ff or 100/100, the slower allele would then be named BB, A_2, ss or, e.g., 70/70 (indicating the percentage of migration in relation to the fastest allele). In the case of a monomeric enzyme and two alleles we would expect the following three electrophoretic phenotypes distributed among the members of the population: AA, AB and BB. AA (BB) is equal to H_o, AB is equivalent to H_e.

The "standard error" of the frequency of an allele (p) is estimated by:

$$\sigma = \sqrt{\frac{p\,(1-p)}{2\,N}} \tag{2}$$

Example 1:

In the examination of 49 trees of Scots pine (*Pinus sylvestris*) in Germany the following numbers of the alleles B1 and B2 at the locus AAT-B (aspartate aminotransferase, locus B) were observed:

$$H_o\,(B1) \quad = 8$$
$$H_o\,(B2) \quad = 15$$
$$H_e\,(B1B2) = 26$$
$$N \qquad\quad = 49$$

$$p_{(B1)} = \frac{(2 \cdot 8 + 26)}{2 \cdot 49} = 0.43$$

$$\sigma_{p\,(B1)} = \sqrt{\frac{0.43\,(1 - 0.43)}{2 \cdot 49}} = 0.050$$

The frequency of allele B2 = $1 - p_{B_1}$ = 0.57. The standard error of B2 is equal to σp (B2) = 0.050. Thus the frequencies of the alleles B1 and B2 at the locus AAT-B are: B1 = 0.43 ± 0.05 and B2 = 0.57 ± 0.05.

7.2.2 Allelic Frequencies with One Dominant Allele and One Recessive Allele

Some genes may have silent (or null) alleles. They are recessively inherited. Therefore, in the case of two alleles, one silent and one active, three phenotypes would exist: AA, AN and NN (A = active, N = non-active). The proportion of the genotypes can be found by applying Hardy-Weinberg's Law [11].

Example 2:

Phenotypes	No.	Genotypes
A	77	AA, AN
N	25	NN
Σ	102	

The frequencies p of the alleles A and N are, Hardy-Weinberg's Law (cf. Sect. 7.2.5.1) applying:

$$(pA)^2 + 2\,(pA \cdot pN) + (pN)^2 = 1 \tag{3}$$

$$p(N) = \sqrt{\frac{25}{102}} = 0.495$$

$$p(A) = 1 - p(N) = 0.505$$

7.2.3 Sex Linked Genes

Some enzyme loci are linked to a sex chromosome. This means that the occurrence of respective bands is limited to one sex only. A precondition of the calculation of allelic frequencies in this case is that the difference between the distributions of the phenotype in the two sexes is statistically significant.

Species may be grouped into three classes, depending on the nature of the sex chromosomes present in the heterogametic sex [11]:

a) organisms where the sex chromosome is present in two copies in the homogametic sex and in only one copy in the heterogametic sex;
b) organisms where the sex chromosomes differ from the autosomes only by the presence of a gene or a group of genes responsible for sex determination. Except for the sex factor all genes exist in two copies in the heterogametic sex and the homogametic sex. The genes carried by this type of sex chromosome normally behave like autosomal genes, except when they are in close proximity to the sex gene. Then significant deviations of phenotypes between males and females will occur;

c)   organisms where individuals of the heterogametic sex possess morphologically different chromosomes (X, Y). Depending on whether or not the X or Y chromosomes have homologous segments, their genes will exist in different numbers of copies:

   α)   in the case where the genes are located on a segment of the X chromosome which has no homologue on the Y chromosome, there will be one copy of genes in the heterogametic sex and two copies in the homogametic sex;

   β)   the genes exist in one copy in the heterogametic sex only. This is the case if the genes are located on a segment of the Y chromosome which has no homologue on the X chromosome;

   γ)   provided the genes are located on homologous parts of the X and Y chromosomes, two gene copies exist in both sexes. The two chromosomes may have the same gene arrangement, and in this case allelic frequencies are calculated as described in Example 3.

   On the other hand, a chromosome rearrangement may have occurred leading to the fact that X and Y carry different alleles [11]:

Example 3:

Calculation of allelic frequencies for the autosomal gene AAT-1 with codominant alleles in the diploid mosquito *Culex pipiens* [11]:

Phenotypes	Genotype	Males	Females	Total
110	110/110	0	0	0
100	100/100	9	14	23
70	70/70	1	2	3
110/100	110/100	2	2	4
110/70	110/70	1	1	2
100/70	100/70	12	10	22
		25	29	54

Allelic frequencies:

   Total number: $2N = 2 \cdot 54 = 108$

110 Allele:

   Number: $2(\overline{110}) + (\overline{110/100}) + (\overline{110/70}) = 2 \cdot 0 + 4 + 2 = 6$
   $p_{110} = 6/108 = 0.055$

100 Allele:

   Number: $2(\overline{100}) + (\overline{110/100}) + (\overline{100/70}) = 2 \cdot 23 + 4 + 22 = 72$
   $p_{100} = 72/108 = 0.667$

70 Allele:

   Number: $2(\overline{70}) + (\overline{110/70}) = + (\overline{100/70}) = 2 \cdot 3 + 2 + 22 = 30$

$$p_{70} = 30/108 = 0.278$$

and:

$$p_{110} + p_{100} + p_{70} = 0.055 + 0.667 + 0.278 = 1$$

The difference between the frequencies in the two sexes is not significant as can be judged by the X^2 test.

Example 4:

Calculation of allelic frequencies for a locus which exists in one copy in one of the sexes (heterogametic sex for diploid animals, haploid sex for haplo-diploid animals): the LDH-2 locus in the hymenopteran *Diadromus pulchellus* [11]:

Phenotypes	Genotypes	Females	Males
A	A	–	43
	AA	16	–
B	B	–	93
	BB	34	–
AN	AB	3	–
		53	136

The females are diploid, while the males are haploid, therefore a total of $2 \cdot 53 + 136 = 242$ genes, 242 individual alleles, were analyzed:

Frequency of allele A:

$$p_A = \frac{2\,(\overline{AA}) + (\overline{A}) + (\overline{AB})}{242} = \frac{32 + 43 + 3}{242} = 0.32$$

Frequency of allele B:

$$p_B = \frac{2\,(\overline{BB}) + (\overline{B}) + (\overline{AB})}{242} = \frac{68 + 93 + 3}{242} = 0.68$$

and: $0.32 + 0.68 = 1.00$

Example 5:

Calculation of allelic frequencies in the case of a sex-linked locus with two alleles; one allele is not present on the Y (or W) chromosome: the MPI locus of the common lizard [11]:

Phenotypes	Genotypes	Males	Females
100	100/100	47	0
100/120	100/120	15	80
120	120/120	0	7

Frequencies for Y (W) chromosomes:

$$p_{100} = 0$$
$$p_{120} = 1$$

Frequencies for Z chromosomes:

$\hat{p}$ (males) = 15 / (15 + (15 + 2 · 47)) = 0.121

$\hat{p}$ (females) = 7 / (80 + 7) = 0.080

$\hat{p}$ (males) = 1 − $p_{(males)}$ = 0.879

$\hat{p}$ (females) = 1 − $p_{(females)}$ = 0.920

On average, there are 90 MPI^{100} alleles for every 10 MPI^{120} alleles.

Diadromus pulchellus (Hymenoptera) has a arrhenotokus parthenogenetic system of reproduction, where the females are produced from fertilised eggs (diploid), and the males from non-fertilised eggs (haploid). Allelic frequencies in a haploid-diploid organism like *Diadromus* are accessible as described in Example 4. The same method of calculating allelic frequencies is applicable to those genes in diploid animals which are situated on a segment of the X chromosome which has no homologue on the Y chromosome. The method should also be used for animals with no Y chromosome [11]. Heterozygote phenotypes are completely absent in the heterogametic sex of haplo-diploid organisms. Therefore, significant differences in genotype distributions between sexes result if genes on one sex chromosome are polymorphic [11].

7.2.4 Analysis of Allelic Structures

The genetic variation of populations may be measured at threee different levels (a) the allelic, (b) the genotypic, and (c) the gene pool level (Table 7.1).

Allelic frequencies can be used to measure the genetic variation within and between populations.

7.2.4.1 Measures of Variation within Populations

The genetic diversity within populations may be quantified by the following measures: (a) percent polymorphic loci, (b) average number of alleles per locus, (c) effective number of alleles per locus n_e (also named diversity (v)), (d) total population differentiation (S_T), (e) evenness (e) and (f) expected heterozygosity (h_e). Usually a locus is considered polymorphic if the frequency of the most common allele is less than 0.99. Percent polymorphic loci for a species is defined as the mean of all loci investigated. The average number of alleles per locus is calculated with an arithmetic mean. The effective number of alleles (n_e) takes into account both the number of alleles and their frequencies. The expected heterozygosity (h_e) is obtained by applying Hardy-Weinberg's Law. This number is compared to the observed heterozygosity (h_i) to identify deviations from a population in equilibrium.

Table 7.1. Characterization of Genetic Structures

---

1   Analysis of allelic structures

1.1 Measures of variation within populations
   - allelic frequencies (p)
   - effective number of alleles (n_e)
   - total population differentiation (δ_T)
   - evenness (e)

1.2 Measures of variation between populations
   - genetic identity (I)
   - genetic distance (D, d_o)
   - subpopulation differentiation (D_j) and (δ)
   - test of homogenety of distributions

2   Analysis of genotypic structures

   - genetic diversity
      - average heterozygosity H_I
      - mean heterozygosity H_e
      - mean heterozygosity H_s
   - test of Hardy-Weinberg-structure (λ^2-test)
   - test of inbreeding
   - test of ordered genotypes

3   Analysis of gene pools

3.1 Measures of variation within populations
   - genepool diversity (v)
   - hypothetic-gametic diversity (v)
   - total population differentiation (δ_T) within gene pool

3.2 Measures of variation between populations
   - genetic distance of gene pools (d)
   - sub population differentiation (δ) within gene pool

Effective number of alleles n_e

It results from:

$$n_e = \frac{1}{(\sum x_i^2)} \tag{4}$$

where x_i equals the frequency of alleles at a gene locus.

Example 6:

If two alleles would occur at a locus with frequencies $p(B1) = 0.43$ and $p(B2) = 0.57$ (Example 1), the effective number of alleles would result as:

$$n_e = \frac{1}{(0.43^2 + 0.57^2)} = 1.962$$

In a german population of *Pinus sylvestris* the effective numbers of alleles at 7 enzyme loci (SKDH-A to LAP-B) were as given below:

Locus	n_e
SKDH-A	1.4988
SKDH-B	1.1271
AAT-A	1.3676
AAT-B	1.2677
AAT-C	1.9928
LAP-A	1
LAP-B	1

The total effective number of alleles would result from the geometric mean of all n_e's:

$$n_e = \sqrt[7]{1.4988 \cdot 1.1271 \cdot 1.3676 \cdot 1.2677 \cdot 1.9928 \cdot 1 \cdot 1} = 1.2866$$

Total population differentiation (δ_T)

The term differentiation denotes the measurement of variation between populations or demes (it may also be applied to variation within a single population by regarding each genetic type as a deme of its own [12]). Consistent extension of the measure δ to this situation yields a measure δ_T of total population differentiation [12]:

$$\delta_T = \frac{N}{N-1} \cdot (1 - \Sigma X_i^2) \tag{5}$$

where N equals population size, and x_i is the relative frequency of alleles at a gene locus (i-th genetic type). If two alleles would accur at a locus at frequencies (x_i) of pB1 = 0.43 and pB2 = 0.57 and the size of the population would be 49 (Example 1) then the total population differentiation would result from:

$$\delta_T = \frac{49}{49-1} \cdot (1 - (0.43^2 + 0.57^2))$$

$$\delta_T = 0.5004$$

When applied to allele frequencies and to large N, δ_T is identical to what Nei [22] called "gene diversity", and only for Hardy-Weinberg proportions is it identical to the actual heterozygosity. If the population would be completely monomorphic δ_T is equal to 0, if each individual would be genetically different from each other then δ_T would be equal to 1.

Population differentiation δ_T and effective number of alleles n_e are related by the equation $\delta_T = N/(N-1) \times (1 - v^{-1})$, which becomes a one-to-one relationship for a large population size. In highly differentiated populations the estimation of n_e from samples may become problematical [16]. In such cases it is safer to rely on δ_T estimates.

Evenness (e)

"Given a distribution of types of individuals in a collection, the evenness (e) of the distribution is considered to measure the degree to which these types are equally represented" [13].

The absolute evenness is given by

$$e = 1 - d_{min} \tag{6}$$

At a uniform distribution $e = 1$ (the "genetic distance" d is defined in Sect. 7.2.4.2.1). As e approaches a lower bound of 0.5, the unevenness increases. The relative evenness of a population varies between 0 and 1 and is defined [13] as:

$$e = 1 - 2 \cdot d_{min}. \tag{7}$$

7.2.4.2 Measures of Variation between Populations

The genetic variation between populations may be quantitated by a number of measures such as (a) the genetic identity (I), (b) the genetic distance (D) (or d_o) or (c) the subpopulation differentiation (D_j and δ).

7.2.4.2.1 Genetic Distance d_o

The genetic distance d_o between populations may be quantified as the difference between two statistical distributions of gene frequencies [14]:

$$d_o = \frac{1}{2} \sum_{i=1}^{n} |X_i - Y_i| \tag{8}$$

The genetic distance can have values ranging from 0 to 1. The value 1 is obtained when no common alleles are present in two populations. The value 0 indicates that two populations have identical allele frequencies at the locus investigated.

Example 7:

The following allele frequencies were observed at the locus AAT-A of two different pine (*Pinus sylvestris*) populations in Germany:

	Populations	
AAT-A	x	y
p (A1)	0.16	0.43
p (A2)	0.84	0.57

The genetic distance d_o of the two populations at this locus results as:

$$d_o = 1/2 \left[(|0.16 - 0.43|) + (|0.84 - 0.57|) \right]$$

$$d_o = 0.27.$$

The distance measure d_o has the advantage that it increases linearly over the whole scale of allelic frequencies. Besides, rare alleles are not underestimated with this measure.

If more than one gene locus is considered, the specifically weighted single distances may be summed up. If each distance is equally weighted ($a_j = 1/m$) then the total genetic distance is equal to the arithmetic mean of all loci where X_{ij} is the frequency of the i-th allele at the j-th locus in population x:

$$d_o = \sum_{j=1}^{n} a_j \cdot \frac{1}{2} \sum_{i=1}^{nj} |X_{ij} - Y_{ij}| \tag{9}$$

Example 8:

Taking the following allelic differencies at six enzyme gene loci observed at two different pine (*Pinus sylvestris*) populations (X, Y) the total genetic distance d_o would result as:

Gene locus	$|X_{ij} - Y_{ij}|$
AAT-A1	0.27
AAT-A2	0.27
AAT-B1	0.02
AAT-B2	0.02
AAT-C1	0.17
AAT-C2	0.17
Σ	0.92

$d_o = 0.92/6 = 0.153$.

Subpopulation differentiation D_j and δ

If a population is divided into subpopulations (demes) its *genetic differentiation D_j* to the remainder of the population is defined as the proportion of genetic elements (alleles, genes at multiple loci, gametes, genotypes) by which it differs from the remainder of the population [15]. This proportion is defined as:

$$D_j = d_o\,(p_j,\, \bar{p}_j) \tag{10}$$

where p_j equals the frequency of elements in the deme while $\bar{p}_j$ is the frequency of elements in the remainder of the population, or:

$$D_j = \frac{1}{2} \sum_{i=1}^{n} |X_i - Y_i| \tag{11}$$

with X_i = frequency of the i-th allele (i = 1, 2, 3, ..., n) in the deme j and Y_i = frequency of the i-th allele (i = 1, 2, 3, ..., n) in the remainder of the population (at a certain gene locus).

The *subpopulation differentiation δ* is defined as [15, 16]:

$$\delta = \sum_{j} c_j \cdot D_j. \tag{12}$$

7.2.4.2.2 Genetic Identity (I) and Genetic Distance (D)

The *genetic identity (I)* between two populations (or taxa) as defined by Nei [14, 17] is given by:

$$I_{xy} = \frac{\sum x_i \cdot y_i}{\sqrt{\sum x_i^2 \cdot \sum y_i^2}} \tag{13}$$

where x_i and y_i are the frequencies of the i-th allele in population X and Y, respectively.

Example 9:

The frequencies (p) of alleles F1 and F2 at locus ACP-F in two beech populations (X, Y) in Germany were:

Allelic frequencies	Population	
ACP-F	X	Y
p F1	0.77	0.73
p F2	0.23	0.27.

Therefore,

$$I = [(0.77 \cdot 0.73) + (0.23 \cdot 0.27)] / [(0.77^2 + 0.23^2) \cdot (0.73^2 + 0.27^2)]^{0.5}$$
$$I = 0.9979.$$

If the frequency of alleles at a locus is $I = 1$ in two taxa, equal allelic frequencies are given. On the other hand, if $I = 0$ then allelic frequencies are completely different.

The mean genetic identity (I) is the arithmetic mean over all loci studied (including monomorphic ones) [10]:

$$I = \frac{I_{xy}}{\sqrt{(I_x \cdot I_y)}} \tag{14}$$

where I_{xy}, I_x and I_y are the arithmetic means over all loci of $\sum x_i y_i$, $\sum x_i^2$ and $\sum y_i^2$, respectively.

The *genetic distance (D)* as defined by Nei [14] is:

$$D = - \ln I. \tag{15}$$

It can be taken as a measure of genetic differentiation among populations.

If $D = 0$, then the allelic frequencies at a locus are the same in two populations; on the other hand, if $D = 1$ then frequencies are completely different.

D may also be interpreted as a measure of the mean number of electrophoretically detectable substitutions of amino acids which occurred since two populations separated from a common ancestor [10]. The actual number of codon substitutions is defined by D/c, where c represents the proportion of codon substitutions which are electrophoretically detectable.

Under certain circumstances, i.e., if two populations are (within an order of magnitude of the same size) in equilibrium with respect to the effects of mutation, selection and genetic drift, then D is proportional to the time of divergence T of the two populations:

$$D = 2 \cdot \alpha \cdot T \tag{16}$$

where α is the rate of electrophoretically detectable codon changes per locus per year [10]. According to Nei [14], $\alpha = 10^{-7}$ which means that

$$T = 5 \cdot 10^6 \cdot D. \tag{17}$$

i.e., when $D = 1$, the populations have been isolated for approximately 5 million years. On the other hand, there are enzymes which accumulate amino-acid substitutions ten times more rapidly than others do. Enzymes not involved in complex metabolic pathways such as non-specific esterases, ribonuclease, lysozyme and carbonic anhydrase "change" more rapidly during evolution than do those enzymes which are normally studied in electrophoretic surveys.

The more correct T-values are [10]:

T (years) = 30 · 10⁶ D for slowly evolving enzyme loci, and
T (years) = 2.4 · 10⁶ D for rapidly evolving loci.

7.2.5 Analysis of Genotypic Structures

7.2.5.1 Hardy-Weinberg Distribution for Autosomal Genes with Codominant Alleles

In many cases the genetic basis of enzyme polymorphism has not been established by breeding experiments. In such cases the agreement of observed results with those expected from the Hardy-Weinberg Law is taken not to reject the genetic hypothesis. The Hardy-Weinberg Law applies to an equilibrium population only. In such a population the frequencies of two alleles (A and B) are given by p and q such that $p + q = 1$. An ideal population with respect to the Hardy-Weinberg Law is outbreeding and sexually reproducing. It produces two types of male gametes (A and B) and two types of female gametes (A and B). These will form three types of zygotes (AA, AB, BB) in the ratio 1:2:1. Since the frequency of allele A is p, the frequency of the genotype $AA = p^2$, that of the genotype $AB = 2\,pq$ and that of $BB = q^2$ or:

$$p^2 + 2pq + q^2 = 1. \tag{18}$$

For a three allele genotype:

$$p^2 + 2pq + 2pr + 2qr + q^2 + r^2 = 1, \tag{19}$$

and so on.

Example 10:

For a two allele system, the number of genotypes which are expected for a population result from [18]:

genotype	h_e
AA	$p^2 \cdot N$
AB	$2pq \cdot N$
BB	$q^2 \cdot N$

where p = frequency of allele A, q = frequency of allele B, and N = number of investigated individuals. In the following example the number of observed genotypes (h_i) at a locus were: F1F1 = 106, F1F2 = 62 and F2F2 = 9; and p(F1) = 0.77 and p(F2) = 0.23.

The expected number of genotypes (h_e) results from:

$$h_e \text{ (F1F1)} = p^2 \cdot N = 0.77^2 \cdot 177 = 105$$
$$h_e \text{ (F1F2)} = 2pq \cdot N = 2 \cdot 0.77 \cdot 0.23 \cdot 177 = 63$$
$$h_e \text{ (F2F2)} = q^2 \cdot N = 0.23^2 \cdot 177 = 9.$$

Therefore, for the example given, the numbers of observed (expected) genotypes are: AA = 106 (105), AB = 62 (63) and BB = 9 (9).

The difference between the observed and expected values can be tested for statistical significance by applying an X^2 test for goodness of fit.

X^2 is defined as the sum of the squared differences of the observed number of genotypes (h_i) and the expected number of genotypes (h_e), divided by the number of expected genotypes:

$$X^2 = \sum_{i=1}^{k} \frac{(h_i - h_e)^2}{h_i} \tag{20}$$

When the expected values are small, the log likelihood X^2 test (G-test) should be used:

$$G = 2\sum h_i \ln\left(\frac{h_i}{h_e}\right). \tag{21}$$

The more the observed values deviate from the expected ones, the more the value of X^2 increases. If the deviation exceeds a certain significance value (α), the difference is non-random and the result is not significant. Each value of X^2 corresponds to a probability value (P). Appropriate tables providing P-values are, e.g., to be found in [19]. The P-value indicates the probability of finding the calculated deviation in a repeated assay.

It is important to use the correct degrees of freedom when establishing P-values [10]. The number of independent genotypes is less than N-1 and the number of degrees of freedom (v) results from:

$$v = \frac{1}{2}(N^2 - N) \tag{22}$$

where N denotes the number of alleles.

In the case of 2-alleles and 3-genotypes the degree of freedom is one (v = 1). However, if the degree of freedom is one, a corrected chi square value (X_c^2) should be used:

$$\chi_c^2 = \sum_{i=1}^{2} \frac{(\,|h_i - h_e| - 0.5)^2}{h_i} \tag{23}$$

The expected number of genotypes (h_e) should not be smaller than one, and not more than 20 % of it should be less than five.

Example 11:

X^2-test for goodness of fit:

genotype	h_i	h_e
F1F1	106	105
F1F2	62	63
F2F2	9	9

Number of degrees of freedom: v = 1/2 ($2^2 - 2$) = 1

Correcting the X^2 value according to [18] (equ. (20)):

$$\chi_c^2 = \frac{(\,|\,106 - 105\,|\, - 0.5)^2}{105} + \frac{(\,|\,62 - 63\,|\, - 0.5)^2}{63} + \frac{(\,|\,9 - 9\,|\, - 0.5)^2}{9} = 0.034$$

The value of $X_c^2 = 0.034$ corresponds to a probability of $P = 0.8 - 0.9$. Therefore, the number of observed genotypes at the locus investigated corresponds, at a significance level of $\alpha = 0.05$ to the number of expected genotypes.

7.2.5.2 Total Genetic Diversity H_T

The total genetic diversity is defined [20] as

$$H_T = 1 - \sum_k X_k^2 \tag{24}$$

where $X_k = \sum X_{ik}/s$, and s = number of populations investigated, and X_{ik} = frequency of the k-th allele at a locus of the i-th population.

Example 12:

Taking the following allelic frequencies at the locus A of shikimate dehydrogenase (SKDH) observed at three different pine (*Pinus sylvestris*) populations in Germany, the total genetic diversity H_T results as:

Allelic frequencies	Populations		
	X	Y	Z
p (A1)	0.04	0.03	0.04
p (A2)	0.80	0.84	0.84
p (A3)	0.16	0.13	0.12

$X_k(A1) = 1/3 \cdot (0.04 + 0.03 + 0.04) = 0.0367$
$X_k(A2) = 1/3 \cdot (0.80 + 0.84 + 0.84) = 0.8267$
$X_k(A3) = 1/3 \cdot (0.16 + 0.13 + 0.12) = 0.1367$
$H_T = 1 - (0.0367^2 + 0.8267^2 + 0.1367^2)$
$H_T = 0.2965$

7.2.5.3 Nei's Coefficient of Genetic Diversity GST

Nei's [21] parameters of genetic diversity (H_T, H_S, D_{ST}) are often applied to investigate the distribution of genetic diversity.

The total genetic diversity H_T of Nei is the sum of the average genetic diversity within populations H_S and between populations D_{ST}:

$$H_T = H_S + D_{ST} \tag{25}$$

The coefficient of genetic diversity G_{ST} [22] is the quotient of D_{ST} ($D_{ST} = H_T - H_S$) and H_T [22]:

$$G_{ST} = \frac{D_{ST}}{H_T}. \tag{26}$$

Example 13:

Taking the total genetic diversity H_T as calculated in Example 12 ($H_T = 0.2965$) and the genetic diversity within populations H_S as given in Example 15 ($H_S = 0.2959$), the genetic diversity between populations (D_{ST}) would result as:

$D_{ST} = 0.2965 - 0.2959$

$D_{ST} = 0.0006$

Then the coefficient of genetic diversity (G_{ST}) results as:

$G_{ST} = 0.0006/0.2965$

$G_{ST} = 0.0020$

This means that related to the gene locus SKDH-A 0.2% of the genetic diversity is found between the investigated populations; the gentic diversity within populations is therefore 99.8%.

7.2.5.4 Gene Flow N_m

The parameter of gene flow is N_m, where N is the effective population size and m is the proportion of migrants exchanged between populations per generation. Mostly, N and m are not known and gene flow is reported as N_m, the number of migrants per generation. It was shown [23] that there is a linear relationship between ln (N_m) and ln [p(I)], the natural logarithm of the average frequency of private alleles (those found in only one population). Low mean private allele frequencies in a species indicate a high rate of gene flow, while high mean frequency of rare alleles indicate low gene flow. N_m-values greater than one are considered high [23]. N_m is related to G_{ST} as

$$G_{ST} = \frac{1}{4\,N_m + 1} \, . \tag{27}$$

7.2.6 Analysis of Gene Pools

7.2.6.1 Measures of Variation within Populations

The *gene pool diversity v* as calculated from allelic frequencies is defined as [12]:

$$v = \left(\frac{1}{L} \cdot \sum_{l=1}^{L} \frac{1}{v_{(l)}} \right)^{-1} \tag{28}$$

or

$$v = \left(\frac{1}{L} \left(\frac{1}{v_1} + \frac{1}{v_2} + \frac{1}{v_3} + \dots \frac{1}{v_L} \right) \right)^{-1} \tag{29}$$

where L denotes the number of loci and $v_1, v_2, v_3, \dots, v_L$ is the allelic diversity at the loci 1, 2, 3, …, L. The allelic diversity results from

$$v_{(l)} = \left(\sum_{i=1}^{n} X_{il}^2 \right)^{-1} \tag{30}$$

with l = locus 1, 2, 3, ..., L and x_{il} = frequency at the i-th allele (i = 1, 2, 3, ..., n) at the l-th gene locus.

The *hypothetical genetic diversity* v is defined as [24]:

$$v = \prod_{l=1}^{L} v_{(l)} \tag{31}$$

or

$$v = v_1 \cdot v_2 \cdot v_3 \cdot ... v_L \tag{32}$$

with

$$v_{(l)} = \left(\sum_{l=1}^{n} X_{il}^2 \right)^{-1} \tag{33}$$

The *total population differentiation* δ_T of the gene pool is defined as the arithmetic mean of the total population differentiation at each locus, that is [12]:

$$\delta_T = \frac{1}{L} \cdot \sum_{l=1}^{L} \delta_{T(l)} \tag{34}$$

or

$$\delta_T = \frac{1}{L} (\delta_{T_1} + \delta_{T_2} + \delta_{T_3} + ... + \delta_{T_L}) \tag{35}$$

7.2.6.2 Measures of Variation between Populations

The gene pool genetic distance d between two populations may be defined as [16]:

$$d = \frac{1}{L} \cdot \left(\frac{1}{2} \Sigma \, | X_{i1} - Y_{i1} | + \frac{1}{2} \Sigma \, | X_{i2} - Y_{i2} | + \frac{1}{2} \Sigma \, | X_{i3} - Y_{i3} | + ... \frac{1}{2} \Sigma \, | X_{iL} - Y_{iL} | \right) \tag{36}$$

with 1, 2, 3, ..., L representing the gene loci investigated, with x_{il} equal to the frequency of the i-th allele at the gene locus l of one population, and y_{il} representing the frequency of the i-th allele at the gene locus l of another population.

The *subpopulation differentiation* δ of the gene pool can be defined as the arithmetic mean of the subpopulation differentiation at each locus [16]:

$$\delta = \frac{1}{L} \cdot \sum_{l=1}^{L} \delta_{(l)} \tag{37}$$

It results if the subpopulation differentiations at the loci 1 to L are summed up and divided by the number of gene loci.

7.2.7 Heterozygosity

7.2.7.1 Average Heterozygosity H_I

The genetic diversity of an individual can be defined as the number of different alleles with respect to the total number of polymorphic gene loci [25]. An increase in the number of homozygous loci would cause a loss in genetic diversity. Genetic diversity may consequently be understood as the average heterozygosity of populations.

H_I denotes the average heterozygosity at a gene locus as observed within a given population [20]:

$$H_I = 1 - \sum_k X_{ik}^2 \tag{38}$$

where X_{ik} is the frequency of the k-th allele at a locus of population i.

Taking the allelic frequencies given above (Example 1: $p(B1) = 0.43$ and $p(B2) = 0.57$) the average heterozygosity at the locus AAT-B would result from:

$$H_I = 1 - (0.43^2 + 0.57^2) = 0.49.$$

7.2.5 Mean Heterozygosity H_e

The mean heterozygosity per locus (H_e) is a quantification of the width of genetic variation in a population (including monomorphic loci where $H_I = 0$):

$$H_e = \sum \frac{H_I}{r} \tag{39}$$

where $H_I = 1 - \sum X_i^2$ for the first locus, r = number of the examined loci and X_i = frequency of i-th allele at a locus.

Example 14:

Average heterozygosities (H_I) at seven enzyme loci of a german pine (*Pinus sylvestris*) population were estimated to be:

Gene locus	H_I
SKDH-A	0.3328
SKDH-B	0.1128
AAT-A	0.2688
AAT-B	0.2112
AAT-C	0.4982
LAP-A	0
LAP-B	0
$\sum$	1.4238

Therefore, H_e is equal to:

$H_e = 1.4238/7$

$H_e = 0.2034$

The mean heterozygosity is the same as the average probability of two genes selected at random not being identical. The figure is independent of the frequency of observed heterozygotes, since it does not take non-random crossings and selection into account. It is based on the observed gene frequencies [26].

The variance (v) of H_e for r loci is:

$$v = \sum \frac{(H_i - H_e)^2}{r(r-1)} \tag{40}$$

where H_i is the heterozygosity at the i-th locus. The standard error of $H_e = \sqrt{v}$.

7.2.6 Mean Heterozygosity H_s

The mean heterozygosity H_s denotes the arithmetic mean of H_1 over all the populations studied:

$$H_S = \sum \frac{H_I}{n} \tag{41}$$

where H_s = average heterozygosity at a locus and n = number of populations.

Example 15:

In three different pine (*Pinus sylvestris*) populations in Germany the following average heterozygosities at the gene locus SKDH-A were observed:

population X $H_I = 0.3328$,
population Y $H_I = 0.2766$ and
population Z $H_I = 0.2784$.
Therefore, $H_s = (0.3328 + 0.2766 + 0.2784)/3 = 0.2959$.

The observed heterozygosity is the fraction of heterozygous individuals [11]. For example:

$H_I \text{ (obs)} = 26/100 = 0.26$.

If the population is in the Hardy-Weinberg equilibrium, then the calculated and observed heterozygosities will be very similar [11].

The calculated heterozygosity can also be used for asexual organisms, but then it should be called gene diversity and not heterozygosity because it is not related to the number of heterozygotes [17].

7.3 Calculation of Dendrograms

Measurements of genetic similarity may be used to reconstruct the genetic links of operational taxonomic units (OTUs) which may be populations of one species, different species or taxa. The values of genetic similarity for all possible OTUs are usually presented in the form of a matrix and a dendrogram is used for a visual display of the results.

Based on the various levels of information, three types of dendrograms can be constructed: (a) a phenogram, based on phenetic information, (b) a cladogram, constructed from cladistic information and (c) a phylogram or phylogenetic tree which results from phenetic and phylogenetic data [10, 27]. A phenogram shows the present-day similarities of OTUs without indicating the probable lines of descent. A cladogram demonstrates the sequence of origin of the various lineages and may indicate the times of debranching OTUs. A phylogram indicates the cladistic branching and the length of each branch.

Two often used methods of dendrogram construction are (a) the "unweighted pair-group arithmetic average clustering method" [10], and the phylogenetic tree construction procedure of Fitch and Margoliash (cf. [10]). The latter method does not assume uniform evolutionary rates, whereas the former does.

7.3.1 Unweighted Pair-Group Arithmetic Average Cluster Analysis

Example 16 [10]:

Values of genetic identity (I) or genetic distance (D) are usually presented in form of a matrix like the one shown in Table 7.2. Unweighted pair-group arithmetic average cluster analysis starts with depicting those two OTUs with the highest genetic identity values (or lowest distance value) between them. In Table 7.2 the two most similar OTUs are OTU2 and OTU3, their I-value is 0.83. Dendrogram construction is started by drawing an appropriate scale covering the range of similarity values found in the OTUs to be clustered. Afterwards OTU2 and OTU3 are joined by a vertical line at an I-value of 0.83 (Fig. 7.4). The distance between the two OTUs is arbitrarily chosen. In the next step of calculation the OTU2 and OTU3 are combined as a single OTU2 – 3 and a new matrix is calculated. The similarity value of OTU2 – 3 to any other OTU is the arithmetic mean of OTU2 to that OTU and the arithmetic mean of OTU3 to that OTU. Thus the I-value of OTU2 – 3 to OTU1 is

$$(0.48+0.36)/2 = 0.42.$$

The new matrix is given in Table 7.3.

Now the next most similar pair of OTUs is joined in the dendrogram; this is OTU4 and OTU5 which have an $\bar{I}$-value of 0.53. These two are joined by a line at $\bar{I} = 0.53$ (Fig. 7.4).

A new matrix taking OTU4-5 as a group is now calculated as demonstrated before (Table 7.4). Calculations are continued until all OTUs are joined together (Fig. 7.4).

Table 7.2. Values of genetic identity (I) as reported ([10]) for five different operational taxonomic units (OTUs)

	OTU 2	OTU 3	OTU 4	OTU 5
OTU 1	0.48	0.36	0.35	0.27
OTU 2		0.83	0.12	0.03
OTU 3			0.05	0.01
OTU 4				0.53

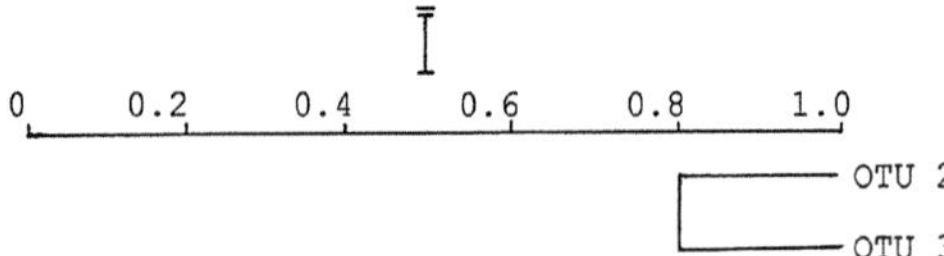

(a) First step in dendrogram construction. The two OTUs 2 and 3 are those with the highest I-value and therefore clustered together

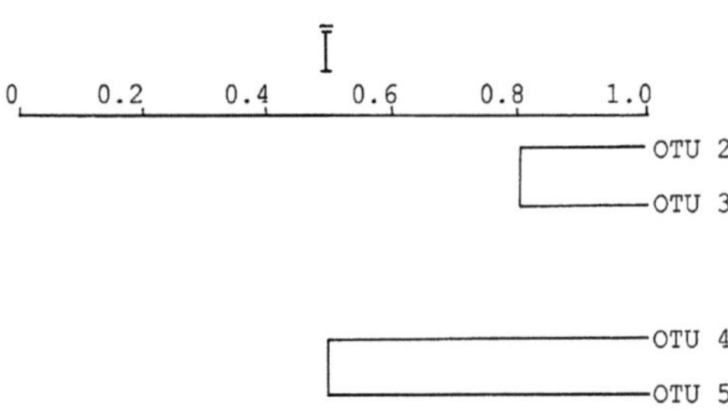

(b) Second step in dendrogram construction

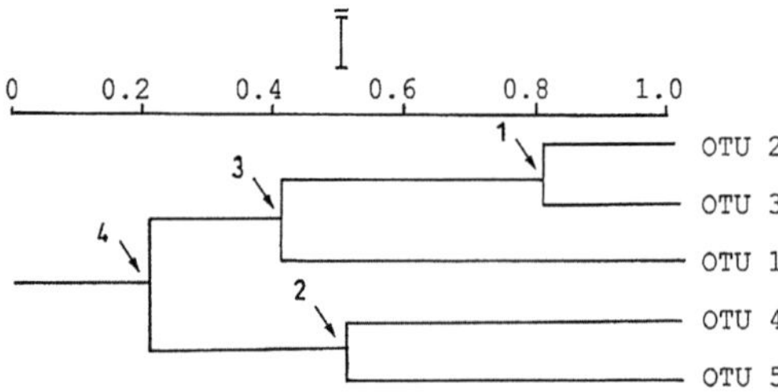

(c) Last step in dendrogram construction

Fig. 7.4. Stages in dendrogram construction. (a) first step in dendrogram construction. The two OTUs 2 and 3 are those with the highest T-value and therefore clustered together. (b) second step in dendrogram construction. (c) last step in dendrogram construction

Table 7.3. Recalculated matrix after clustering OTU 2 and OTU 3 of Table 7.2

	OTU 2–3	OTU 4	OTU 5
OTU 1	0.42	0.35	0.27
OTU 2–3		0.09	0.02
OTU 4			0.53

Table 7.4. Recalculated matrix after clustering OTU 4 and OTU 5 of Table 7.3

	OTU 2–3	OTU 4–5
OTU 1	0.42	0.31
OTU 2–3		0.05

7.3.2 Cladogram Construction

Example 17 [10]:

Three OTUs (A, B and C) may have the following genetic distance (D) values

	B	C
A	0.08	0.19
B		0.17.

A phylogenetic tree can be constructed by joining the pair of highest similarity (lowest D-value) which is A and B and then joining C to this group.

The length of the branches a, b and c which link A to B, A to C and B to C can be calculated from the given matrix values

$$a + b = 0.08 \qquad (a = 0.08\text{-}b)$$
$$a + c = 0.19 \qquad (b = 0.17\text{-}c)$$
$$b + c = 0.17 \qquad (c = 0.19\text{-}a)$$

and substituting values (2a = 0.08 − 0.17 + 0.19 = 0.10, a = 0.05; c = 0.19 − 0.05, c = 0.14 and b = 0.17 − 0.14, b = 0.03).

A, B and C can now be joined by branches of the calculated length:

The phyllogenetic tree illustrates that more changes took place in the line from a common ancestor to A or C than to B.

If more than three OTUs are given, the problem of deciding which two should be joined first can be solved as follows [10]. All possible pairwise combinations are used as A and B, with the remaining OTUs set to C. If there are six OTUs, for example, one is taken as A, one as B and the remaining four as C; this procedure is carried out for all fifteen possible alternative combinations. Finally, all OTUs are in each alternative part of one of three sets. Each set is then treated as for A, B and C above, except that now similarity values are means of A to each member of OTU of C, and likewise for B. The lowest distance from A to B is used to join these two. Henceforward A and B are treated as a single OTU and the calculation procedure is repeated. After each cycle of calculations the number of OTUs is reduced by one. Further explanations are to be found in Ref. [10]. A number of computer programs are available if many cyclical calculations need to be carried out.

7.3.3 Evolution of Gene Loci

Starch gel electrophoresis is mostly taken to compare allozyme frequencies within species. But it may also be used to study the number of loci coding for a special en-

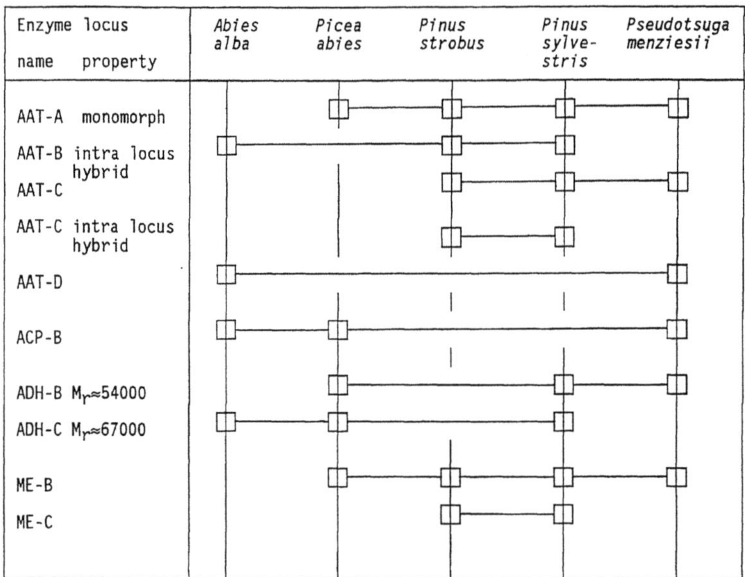

Fig. 7.5. Apomorphic conformity of some enzyme gene loci as observed in five different *Pinaceae*. Common locus traits (monomorphic loci, formation of intra locus hybrid bands or loci coding for enzymes of distinct size (M_r) are indicated by *linkage lines*)

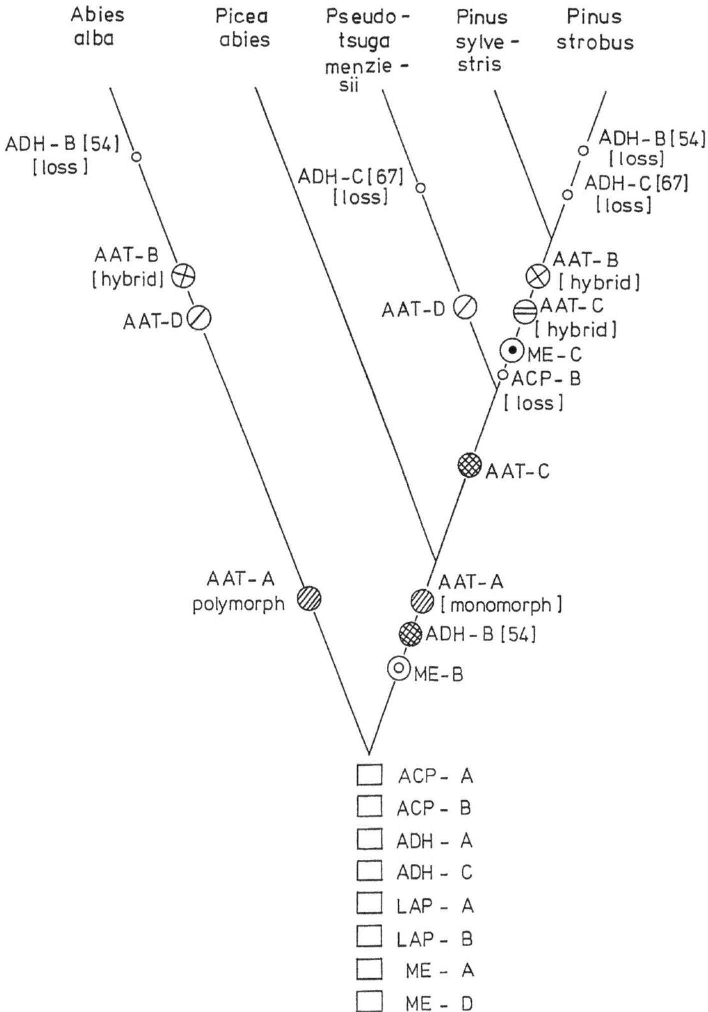

Fig. 7.6. Hypothetical scheme of the evolution of some enzyme loci in five *Pinaceae* based on Fig. 7.5 and unweighted pair-group arithmetic average cluster analysis

zyme system in related species. Different numbers of loci can be used to rank species phylogenetically. The conifer *Picea mariana*, which has only two 6-phosphogluconate dehydrogenase loci, may be considered more primitive than five other *Picea* species (*P. abies, P. engelmannii, P. glauca, P. pungens* and *P. sitchensis*) which have three PGD loci [28].

Pore gradient gel electrophoresis however provides more possibilities to compare enzyme patterns among species. It allows species to be grouped according to common enzyme traits such as : (a) size (molecular weight), (b) locus membership, (c) number of alleles and (d) ability to form intra- (and inter-) locus hybrid forms. The relatedness of species is elucidated from the occurence of derived ("apomorphic") locus traits. Traits are defined as derived if they occur in at least one but never in all of the studied species pairs [27]. An example is given in Fig. 7.6 where the apomorphic conformity of nine enzyme loci among five *Pinaceae* is shown. Conformity may also be expressed as a number and "unweighted pair-group arithmetic average cluster analysis" may be used to construct a dendrogram of genetic similarity.

The dendrogram and the expression of loci within the ten species pairs (Fig. 7.5) can be used to reconstruct the evolution of the studied enzyme loci (Fig. 7.6). Eight loci occur in all species pairs and they are considered to be ancient genes. The derived loci do not occur in every species pair. There are only a few conflicting situations within the suggested evolution of enzyme loci and enzyme traits. These concern the evolution of the loci AAT–B (hybrid), AAT–D, ADH–B (54) ($M_r = 54\,000$) and ADH–C (67) ($M_r = 67\,000$). AAT–B (hybrid) and AAT–D may have evolved indepently in *Abies alba* and one or two other of the investigated species. The loci ADH–B (54) ($M_r = 54\,000$) and ADH–C (67) ($M_r = 67\,000$) may have been lost in *Pinus strobus* and other *Pinaceae* species. However, these uncertainties do not interfere with the main line of evolution.

7.4 References

1. Rothe GM (1991) Efficiency and limitations of isozyme studies in forest tree genetics. In: Forest Genetic Resources, Information-No. 18, Food and Agriculture Organization of the United Nations, Rome, pp 2 – 15
2. Kephart SR (1990) Amer J Bot 77: 693 – 712
3. Weeden N, Gottlieb L (1979) Biochem Genet 17: 287 – 296
4. Weeden N, Gottlieb L (1980) Plant Physiol 66: 400 – 403
5. Gottlieb L (1982) Science 216: 373 – 380
6. Gottlieb L (1983) Isozyme number and phylogeny. In: Jensen U, Fairbrothers D (eds) Proteins and nucleic acids in plant systematics, Springer-Verlag Berlin Heidelberg, pp 209 – 221
7. Lewis D (1980) Polyploidy in species populations. In: Lewis W (ed) Polyploidy, biological relevance, Plenum Press, New York London, pp 103 – 144
8. Weeden N, Wendel J (1989) Genetics of plant isozymes. In: Soltis D, Soltis P (eds), Isozymes in plant biology, Dioscorides Press, Portland, pp 46 – 72
9. Jackson R, Hauber D (1983) Polyploidy, Benchmark Papers in Genetics 12, Hutchinson Ross, Stroudsburg, PA
10. Ferguson A (1980) Biochemical systematics and evolution, Blackie, Glasgow
11. Pasteur N, Pasteur G, Bonhomme F, Catalan J, Britton-Davidian J (1988) Practical isozyme genetics. Ellis Horwood series in gene technology, John Wiley & sons, Chichester
12. Gregorius H-R (1987) Theor Appl Genet 74: 397 – 401
13. Gregorius H-R (1990) Amer Natur 136: 701 – 711
14. Nei M (1972) Amer Natur 106: 283 – 291
15. Gregorius H-R (1984) Measurement of genetic differentiation in plant populations. In: Gregorius H-R (ed) Population genetics in forestry. Springer Verlag Berlin Heidelberg New York Tokyo, pp 276 – 285
16. Gregorius H-R (1986) Theor Appl Genet 71: 826 – 834

17. Nei M (1975) Molecular population genetics and evolution; North-Holland Publishing Company, Amsterdam
18. Bartels H (1971) Planta (Berl) 99: 283 – 289
19. Zar JH (1984) Biostatistical analysis, 2nd edition Prentice-Hall International Editions
20. Falkenhagen ER (1985) Theor Appl Genet 69: 335 – 347
21. Nei M (1977) Ann Hum Genet 41: 225 – 233
22. Nei M (1973) Proc Nat Acad Science USA 70: 3321 – 3323
23. Slatkin M (1985) Evolution (Lawrence, Kans.) 39: 53 – 65
24. Gregorius H-R (1978) Math Biosciences 41: 253 – 271
25. Gregorius H-R (1977) Math Biosciences 34: 267 – 277
26. Lundkvist K, Rudin D (1977) Hereditas 85: 67 – 74
27. Ax P (1984) Das phyllogenetische System, Systematisierung der lebenden Natur aufgrund ihrer Phyllogenese. Gustav Fischer Verlag, Stuttgart
28. Giannini R, Morgante M, Vendramin GG (1991) A putative gene duplication in Norway Spruce for 6-PGD and its phyllogenetic implications. In: Fineschi S, Malvolti ME, Cannata F, Hattemer HH (eds) Biochemical markers in the population genetics of forest trees. SPB Academic Publishing bv, The Hage, The Netherlands, pp. 23 – 29

Enzyme Handbook

The **Enzyme Handbook** provides in a concise form data on enzymes sufficiently well characterized. Data of about 3000 enzymes are presently known and their data sheets will be published at a frequency of 200 per quarter. The data sheets are arranged in their EC Number sequence. Each data sheet is divided into seven sections:

- Nomenclature
- Reaction and specificity
- Enzyme structure
- Isolation/Preparation
- Stability
- Cross references
- Literature references

This collection is an indispensable source of information for researchers applying enzymes in analysis, synthesis and biotechnology.

Also available:

Enzyme Handbook 2. Class 5: Isomerases; Class 6: Ligases. 1990.
DM 257,- ISBN 3-540-52580-7

Enzyme Handbook 3. Class 3: Hydrolases. 1991.
DM 257,- ISBN 3-540-53729-5

Enzyme Handbook 4. Class 3: Hydrolases. 1991.
DM 257,- ISBN 3-540-53730-9

Enzyme Handbook 5. Class 3: Hydrolases. 1991.
DM 248,- ISBN 3-540-54209-4

Enzyme Handbook 6. Class 1.2–1.4: Oxidoreductases. 1993.
DM 248,- ISBN 3-540-56435-7

Enzyme Handbook 7. Class. 1.5 - 1.12: Oxidoreductases. 1994.
DM 257,- ISBN 3-540-56435-7

Tm.BA.94.8.19

If you have any concerns about our products,
you can contact us on
ProductSafety@springernature.com

In case Publisher is established outside the EU,
the EU authorized representative is:
Springer Nature Customer Service Center GmbH
Europaplatz 3, 69115 Heidelberg, Germany

Printed by Libri Plureos GmbH
in Hamburg, Germany